Security First

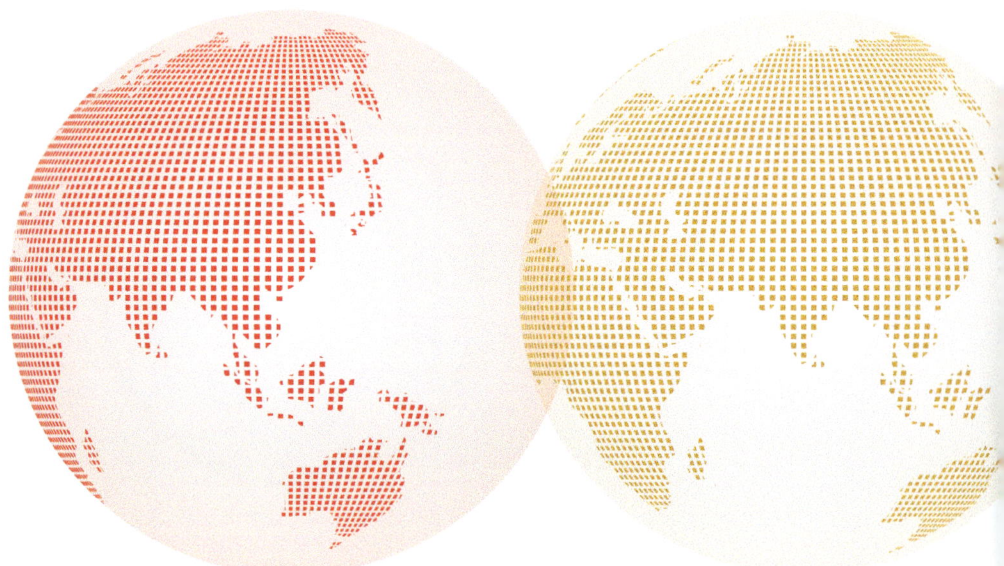

Security First

Geospatial Workflows for a Safe and Equitable World

Edited by

Darren Martin Ruddell

Diana Ter-Ghazaryan

Esri Press

Redlands, California

Esri Press, 380 New York Street, Redlands, California 92373-8100

Printed in the United States of America.
29 28 27 26 25 1 2 3 4 5 6 7 8 9 10

ISBN: 9781589487857
Library of Congress Control Number: 2025940735

For purchasing and distribution options (both domestic and international), please visit esripress.esri.com.

Contents

Foreword

| *By Ronda Schrenk*

The field of human security and geospatial intelligence (HSGI) is vast, varied, and vital to life on Earth. HSGI addresses some of the world's most pressing challenges arising from global issues as varied as natural disasters, infectious diseases, famine, poverty, and war. Confronting these challenges calls for a sophisticated combination of technology and technique.

Security First: Geospatial Workflows for a Safe and Equitable World will improve knowledge of today's cutting-edge geospatial intelligence (GEOINT) tools and how to apply them using the tradecraft of the US intelligence community to mitigate a range of threats to human security.

With this workbook, you will embark on a journey of learning that will equip you with the technical skills, reasoning, and analytic ability needed to succeed as an HSGI professional. It will prepare you to provide information to high-level decision-makers that may affect the course of humankind far into the future.

The evolving landscape of security

The concept of security has transcended traditional notions of military defense. Today, human security encompasses a wide range of threats, including economic instability, water scarcity, pandemics, environmental degradation, violence, and social injustice. Addressing these complex issues requires a holistic approach that uses the power of geospatial intelligence.

HSGI: A powerful solution

HSGI stands at the intersection of geospatial intelligence technologies and human security. It harnesses the power of various sources—satellites, drones, advanced software platforms—to gather and analyze data across vast spatial and temporal scales. This data provides invaluable insights that inform decision-making and guide interventions to safeguard people and promote a more just and equitable world.

This book is designed to be a comprehensive resource for college and university students, faculty, and early-career professionals in HSGI. Whether you envision working in government, the commercial sector, or academia, the knowledge you gain here will be instrumental. Every chapter includes learning objectives that lay out the specific skills you'll acquire. In each case, those skills are built through a step-by-step tutorial process that's crystal clear.

What sets this book apart?

This book offers a unique learning experience consisting of the following features:

- **Topical content:** The book's 20 chapters delve into a diverse range of human security concerns. You will explore such topics as food scarcity, human rights violations, wildfire preparedness, and air pollution monitoring.
- **Technical breadth:** The material will equip you with a broad suite of geospatial methods. You will learn about remote sensing, imagery analysis, machine learning, change detection, and more.
- **Data-driven approach:** The book uses an extensive array of authoritative and nonauthoritative data sources, including satellite imagery, census data, and environmental datasets.
- **Software mastery:** The chapters employ a variety of software platforms to empower your analyses. You will gain proficiency in ArcGIS® Online, ArcGIS Pro, ArcGIS AllSource™, ArcGIS Drone2Map®, Python, and other industry-standard tools.
- **Three-part learning model:** The tutorials provide a structured learning experience. You will gain hands-on skills through guided technical training, followed by independent analysis, and the creation of an intelligence brief that meets the standards of the US intelligence community. Each chapter concludes with additional resources for deeper exploration.

A collaborative effort

This workbook is a culmination of expertise from leading figures across government, industry, and higher education. The editors are leading experts with the University of Southern California's Spatial Sciences Institute, a world-class academic program accredited by the United States Geospatial Intelligence Foundation (USGIF). Together, they have assembled contributions from renowned specialists to ensure the highest quality and relevance of the content.

Filling a critical gap

HSGI is a burgeoning field, yet training resources are struggling to keep pace with its growth. This book aims to address that critical gap by providing a comprehensive learning platform with technical, spatial, and critical thinking resources designed to develop and advance the HSGI community. By equipping you with the necessary skills and knowledge, *Security First* empowers you to become an active participant in this community, helping shape a safer and more equitable world.

Ronda Schrenk *is the chief executive officer of USGIF. She leads the foundation in its mission to promote geospatial intelligence tradecraft and develop a stronger GEOINT community among government, industry, academia, professional organizations, and individuals whose mission focus is the development and application of geospatial intelligence to address national security objectives.*

Introduction

HSGI is an interdisciplinary field that combines geospatial technologies and data analysis with human security concerns to address urgent and complex global challenges. Its applications are wide-ranging, from military to humanitarian, and include strategies for disaster planning, threat response, and risk mitigation. As you will see in the workflows presented in *Security First: Geospatial Workflows for a Safe and Equitable World*, HSGI represents a rapidly growing and increasingly diverse field of application areas in local and global settings. HSGI professionals deploy various geospatial tools for monitoring and assessment to provide actionable information for promoting human security and safety.

The purpose of this book is to provide foundational resources and materials to the growing HSGI community, including (1) students, (2) working professionals who want to gain new skills or deepen their skills in the HSGI space, and (3) educators looking for resources to teach HSGI courses. Currently, within the US intelligence community, the value of professionals who have training in intelligence analysis and applications of geospatial tools cannot be overstated. With the rapid expansion and availability of geospatial data and tools and their integration with high-performance computing, this skill set will continue to be a high-value asset for students and professionals at all career stages. This book seeks to ease the current shortage of training resources and curricular materials in support of HSGI learning.

As the editors of this volume, we have extensive expertise and experience in geospatial technologies and HSGI:

- Darren Martin Ruddell, PhD and GISP, is a professor, director of graduate studies, and codirector of the Human Security and Geospatial Intelligence Lab at the University of Southern California's Spatial Sciences Institute, where he teaches and conducts research on HSGI issues. Ruddell earned his PhD in geography from the School of Geographical Sciences and Urban Planning at Arizona State University and is a geographic information systems professional (GISP) certified by the GIS Certification Institute (GISCI). He has served in multiple leadership positions at USC and beyond, working to advance geospatial education, promote interdisciplinary research, and facilitate partnerships among industry, government, and academia.
- Diana Ter-Ghazaryan, PhD and GISP, is an associate professor of spatial sciences, director of undergraduate studies, and codirector of the Human Security and Geospatial Intelligence Lab at the University of Southern California's Spatial Sciences Institute, where she teaches and

conducts research on issues of geopolitics in the context of HSGI. Ter-Ghazaryan earned her PhD in international relations and geography, with a specialization in GIS and remote sensing, from Florida International University and is a GISP certified by the GISCI. Before joining USC, Ter-Ghazaryan was at Florida International University, where she worked on various GEOINT research projects focusing on Latin America.

In our combined experience teaching a variety of HSGI courses, we have encountered a lack of resources for technical training using geospatial technologies on HSGI topics—hence this book.

This book is structured as a series of exercise workflows designed to give readers a solid grounding in applied tools and concepts on issues of human security. The workflows cover a range of topics—including disaster management, social justice and human rights, climate challenges, safer seas, and global security—that reflect the scope of HSGI today. At the end of each workflow, you will be prompted to produce a three- to five-page intelligence brief.

Intelligence writing

There are major differences between academic writing and intelligence writing. One of the goals of this book is to introduce and help learners gain an understanding of intelligence writing and then implement best practices to develop experience and confidence in intelligence writing.

For comparison, academic writing generally follows a logical flow and progresses through the following sections: introduction, background, methods and data, results, and conclusions. In academic writing, the key findings and conclusions are presented at the end of the document. In contrast, intelligence writing begins with the most important concept and then progresses to supporting content. Intelligence writing generally uses the following format: bottom line up front (BLUF), a communication technique that gives the main idea first; key judgements; source summary; technical analysis; strategic analysis; and conclusions.

To summarize, intelligence writing is characterized as follows:

- Written for generalists facing real problems
- Provides meaningful insights to help inform and guide decision-makers
- Begins with the bottom line and then explores the implications
- Generally shorter than five pages because anything longer is unlikely to be read

Intelligence products and delivery formats

Producing an intelligence product—whether for a local emergency response scenario, a national security response scenario, or a scenario at an international scale—requires the investigation of specific spatial and temporal data to understand the challenge and potential threats at hand. Intelligence products almost always incorporate maps, charts, infographics, or other visualization techniques that help present the analysis and provide context. The geospatial intelligence (GEOINT) product can take a variety of forms. The most fundamental GEOINT product is a single static map. In contrast, dynamic web maps or dashboards incorporating real-time data display and analytics represent the cutting edge of GEOINT products.

Regardless of the GEOINT product form, it can be delivered through an oral presentation (that is, an intelligence brief) or in the form of a written report (that is, an intelligence report). The preferred GEOINT product and delivery format will often be determined by the client or decision-maker leading the project. The intelligence producer (you) should consider the audience's experience, knowledge, or familiarity with the subject of investigation and their experience or knowledge of geospatial technologies and principles. Their level of knowledge will determine how to best communicate your analysis for their understanding. This will help determine how technically explicit your explanation needs to be regarding findings, recommendations, or analytic methodologies.

Intelligence briefing type

Three broad categories of analytic insight can be derived and delivered in an intelligence brief or intelligence report:

- **Information:** The informational briefing type aims to provide situational context and understanding of a given scenario.
- **Guidance:** This analysis examines the present and future forecast to project the most likely path for a scenario to develop.
- **Recommendation:** This briefing type offers recommendations in the form of a solution or mitigation effort to alter developments to the most favorable conditions.

BLUF

Your task is to deliver scientifically sound, actionable information in a timely and accessible way so that decision-makers can make the most informed decision possible, considering the current situational understanding. Grounded in concision and precision of writing, the BLUF format helps ensure that the customer receives the most crucial information first before diving into evidence and deeper analytic underpinnings to support the conclusions.

The following best practices provide an effective BLUF first paragraph:

- Do not exceed five sentences.
- Include estimative language in your assessments.
- Briefly answer who, what, when, where, why, and how.
- Summarize key points made throughout the GEOINT product.
- Provide recommendations.
- Expand on the information in subsequent sections.

Confidence reporting

An essential aspect of an intelligence brief or intelligence report is a statement on the confidence level that analysts have in their findings. Confidence reporting can be performed on a three-point scale (low, medium, high) or using a numeric rating (5–95 percent).

Confidence reporting is an objective way for an analyst to describe any known limitations in the data or methods employed in the analysis. Limitations could include weaknesses in data availability or limitations in spatial, spectral, or temporal resolutions.

How to use this book

Security First: Geospatial Workflows for a Safe and Equitable World has been tested for compatibility with ArcGIS Pro 3.4.

Chapters vary in their reach and scope. In general, at least an introductory knowledge of geographic information systems (GIS) is helpful, and some chapters benefit from an intermediate knowledge of GIS and some Python programming.

Hardware and software requirements

To perform the tutorials in this book, you'll need the following: ArcGIS Pro 3.4 installed on a computer that's running the Windows operating system, an internet connection, Microsoft Excel, and a web browser to access ArcGIS Online. Earlier software versions may not be fully compatible with the data and may not operate as described in the tutorials. The hardware requirements for ArcGIS Pro are listed at links.esri.com/SysReqs.

Information on software trial options, as well as Personal Use and Student Use licensing, can be found at esri.com.

Licensing

Use an existing license

If you have existing credentials or can obtain credentials from your educational institution or organization, you can use those credentials to obtain the required software listed in this section.

Purchase new credentials

If you don't have credentials, purchase an ArcGIS for Personal Use license at links.esri.com/PersonalUseLicense.

Downloading the exercise data

The tutorial data for this book is available on ArcGIS Online.

1. Go to links.esri.com/SecurityFirstData.
2. Download the tutorial data and store it on your computer.

   All the tutorial data that accompanies this book is covered by a license agreement that stipulates the terms of use. To view the license agreement, go to links.esri.com/LicenseAgreement.

Special font styles used

In this book, student input for a work step is designated by colored highlighted text. The computer user interface elements are styled in **bold font**.

Resources

Visit links.esri.com/Help for Esri-curated ArcGIS Pro Help documentation that can answer many of your questions and provide in-depth details and troubleshooting tips.

Esri Community brings together thousands of Esri® users worldwide. Users can find or leave feedback, obtain updates, and collaborate with others at links.esri.com/EsriPressCommunity. Feel free to share your experiences or questions about this book.

Additional information is available on the book's web page at links.esri.com/SecurityFirst.

Part 1

Disaster management

Chapter 1

Applying the US National Grid in emergency response

Justice Batiste, *University of Southern California, Spatial Sciences Institute*
Joseph Cañas, *University of Southern California, Spatial Sciences Institute*
Tristan Pekron, *University of Southern California, Spatial Sciences Institute*

Overview

In this chapter, you will learn how to implement geographic information science and technology (GIST) tools and methods to meet the needs of emergency first responders and incident commanders during natural disasters or public safety incidents. You will learn how the US National Grid (USNG) system can be applied as a systematic approach to assist in imagery tasking, collection, processing, exploitation, and dissemination (TCPED) specifically during these emergency scenarios. The workflow presented here provides an objective approach to triage, using the USNG and other prioritization parameters to identify which geographic locations are most in need of emergency support, making emergency response more efficient and fostering community resilience.

Learning objectives

- Integrate geospatial data into a gridded system to identify at-risk areas during natural disasters.
- Systematically triage geographic areas of interest for emergency response operations.
- Identify critical information needs to support emergency support functions and develop spatially informed and data-driven collection plans to gather this information.

Technical components

- Software: ArcGIS Pro 3.4
- Datasets:
  - US National Grid
  - WorldPop United States Population 2018 100 m
  - California Historical Fire Perimeters
  - PopInButteAndPlumas

Prerequisites

- Some experience with GIS software is recommended but not required.
- Technical skill level: Introductory GIS.
- Time to complete chapter: Two to three hours without intelligence brief, three to five hours with intelligence brief.

Background and situational context

In emergency management, the time-sensitive nature of interventions is crucial. Such imperatives underscore the importance of emergency support functions (ESFs), mechanisms that methodically organize and facilitate assistance during national incidents by classifying capabilities and resources into categorical units. Central to these ESFs is the rapid assessment of affected areas following a catastrophic event.

Traditionally, damage assessments were conducted by first responders and augmented by crewed aircraft and satellite imagery. However, advancements in technological capabilities have ushered in tools such as uncrewed aerial systems (UAS). Although aircraft and satellite imagery have a well-established history in disaster management, encompassing mitigation, response, and recovery, UAS-derived imagery represents a nascent yet transformative evolution in this field.

The commercial accessibility of UAS technology offers government entities an opportunity to employ a cost-efficient methodology for surveying disaster-afflicted zones. In addition, the nimbleness of UAS deployment augments real-time situational awareness, positioning them as valuable assets within the national disaster response framework.

State-led integration of UAS within these frameworks holds considerable promise, especially within the overlapping dimensions of the response and recovery phases of emergency management. For instance, urban search and rescue (US&R) units, when equipped with UAS support, have the potential not only for expedited lifesaving interventions but also for the simultaneous collection of crucial data regarding the nature and scale of the damages. This enhances prioritization strategies for subsequent assessments by specialized personnel. Nonetheless, the integration of UAS introduces complexities, especially when considering potential airspace conflicts with other resources, such as helicopters. Mitigating such risks requires the implementation of structured UAS flight coordination protocols.

This workflow presents a methodology to use the US National Grid (USNG) as a foundational tool to facilitate the coordinated deployment of UAS, crewed aircraft, and satellites during disaster response operations. Similar to the Military Grid Reference System (MGRS), the USNG system offers a standardized location lexicon, having been adopted by the Federal Emergency Management Agency (FEMA) as the preferred grid reference system for operational maps. Beyond serving as a universal location reference system, the segmentation inherent to the grid offers a viable framework for UAS mission planning in addition to airspace deconfliction with crewed aircraft. In tandem, the USNG can also be used to improve the storage of the vast quantities of imagery collected during disaster response. For example, the grid references could serve as nonunique indexes within mosaic datasets. This exploration also extends into dasymetric population mapping—a technique that, when combined with the USNG, could support the decision-making behind imagery collection and the efficient allocation of resources in disaster-affected areas.

Exercise workflow 1

1 Visualize the extent of a disaster

In this step, you will investigate the 2018 Camp Fire in Butte County, California.

1. In your browser, navigate to www.livingatlas.arcgis.com and search for California Historical Fire Perimeters.

2. Click **California Historical Fire Perimeters** to view the item details.

3. Scroll down to find the **URL** section on the right side of the page. Click the **Copy** button next to the **URL**.

4. Open ArcGIS Pro. Under **New Project**, click **Map**.
5. In the **New Project** dialog box, apply the following settings:
   - **Name**: Chapter_01
   - **Location**: C:\SecurityFirst

   Note: This workflow will use C:\SecurityFirst.

6. On the ribbon, click the **Map** tab. In the **Layer** group, click **Add Data From Path.**
7. Under **Path**, paste the URL that you copied earlier. Click **Add.**
8. In the **Selection** group, click the **Select By Attributes** tool.
9. In the **Select By Attributes** window, enter the following:
   - **Input Rows:** California Fire Perimeters (all)
   - **Selection Type**: New selection
   - **Expression**: Where Year is equal to 2018
   - **Expression:** And Fire Name is equal to CAMP

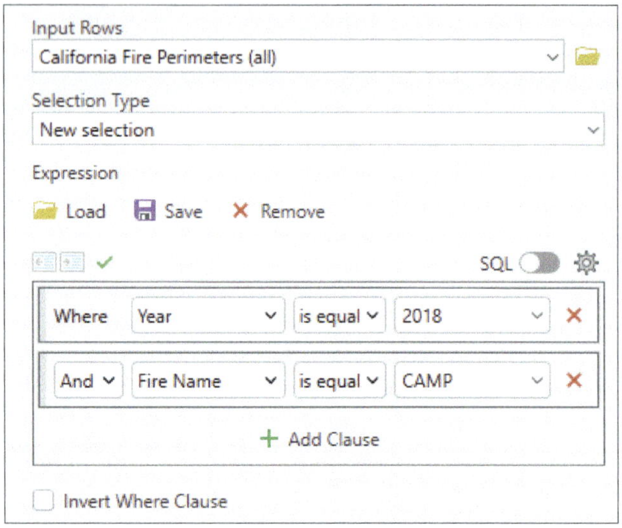

10. Click **OK.**

    The Camp Fire boundary is highlighted on the map.

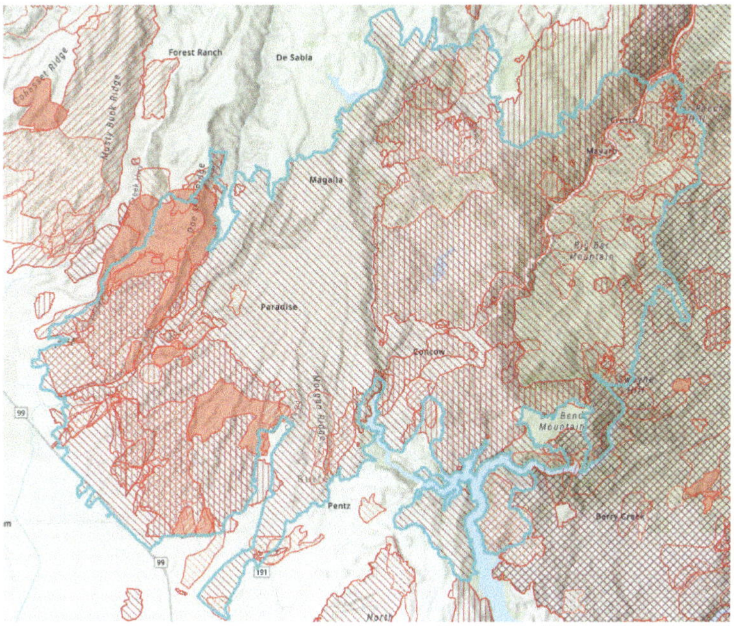

11. In the **Contents** pane, expand the **California_Historic_Fire_Perimeters** layer and right-click **California Fire Perimeters (all)**.
12. Click **Selection** > **Make Layer From Selected Features**.
13. On the ribbon, click the **Analysis** tab. In the **Geoprocessing** group, click **Tools**.
14. In the **Geoprocessing** pane, search for Copy Features and select the tool. On the **Parameters** tab, specify the following settings:
    - **Input Features:** California Fire Perimeters (all) selection
    - **Output Feature Class:** CampFireBoundary
15. Click the **Environments** tab. Next to **Output Coordinate System**, click the **Select coordinate system** button.
16. In the **Output Coordinate System** window, navigate to **Projected Coordinate System** > **State Plane** > **NAD 1983 (2011) (US Feet)** > **NAD 1983 (2011) StatePlane California II FIPS 0402 (US Feet)**.
17. Select the coordinate system and click **OK**.

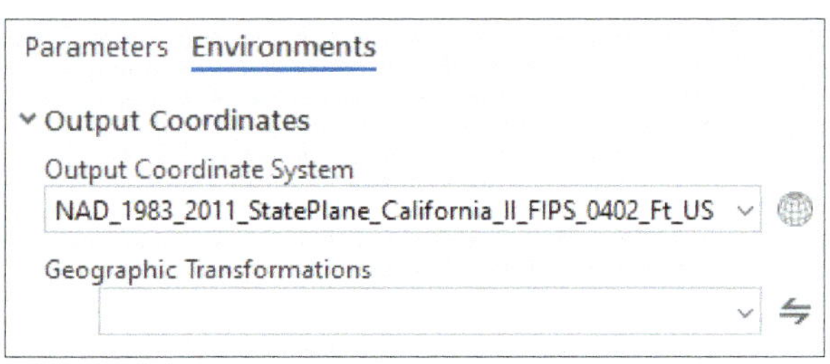

18. Click **Run**.

Tip: Using the state plane coordinate system minimizes distortions. This step may not be crucial in time-restricted workflows but can affect analytic results. You will use Zone II because the vast majority of the disaster extent occurred in Butte County.

2 Obtain the grid system

In this step, you will use the USNG. The purpose of the grid system is to partition the disaster extent into focus areas that can be prioritized based on criteria such as population density, critical infrastructure, natural resources, and so on.

1. In the **Catalog** pane, click the **Portal** tab and click the **Living Atlas** cloud icon to search ArcGIS Living Atlas of the World.
2. In the search bar, type U.S. National Grid. Right-click **U.S. National Grid** and click **Add To Current Map.**

   The grid sizes range from 100 meters to 100,000 meters and beyond. Based on the extent of the fire, the 1,000-meter grid size is the most useful for partitioning the disaster extent into priority areas.

The 10,000-meter grid covers too large an area to be useful for prioritization and rescue. The 1,000-meter grid allows you to prioritize areas based on rescue or intelligence criteria. As you did with the fire boundary data, you will filter and export the data from ArcGIS Living Atlas of the World and add it to your local geodatabase. You will filter the grids by selecting the ones that intersect the Camp Fire boundary.

3. In the **Contents** pane, expand the **U.S. National Grid** layer and select **USNG 1000m.**
4. On the **Feature Layer** tab, in the **Visibility Range** group, change the **Minimum Scale** to **1:500,000.**
5. On the **Map** tab, in the **Selection** group, click **Select By Location** and enter the following settings:
   - **Input Features:** USNG 1000m
   - **Relationship:** Intersect
   - **Selecting Features:** CampFireBoundary
6. Click **OK.**
7. Right-click **USNG 1000m** and click **Zoom To Make Visible.** Pan around the fire boundary to confirm that the layer is properly selected.

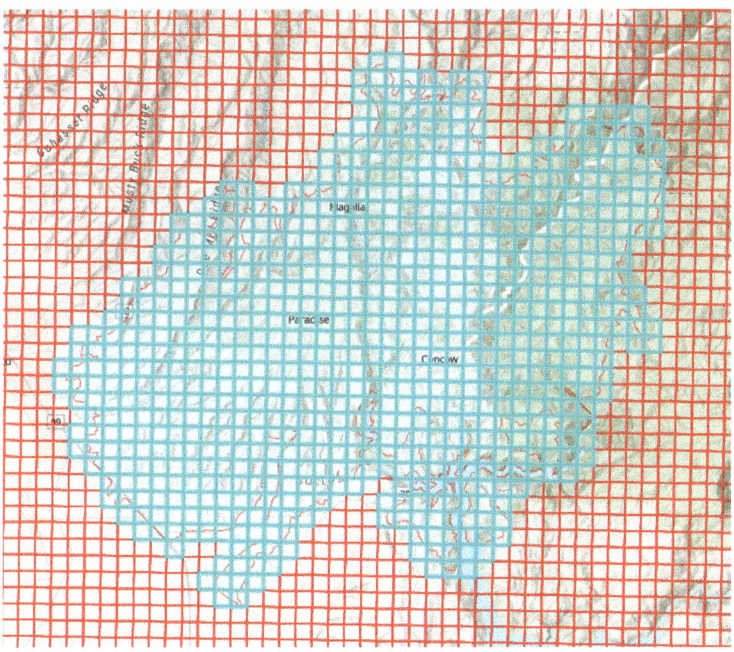

8. Using the **Export Features** tool, export the selected data to your geodatabase using the following settings:
   - **Input Features:** USNG 1000m
   - **Output Feature Class:** USNG1000mExport
   - **Output Coordinate System (Environments):** NAD_1983_2011_StatePlane_California_II_FIPS_0402_Ft_US

9. Run the tool.
10. In the **Contents** pane, turn off all layers except **USNG1000mExport** and **CampFireBoundary**.

    Now you have only the pertinent 1,000-meter USNGs that are coincident with the Camp Fire perimeter.

3 Obtain population data

In this step, you will analyze population within the Camp Fire footprint to prioritize areas within the disaster extent for rescue.

Because the population data spans multiple counties, you will trim it down to the focus area (that is, the disaster extent). Using the **Dissolve** tool to dissolve the individual grids while retaining the border of the grid polygon, you will create one large polygon that encompasses all the 1,000-meter grids.

1. In the **Geoprocessing** pane, search for Dissolve. Click the **Dissolve** tool and enter the following settings:
    - **Input Features**: USNG1000mExport
    - **Output Feature Class**: USNG1000mDissolve
    - **Dissolve Fields**: Universal Transverse Mercator (UTM) Zone
2. Check the box for **Create multipart features**.

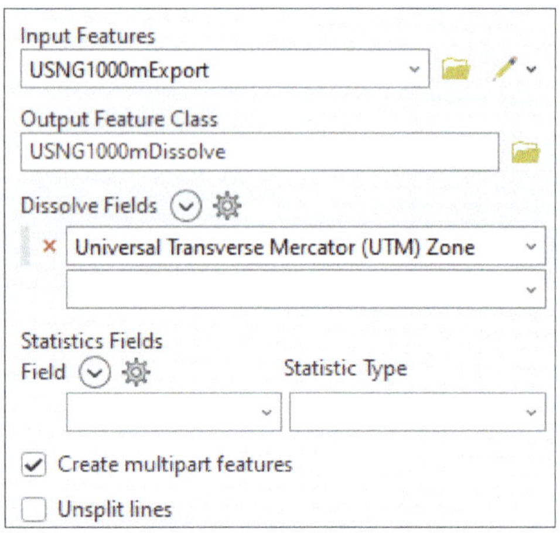

3. Click **Run.**

    For **Dissolve Fields**, you should choose fields in which all the values are the same. In this case, every grid square is in UTM Zone 10. So, when you dissolve by this field, you will have one large polygon dissolved by the UTM Zone field. Alternatively, you could dissolve by the Grid 100,000m or Grid 1,000,000m fields because they are all the same.

    Before running the **Dissolve** tool, each grid is shown.

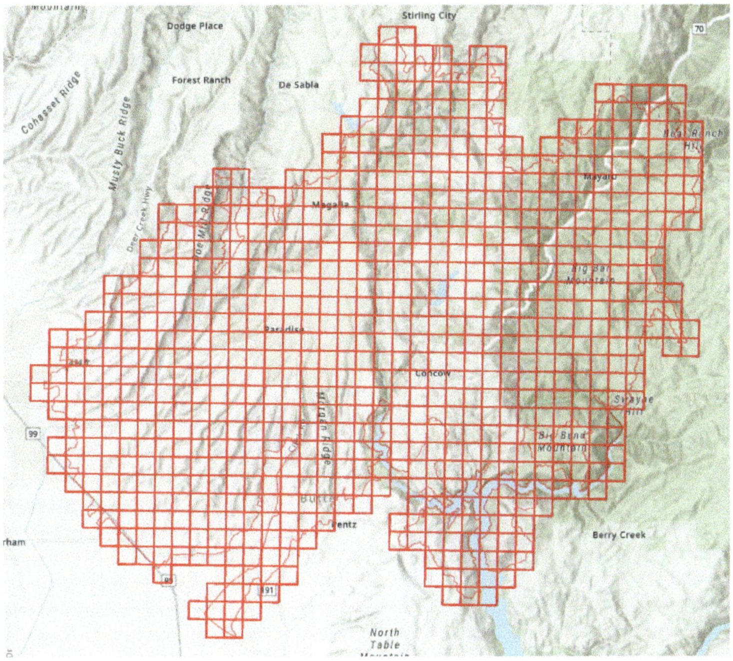

After running the **Dissolve** tool, you have a new layer that keeps only the outline of the grids. You have dissolved each of the individual grids into one large polygon.

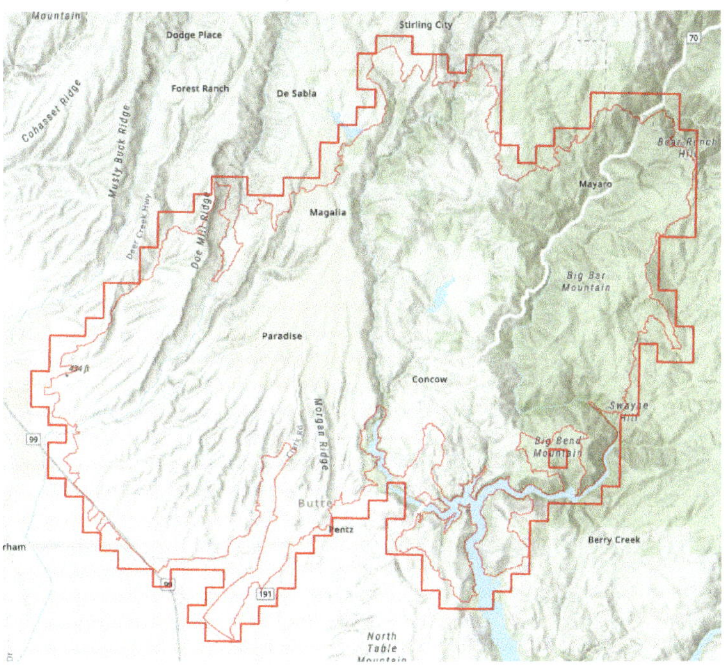

This dissolved USNG polygon will serve as the extent to which you clip the population raster.

The population raster dataset used in this exercise is located in the **SecurityFirst_ExerciseData** folder.

4. In the **Catalog** pane, click the **Computer** tab and navigate to the location of the **SecurityFirst_ExerciseData** folder. Expand this folder and open the **Chapter01** folder. Right-click **PopInButteAndPlumas.tif** and click **Add To Current Map**.

   This population dataset is a raster representing the number of people living in Butte and Plumas Counties, where the fire occurred. The resolution of this raster is 100 meters, which means that it is an estimate of the people residing in a 100- by-100-meter pixel.

   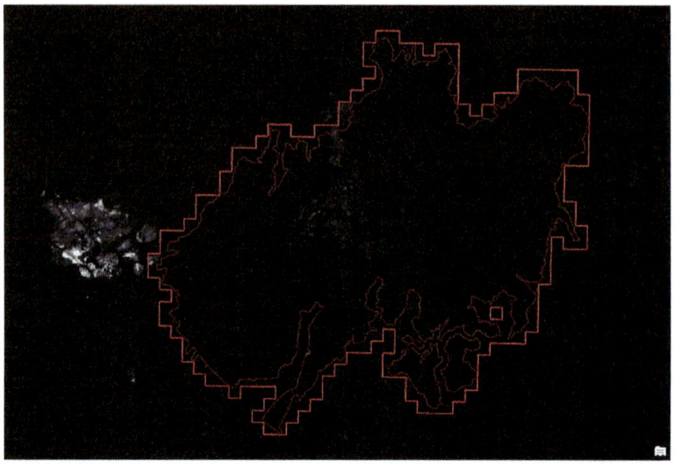

   The town of Paradise and the city of Chico are coincident with high population value pixels.

5. In the **Geoprocessing** pane, search for Extract by Mask. Click the tool and enter the following settings:

   - **Input raster:** PopInButteAndPlumas.tif
   - **Input raster or feature mask data**: USNG1000mDissolve
   - **Output raster:** PopInCampFire
   - **Extraction Area**: Inside

6. On the **Environments** tab, set the **Output Coordinate System** to NAD_1983_2011_StatePlane_California_II_FIPS_0402_Ft_US.

7. Click **Run**.

8. In the **Contents** pane, turn off the **PopInButteAndPlumas.tif** layer. Right-click the **PopInCampFire** layer and click **Symbology**.

9. In the **Symbology** pane, open the **Color scheme** list and check the box for **Show names**. Scroll down and select the **Shadow to Sunshine** theme.

   You have now clipped the **PopInButteAndPlumas** population dataset to the USNG boundaries. The higher population values indicated by yellow and green coincide with the town of Paradise.

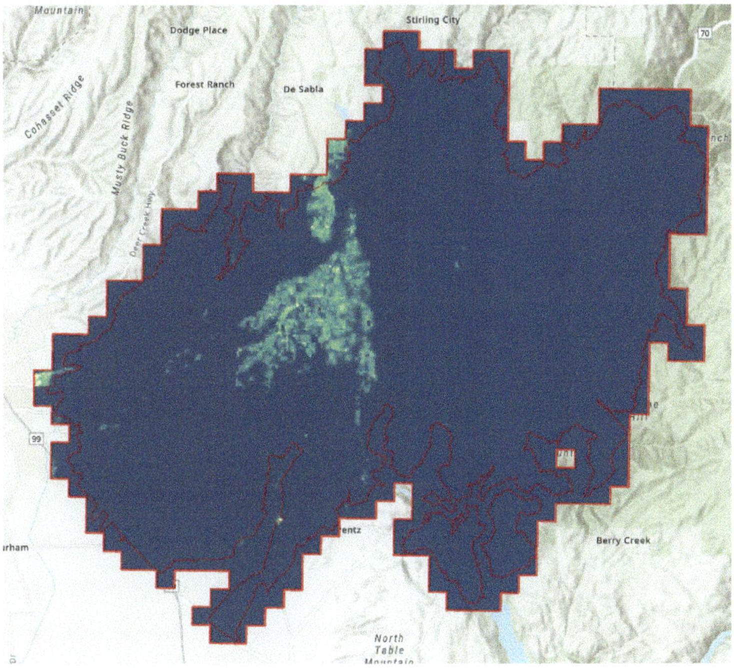

4 Perform the analysis

Now that you have all the data (the grid system and the population raster, both clipped to the disaster extent), you are ready to perform the analysis. This entails calculating the sum of the population that exists within each USNG pixel. To clarify, you will use ArcGIS Pro to count all the 100-meter population pixels that exist within each 1,000-meter USNG.

You will sum the population pixels using the **Zonal Statistics** tool.

> **Tip:** The Zonal Statistics tool works by calculating statistics for an input raster within specific zones.

1. In the **Geoprocessing** pane, search for Zonal Statistics (Spatial Analyst Tools). Click the tool and apply the following settings:
   - **Input Raster or Feature Zone Data:** USNG1000mExport
   - **Zone Field:** United States National Grid (USNG)
   - **Input Value Raster:** PopInCampFire
   - **Output Raster:** ZonalSt_USNG1
   - **Statistics Type:** Sum
2. Click **Run**.

   The next step will be to update the symbology to something intuitive—for example, making the most populated grids a deep red.

3. For the new **ZonalSt_USNG1** layer, open the **Symbology** pane. Change the **Primary symbology** to **Classify**. Change the **Color scheme** to **Yellow to Red**.

   The output raster now displays the USNG according to population, with red to yellow having high to low population values, respectively.

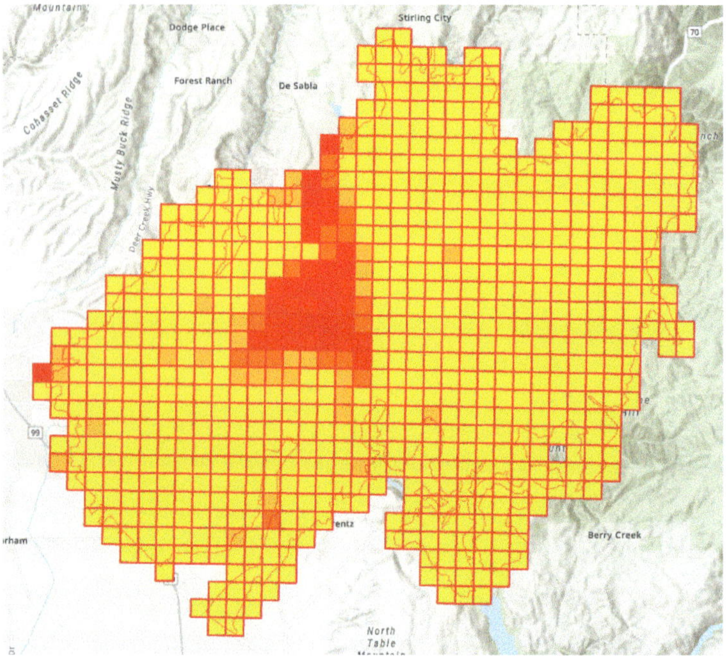

4. In the **Contents** pane, double-click the **ZonalSt_USNG1** layer and rename it Population per 1000m grid.
5. Under **USNG1000mExport**, click the red box outline. In the **Symbology** pane, change the **Polygon Symbol** to **Black Outline (1pt)**.

   First responders can prioritize, by population, sections in the disaster area for search and rescue. Zooming to the town of Paradise reveals the coincident USNG squares with their individual nomenclature. Communicating these areas through the unique USNG identifiers can expedite the federal response from agencies such as FEMA.

Interpretation of analysis

After completing this exercise, answer the following questions in detail. The answers to these questions, as well as the analysis results from the exercise workflow, will inform the intelligence brief.

1. The workflow used three datasets from ArcGIS Living Atlas of the World to prioritize areas for search, rescue, and disaster assessment according to a set of criteria. In this case, population was used as the criterion, where areas with more people are prioritized for disaster response. What are some other relevant criteria that can be used to prioritize search, rescue, and post-disaster assessment activities? Sample answer: critical infrastructure or areas with sensitive natural resources.
2. In the context of postdisaster damage assessment, what are the challenges and limitations of using geospatial data? How can integrating data from multiple sources (for example, satellite, UAV, and crowdsourced data) enhance the accuracy of such assessments?
3. How can machine learning techniques be applied to geospatial data for automating postdisaster damage assessment? Conduct an online search to find the key datasets and models in this realm. Given that many entities developing models for this purpose are private organizations, what are the potential ethical considerations?

Intelligence brief

Based on the analysis from the exercise workflow, produce a three- to five-page intelligence brief by responding to the following question prompts. Incorporate your geospatial outputs from the exercise to enrich and illustrate your conclusions.

- **BLUF:** List three to five primary observations from your analysis in list format that summarize findings of this work.
- **Top-level analysis:** Report the primary data points supporting your BLUF.
- **Secondary analysis:** Provide the data sources and deeper analytic products from this analysis.
- **Impact:** Describe the impacts of your findings on the department or organization.
- **Confidence:** Report the level of confidence of your assessment based on available data, accuracy, and so on.
- **Bias:** Explain any background or information you have that could have affected your analysis.
- **Conclusions:** Determine what type of analytic conclusions there may be and how the data requirements, analysis, and brief writing may change based on what analytic framework you are following (descriptive, predictive, prescriptive).

Applying and extending your skills

Now that you have completed the workflow using the USNG and population density to derive areas that need prioritized imagery collection and produced an intelligence report highlighting the conclusions, you will identify the next steps for helping the town of Paradise cope with the impact of this wildfire and examine how geospatial technologies can be applied to address these tasks.

One important way to expand this workflow is through automation. Using ModelBuilder™ in ArcGIS Pro, you can create a series of steps that will automate the entire workflow when provided with the necessary inputs: boundaries (for example, disaster extent), prioritization criterion (for example, population data), and grid system (for example, USNG).

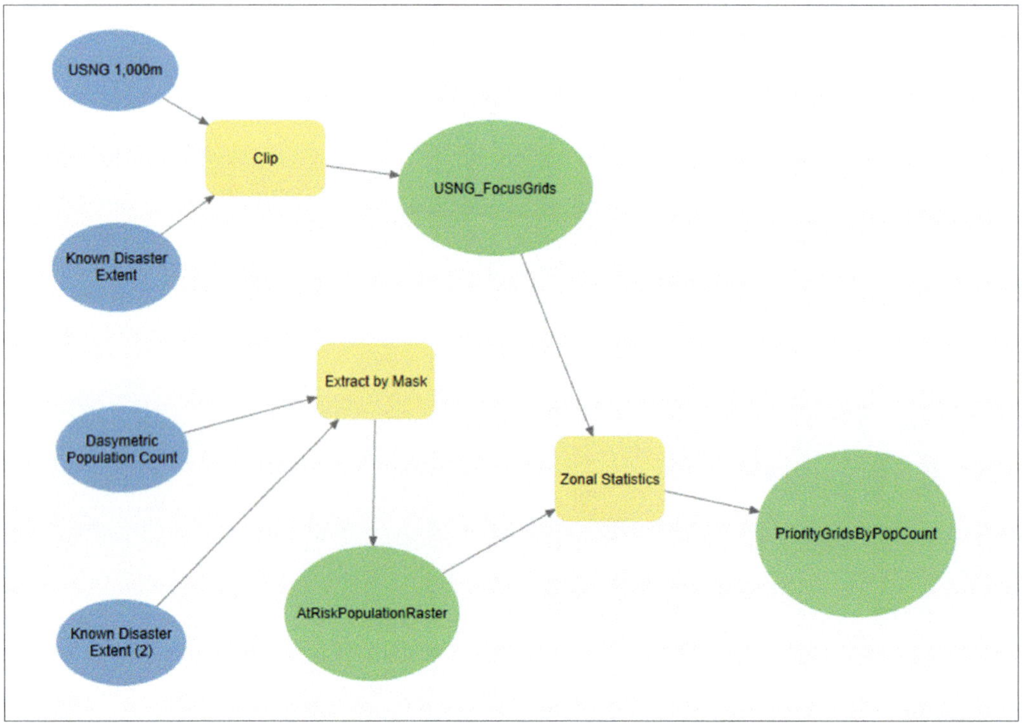

Postdisaster imagery exploitation can help speed the process of damage assessment—pertinent information to homeowners and insurance companies alike. Building footprint data is a critical source of information from which damage assessments can be made and spatially visualized for follow-up rebuilding efforts. However, up-to-date building footprint data may not be available for a given disaster area. When imagery of these areas has been prioritized and collected, artificial intelligence (AI) can help identify building footprints and damage through image feature extraction and detection. These deep learning models are becoming more available and accessible. ArcGIS Living Atlas contains a range of pretrained deep learning models that can assist in this regard.

Building Footprint Extraction—USA (links.esri.com/BldgFootprintExtract) is one such deep learning package available in ArcGIS Living Atlas. It can be combined with National Agriculture Imagery Program (NAIP) imagery of Butte County to create up-to-date building footprint data.

If a previously created building footprint layer is available, comparing this layer with the generated layer from the recent postdisaster imagery may help determine anomalies. If many buildings are damaged to the point that the pretrained model cannot detect the buildings, this layer can act as a sort of preliminary damage assessment.

Damage, and its subsequent appearance through aerial imagery sensors, is not congruent across all disasters. A burned building will have a much different appearance from building damage caused by a hurricane. Having a robust assortment of pretrained damage assessment models that can be used to meet a variety of disasters can help communities respond to and recover from these events more efficiently. Currently, using AI for building damage assessments is not openly available, although there are some proprietary solutions, such as the DamageMap application created by researchers from Stanford and California Polytechnic State University (Galanis et al. 2021). In the future, relying on AI through pretrained models will be fully integrated into the emergency response workflow, as would the USNG for imagery prioritization, tasking, and resource allocation.

How might these tools change your analysis, conclusions, and the way you write your intelligence report? What additional insights or benefits could be gained from integrating AI models into the analysis for emergency response broadly and wildfire response specifically?

Additional resources

- U.S. National Grid Information Center (https://usngcenter.org). This website contains information about the USNG, training resources, and so on.
- GeoCONOPS (https://geoconops-geoplatform.hub.arcgis.com). The Department of Homeland Security Geospatial Concept of Operations is a strategic road map for how geospatial technology and data can be implemented for disaster scenarios. The sixth version of the document can be found at https://publicintelligence.net/dhs-geoconops-v6/.
- NAPSG Foundation (www.napsgfoundation.org). This organization promotes cross-agency adoption of GIS for public safety.
- Wildland Urban Interface (WUI) Look-up Tool (links.esri.com/WUILookUpTool). This web map shows the wildland urban interface in and around Butte County.

References

- CAL FIRE. “Remembering the Camp Fire: The Deadliest and Most Destructive Wildfire in California’s History.” www.fire.ca.gov/our-impact/remembering-the-camp-fire.
- Galanis, Marios, Krishna Rao, Xinle Yao, Yi-Lin Tsai, Jonathan Ventura, and G. Andrew Fricker. 2021. “DamageMap: A Post-Wildfire Damaged Buildings Classifier.” *International Journal of Disaster Risk Reduction* 65 (November). doi.org/10.1016/j.ijdrr.2021.102540.

About the authors

Justice Batiste is a veteran of the US Coast Guard and a solution engineer at Esri. During his time with the Armed Forces, he developed a skill set in the geospatial and emergency response disciplines, which he continues to apply on the job and as a Coast Guard reservist.

Joseph Cañas is a GIS specialist with a master of science in geographic information science and technology from the University of Southern California. He has a background in utility permitting, spatial analysis, and geodatabase management, with experience spanning federal land management and the energy sector. His work emphasizes the use of remote sensing imagery and ArcGIS tools to support infrastructure planning, vegetation management, and wildfire mitigation.

Tristan Pekron recently completed his master's degree in human security and geospatial intelligence at the University of Southern California. He has experience as a geospatial operations analyst and an imagery analyst and enjoys applying the skills and knowledge from his graduate studies to meet real-world challenges.

Chapter 2

Geospatial intelligence for disaster management: Lessons from Hurricane Dorian

Jacob Spear, *US Marine Corps*

Darren Martin Ruddell, *University of Southern California, Spatial Sciences Institute*

Overview

In this chapter, you will learn how to use GIST tools and methods in the context of a human security and geospatial intelligence (HSGI) threat to enhance disaster preparedness and response. The workflow you follow and the final deliverable produced will provide information to help guide strategic decision-making, facilitate timely interventions, and mitigate the impact of tropical cyclones and other related hazards.

Learning objectives

- Explain the role of geospatial intelligence in tracking and monitoring tropical cyclones.
- Explain how geospatial applications and analysis can provide valuable insights for decision-making in disaster management.
- Present practical examples of using geospatial data to support preparedness activities for tropical cyclones.
- Help beginners in the HSGI field understand the potential and importance of geospatial intelligence in disaster management and tropical cyclone tracking.

Technical components

- Software: ArcGIS Pro 3.4, Google Colab (*optional*)
- Datasets:
  - Track, Intensity, and Forecast Advisory 31 Data: National Hurricane Center GIS Archive
  - Population: ArcGIS Online
  - Points of Interest, Transportation, Schools, and Healthcare Facilities: OCHA Humanitarian Data Exchange (HDX)

Prerequisites

- Some experience with GIS software or spatial data is helpful but not required.
- Technical skill level: Introductory GIS.
- Time to complete chapter: One to two hours without intelligence brief, three to five hours with intelligence brief.

Background and situational context

In 2019, Hurricane Dorian developed into a powerful Category 5 storm, causing catastrophic damage to the Bahamas and then affecting the Carolina coast as a weaker hurricane.

Dorian caused catastrophic damage mainly in Abaco Island and the eastern Grand Bahama Islands, resulting in about $3.4 billion (USD) in total. The Bahamas asked the Inter-American Development Bank (IDB) to conduct a study following Dorian's trail of destruction. The IDB stated that the hurricane left 29,500 people homeless or jobless. Abaco was hardest hit, suffering 87 percent of the damage. More than 75 percent of all homes on the island were damaged. In the US, coastal and inland flooding affected homes and businesses in the Carolinas, with losses totaling $1.6 billion (USD).

The storm's intensity, trajectory, and magnitude posed a severe human security threat to communities. The destructive force of Hurricane Dorian highlighted the urgent need for effective disaster management strategies to mitigate the impact of such natural disasters in the future.

Geospatial technologies are pivotal in this type of analysis, providing tools and techniques to process and interpret complex geospatial data. With advanced geospatial tools, analysts and data scientists can conduct in-depth analyses of a region's resources to better understand population, transportation, and critical infrastructure.

The primary objective of this chapter is to demonstrate how geospatial analysis can provide actionable intelligence for disaster management in the context of natural hazards (tropical cyclones, in this example). By presenting a detailed, step-by-step workflow, this chapter aims to assist aspiring HSGI students and professionals in conducting their geospatial analysis for tropical cyclone research. Through this analysis, researchers can identify crucial insights into the hurricane's behavior and its impact on different regions, informing more effective preparedness and response strategies to safeguard communities from similar human security threats in the future.

Exercise workflow 2

You will walk through a step-by-step guide to analyze the impacts of Hurricane Dorian. First, you will look at various data sources to identify vulnerable and potential areas for emergency response centers in preparation for the hurricane. You will then conduct postevent analysis to identify who was affected to help determine where resources should go.

The following workflow uses ArcGIS Pro; however, similar workflows may work with various GIS platforms, including ArcGIS Online and QGIS.

1 Add data to a map

You will first add data to a new map in ArcGIS Pro.

1. Open ArcGIS Pro. Under **New Project**, click **Map** to create a new map.
2. In the **New Project** dialog box, apply the following settings:
   - **Name**: Chapter_02
   - **Location**: C:\SecurityFirst

   Note: This tutorial uses the folder C:\SecurityFirst.

3. Click **OK**.
4. In the **Catalog** pane, click the **Computer** tab and navigate to the **SecurityFirst_ExerciseData** folder. Open the folder and right-click the **Chapter02** folder. Click **Add to Project**.
5. Click the **Project** tab and expand **Folders**. Expand **Chapter02**.

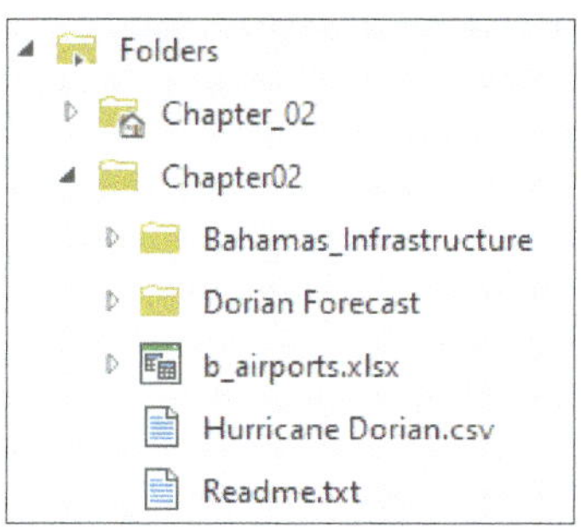

6. Expand the **Dorian Forecast** and **Bahamas_Infrastructure** folders. Add all the shapefiles to your map by either dragging the files to the map or right-clicking each file and selecting **Add To Current Map**.

   Your map should include nine layers, not counting the basemaps.

7. Right-click the **Hurricane Dorian.csv** file and click **Add To Current Map**.
8. Click the **Portal** tab and click the **ArcGIS Online** icon.
9. In the search bar, type bahamas_population.

10. Right-click **bahamas_population** and click **Add To Current Map**.
11. Save your map.

    At this point, your map will look cluttered and chaotic. The next set of steps will focus on symbolizing the data to help make the information on the map more legible.

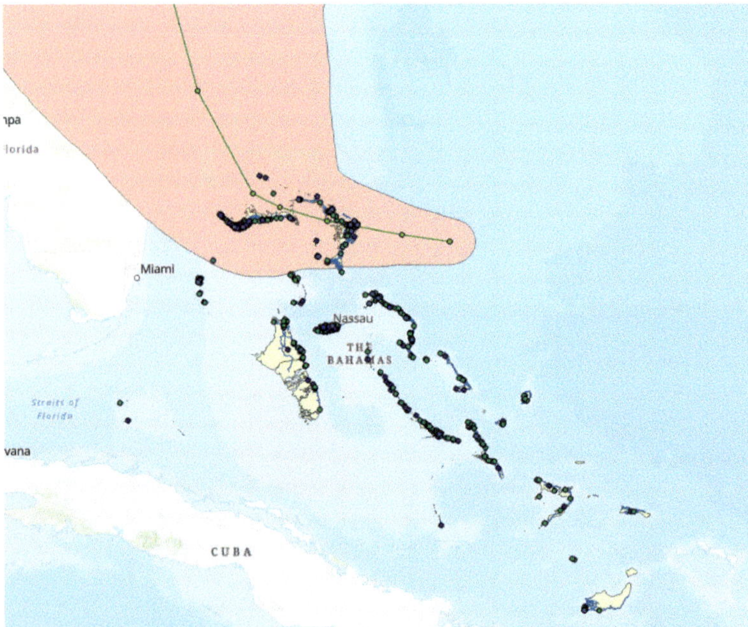

2 Customize symbology

1. In the **Contents** pane, right-click the **positions** layer and click **Symbology**.
2. In the **Symbology** pane, click the **Symbol** icon.

    The **Symbol Gallery** lets you search for predetermined symbols. You will use these preset symbols to symbolize the data in your map. To search, type the term in the search bar and press **Enter**.
3. Set the **Point Symbol** to **Cross 2**.
4. Repeat the preceding steps to symbolize the following layers using the recommended symbols:

Layer	Symbology
sea_ports	Maritime (Medium)
schools	School (Medium)
health_facilities	Hospital (Medium)
airports	Airport (Medium)
track	Limited Access Ramp
cone	Black Outline (2 pts)
roads	0.5 point

Your map will now have symbology that conveys the meaning of each layer, making information more accessible.

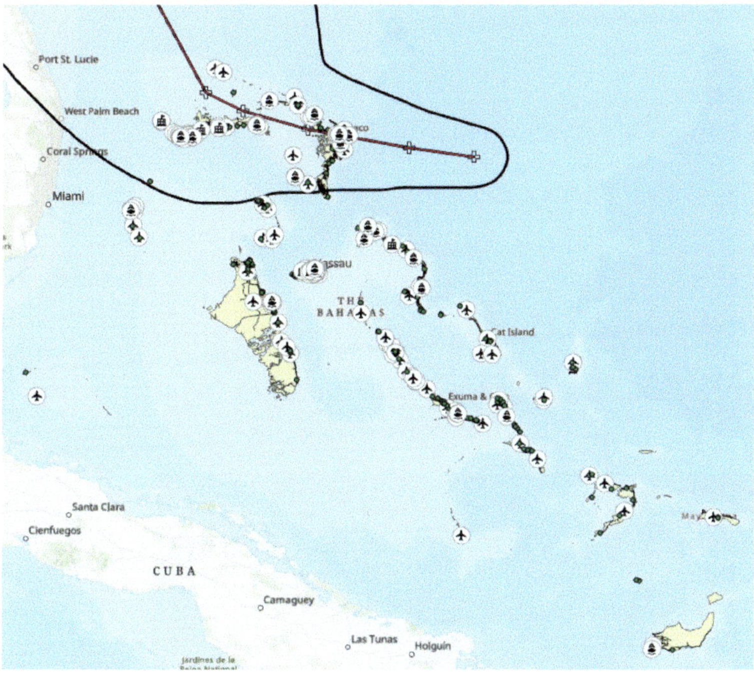

The **points_of_interest** layer will be symbolized differently to highlight features within the layer.

5. Open the **Symbology** pane for the **points_of_interest** layer.
6. In the **Symbology** pane, change the **Primary symbology** to **Unique Values**.
7. For **Field 1**, expand the drop-down list and click **amenity**.
8. Use the following table to set the values for select symbol classes by clicking the format symbol button. Leave all other symbol classes set to the default.

Value/Label	Symbol
bank	Bank Account (Medium)
community_centre	Group (Medium)
fire_station	Fire Station (Medium)
fuel	Gas Station (Medium)
library	Library (Medium)
marketplace	Grocery Store (Medium)
place_of_worship	Place of Worship (Medium)
police	Police Station (Medium)
post_office	Post Office (Medium)

9. In the **Contents** pane, expand the **bahamas_population** layer and click the box symbol to open the **Symbology** pane. Apply the following changes:
   - **Primary symbology:** Graduated Colors
   - **Field:** Pop_2010
   - **Method:** Geometric Interval
   - **Color scheme:** Yellow-Orange-Red (Continuous)

   Note: If your Symbology pane shows you the gallery, click the back arrow in the top left.

   > **Tip:** In the color scheme list, check the box for Show names to see the name of each color ramp.

10. On the ribbon, click the **Map** tab. In the **Layer** group, open the **Basemap** menu and select **Imagery Hybrid.**

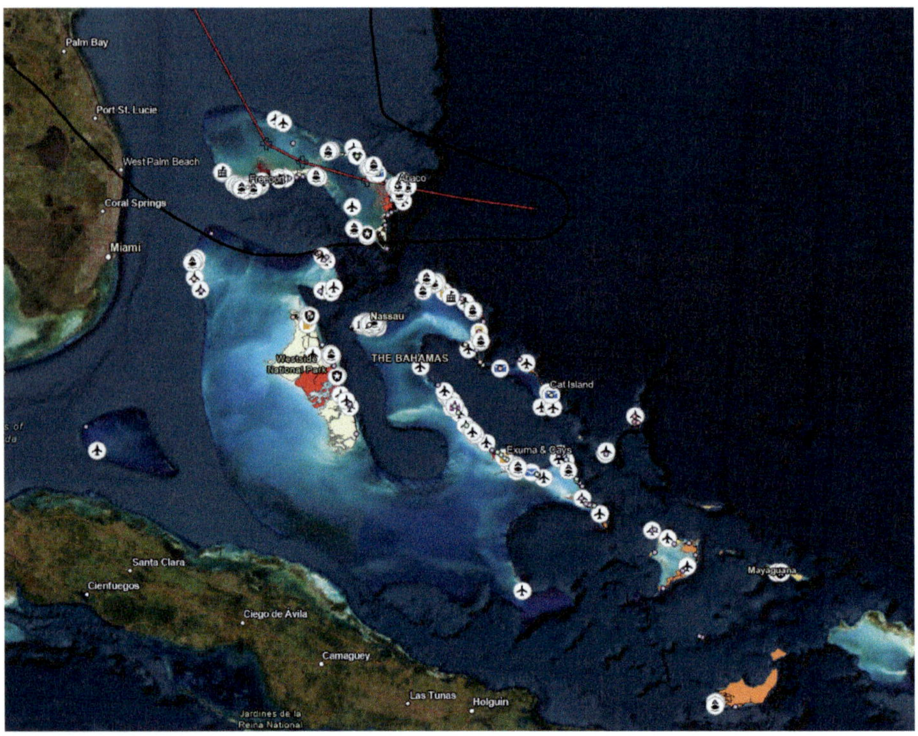

11. Save your project.

3 Perform a prehurricane analysis

Even after updating the symbology, the map is still a bit cluttered, and it is hard to identify observations or draw conclusions from the content provided. The next steps will focus on filtering, sorting, and analyzing the data to provide first responders and emergency planners with information that will help prepare the Bahamas for Hurricane Dorian.

1. In the **Contents** pane, right-click on the **points_of_interest** layer and click **Properties.**
2. In the **Layer Properties** window, click the **Definition Query** tab. Click **New definition query**.

   In the next steps, you will set a definition query. You can use definition queries to filter data that is visualized on the map while preserving the base layer. This is a useful technique when creating tailored products because it allows you to show what you need without deleting data unnecessarily.
3. For the first field, select **amenity**. For the second field, select **includes the value(s)**. For the third field, check the boxes for the following values:
   - **bank**
   - **community_centre**
   - **fire_station**
   - **fuel**
   - **library**
   - **marketplace**
   - **place_of_worship**
   - **police**
   - **post_office**

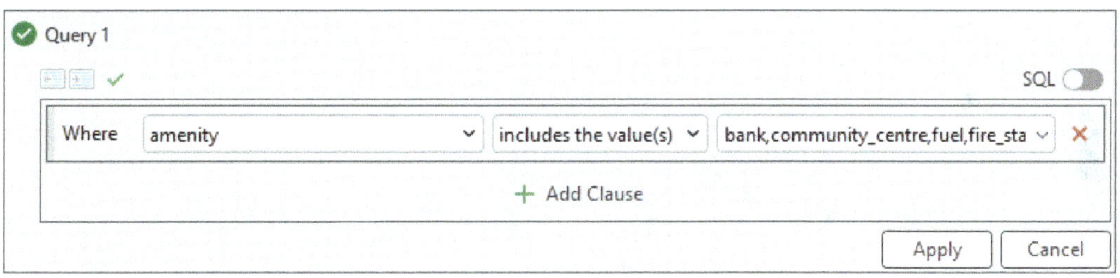

4. Click **Apply** and then click **OK.**
5. Repeat the preceding steps to set definition queries for the **roads** layer using the following query expression: **Where highway includes the value(s) primary, secondary.**
6. Save the project.
7. In the **Contents** pane, expand the **bahamas_population** layer. Right-click the **bahamas_population** sublayer and click **Attribute Table.** Explore the attribute table.

   The attribute table has information on place-names and other important census data to include estimated population counts. You will use this information to identify how many people may need assistance after the hurricane to help emergency services prepare.
8. On the **Map** tab, in the **Selection** group, click **Select By Location.**
9. In the **Select by Location** window, apply the following settings:
   - **Input Features**: bahamas_population
   - **Relationship**: Have their center in
   - **Selecting Features**: cone

10. Click **Apply** and then click **OK**.
11. On the **Analysis** tab, in the **Geoprocessing** group, click **Tools**.
12. In the **Geoprocessing** pane, search for Summary Statistics and open the tool. Enter the following settings:
    - **Input Table**: bahamas_population
    - **Output Table**: Estimated_Population_Affected
    - **Statistics Fields** > **Field**: Pop_2010
    - **Statistics Fields** > **Statistic Type**: Sum

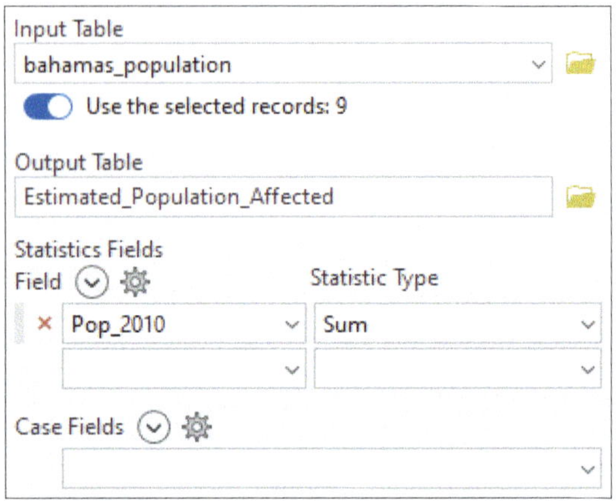

13. Click **Run**.
14. In the **Contents** pane, under **Standalone Tables**, right-click the **Estimated_Population_Affected** table and click **Open**.

    You had applied a selection to the **bahamas_population** layer, so you calculated the sum of only the select polygon's population attribute. This gives you an estimate of the number of people that may be affected by Hurricane Dorian within the study area.

On your own

Explore other datasets in the ArcGIS portal or ArcGIS Living Atlas in the Catalog pane that you find interesting or believe would provide insight. Keep in mind that the data should assist you in helping prepare the Bahamas for Hurricane Dorian.

Explore the map and the data you have visualized. Identify areas where you may need to provide additional resources for people to prepare for the hurricane. Identify potential locations that may serve as disaster recovery sites where emergency responders can stage and organize. Consider the population densities in these areas; it may affect how and where resources need to be directed.

4 Perform a posthurricane analysis

The next steps will walk through various geoprocessing tools to provide insight that will help emergency responders identify areas that are likely to be affected by the hurricane and to help develop a response strategy.

1. In the **Geoprocessing** pane, search for XY Table to Point. Open the tool and enter the following settings:
    - **Input Table:** Hurricane Dorian.csv
    - **Output Feature Class:** HurricaneDorian_Actual_Points
2. Click **Run.**
3. In the **Contents** pane, right-click **HurricaneDorian_Actual_Points** and click **Symbology.**
4. Set the **Point Symbol** to **Cross 3.**
5. Adjust the map so that you can see the full extent of the **HurricaneDorian_Actual_Points** layer.

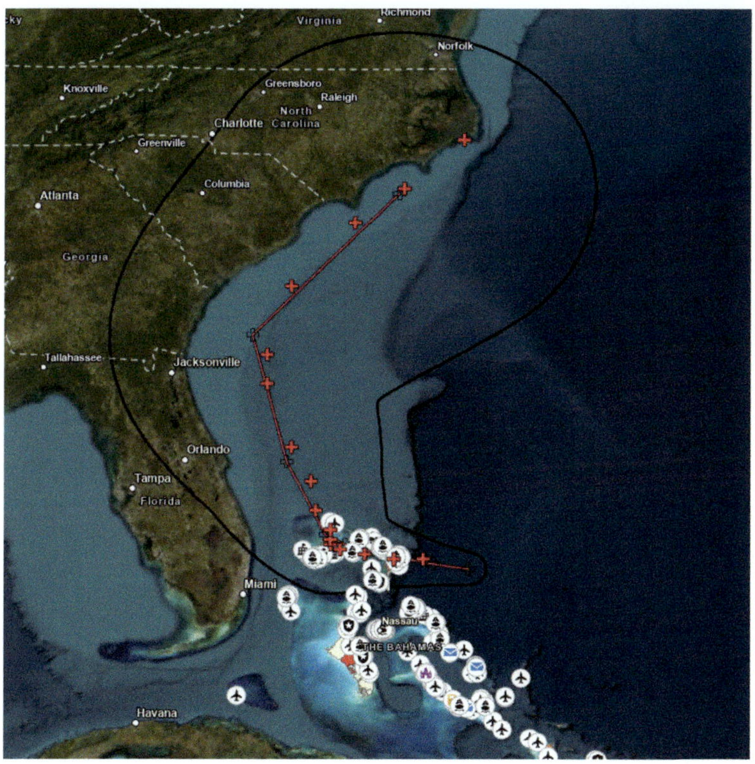

6. Save the project.

    The newly created **HurricaneDorian_Actual_Points** layer represents the observed coordinates of Hurricane Dorian, and the **positions** layer represents the forecast position points. You will draw the actual hurricane track using the points layer you just created.

7. On the **Edit** tab, in the **Features** group, click **Create.**

8. In the **Create features** pane, find the **track** layer and click the line icon below it.
9. Confirm that the **Line** feature is selected. On the map, click each segment of the **HurricaneDorian_Actual_Points** layer. When you finish tracing the points, double-click the last point to complete the line.

> **Tip:** To make this step easier, turn off all layers except the track and HurricaneDorian_Actual_Points layer.

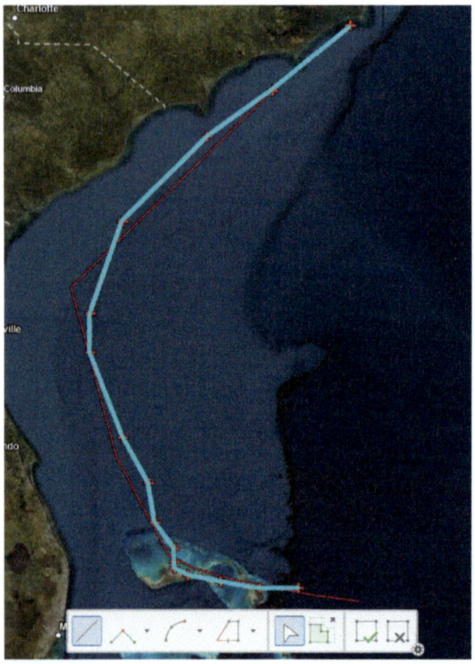

10. In the **Manage Edits** group, click **Save**. Save the project.

    Now that you have added a new line to the track layer, you must update the attribute data to ensure that you can delineate between the observed hurricane track and the projected hurricane track.
11. In the **Contents** pane, right-click the **track** layer and click **Attribute Table.**

    The attribute table has two rows. You will add information to the empty row.
12. In the table, double-click the blank cell under **STORMNAME** and set this to Dorian.
13. Double-click the blank cell under **STORMTYPE** and set this to AC.
14. In the **Manage Edits** group, click **Save**. Save the project.

    Finally, you will update the symbology to ensure that both lines have different symbols.
15. Open the **Symbology** pane for the **track** layer.

16. Change the **Primary symbology** to **Unique Values**.
17. Change the value for **Field 1** to **STORMTYPE**.
18. Symbolize each unique line value with different colors. Adjust the size of the lines to make them easily legible on the map.

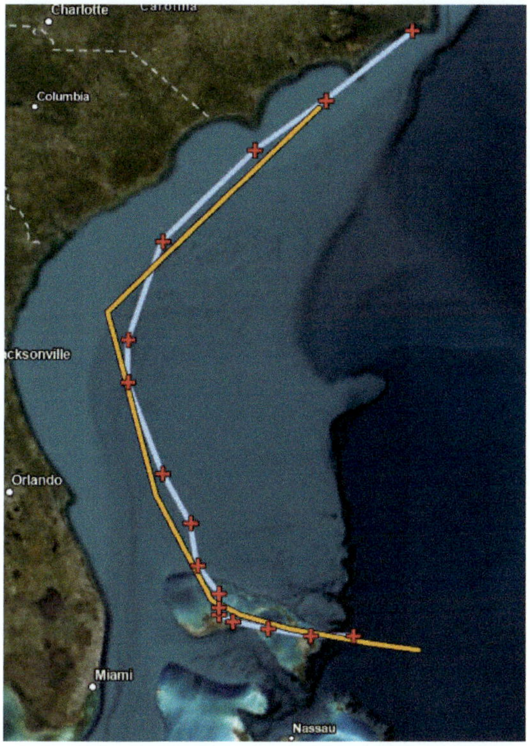

Now you have created a new track for the actual hurricane path. You can use this new data to help identify where Hurricane Dorian traveled through the Bahamas. Hurricane Dorian reached maximum strength with hurricane-force winds encompassing a 45-mile radius from its center. You will use this information as a basis to create a buffer that can be used to identify the areas where Hurricane Dorian had the strongest impact.

19. On the map, confirm that the new hurricane path that you created based on the actual Hurricane Dorian points is selected.
20. In the **Geoprocessing** pane, search for Pairwise Buffer. Open the tool and apply the following settings:
    - **Input Features**: track
    - **Output Feature Class**: Hurricane_Dorian_Wind_Range
    - **Distance**: 45 US Survey Miles
21. Click **Run**.
22. In the **Contents** pane, turn on the **cone** layer.

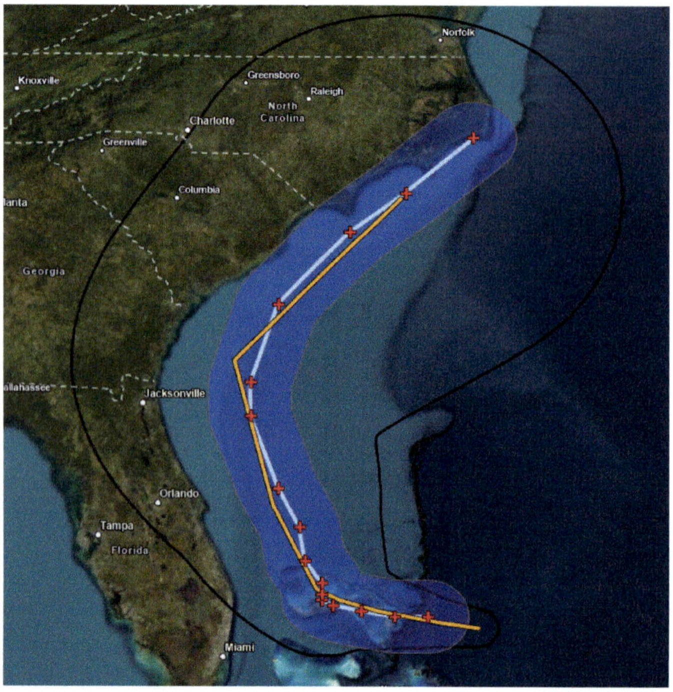

23. Save your project.

    Now, examine the map and analyze where Hurricane Dorian had the most impact.

> **On your own**
>
> At this point, you can continue to add various data sources and layers, such as recent imagery, local reporting, and geolocated social-media feeds to add further depth to your analysis.

5 Disseminate analytic findings

You have completed the analysis of Hurricane Dorian for the Bahama Islands. The next step is to display and share the results with stakeholders.

1. On the ribbon, click the **Insert** tab. In the **Project** group, open the **New Layout** list and click **ANSI–Landscape Letter 8.5x11**.
2. In the **Map Frames** group, open the **Map Frame** list and click **Map**.
3. In the **Layout** window, right-click and drag the cursor across the blank layout to insert a map. Adjust the size so that the **Map Frame** takes up the full extent of the layout.
4. In the **Contents** pane, turn on all the layers.
5. Right-click the **Map Frame** and click **Activate**.
6. Reposition your map extent to show the key areas in the Bahamas.
7. Click the exit button in the top right of the map frame to exit **Activated** mode.

8. Add other map elements such as a legend, north arrow, and scale bar until you are happy with the product.
9. On the **Share** tab, in the **Output** group, click **Export Layout**.
10. In the **Export Layout** pane, update the **Name** to read Hurricane Dorian Map.
11. Leave all other settings with their default values and click **Export**.
12. Save your project.

Interpretation of analysis

After completing this exercise, answer the following questions in detail. The answers to these questions, as well as the analysis results from the exercise workflow, will inform the intelligence brief.

1. Based on the analysis, describe the observed path of Hurricane Dorian. What areas were affected? In the Bahamas, which island and cities did the storm pass through? What was the total number of people in the Bahamas thought to be affected by the hurricane?
2. What information can be derived from this analysis? How might the Bahamas Emergency Management Center use this information to prepare for disaster assessment, coordination, and response?
3. Emergency response units usually establish an emergency operations center (EOC) to serve as the central hub for all coordination and logistics during a crisis. EOCs are often near major transportation centers. Based on the workflow, recommend three locations that would be ideal for the EOC. (*Note*: EOCs are not always located in disaster zones.)
4. Compare the tracks between the projected storm path and the actual storm path. In your opinion, did the climate model perform well in forecasting the storm path?
5. What other factors might help others understand the location and impacts of Hurricane Dorian?

Intelligence brief

Based on the analysis from the exercise workflow, produce a three- to five-page intelligence brief by responding to the following question prompts. Incorporate your geospatial outputs from the exercise to enrich and illustrate your conclusions.

- **BLUF:** List three to five primary observations from your analysis in list format that summarize findings of this work.
- **Top-level analysis:** Report the primary data points supporting your BLUF.
- **Secondary analysis:** Provide the data sources and deeper analytic products from this analysis.
- **Impact:** Describe the impacts of your findings on the department or organization.
- **Confidence:** Report the level of confidence of your assessment based on available data, accuracy, and so on.

- **Bias:** Explain any background or information you have that could have affected your analysis.
- **Conclusions:** Determine what type of analytic conclusions there may be and how the data requirements, analysis, and brief writing may change based on what analytic framework you are following (descriptive, predictive, prescriptive).

Applying and extending your skills

See the following resources for additional information about emergency response and disaster management:

- Office for the Coordination of Humanitarian Affairs—Humanitarian Data Exchange (OCHA HDX) plays a vital role in disaster management and geospatial intelligence (https://data.humdata.org).
- Activities-based intelligence is a method that builds on the workflow presented in this chapter by taking numerous data sources to generate conclusions through finding spatial relationships in various data sources. See "Activity-Based Intelligence: Understanding Patterns-of-Life" (Biltgen et al. 2017) at https://medium.com/the-state-and-future-of-geoint-2017-report/activity-based-intelligence-understanding-patterns-of-life-481c78b7d5ae.

Additional resources

- National Hurricane Center, Tropical Cyclone Report: *Hurricane Dorian*, August–September 2019. www.nhc.noaa.gov/data/tcr/AL052019_Dorian.pdf.
- Esri Industry Blogs, "Staying Ahead of the Next Disaster—How GIS Is Advancing in Disaster Management" (May 11, 2021). www.esri.com/en-us/industries/blog/articles/staying-ahead-of-the-next-disaster-how-gis-is-advancing-in-disaster-management.
- "NGA Deploys to Assist Hurricane Ian Search-and-Rescue Efforts" (October 5, 2022). www.hstoday.us/subject-matter-areas/emergency-preparedness/nga-deploys-to-assist-hurricane-ian-search-and-rescue-efforts.
- "Reviews of Geospatial Information Technology and Collaborative Data Delivery for Disaster Risk Management" (2015). www.preventionweb.net/publication/reviews-geospatial-information-technology-and-collaborative-data-delivery-disaster-risk.
- Pacific Disaster Center (PDC) National Disaster Preparedness Baseline Assessment. www.pdc.org/ndpba.
- "Geospatial Intelligence Revolution in Insurance and Risk Management" (April 2020), United States Geospatial Intelligence Foundation. www.preventionweb.net/news/geospatial-intelligence-revolution-insurance-and-risk-management.

Acknowledgments

The authors would like to thank Eric Allen Jr. for his insights and contributions on Hurricane Dorian.

About the authors

Jacob Spear is an imagery analyst in the US Marine Corps. In that capacity, he has supported national defense, humanitarian response, and civil works missions. He is also an instructor with the National Geospatial-Intelligence Agency College, where he teaches imagery analysis, sensor phenomenology, and GEOINT collections. He has a master's degree in human security and geospatial intelligence and holds the NGIA GEOINT Professional certifications for imagery analysis, geospatial analysis, cartography, and human geography.

Darren Martin Ruddell is a professor and director of graduate studies at the University of Southern California Spatial Sciences Institute, where he directs academic programs in HSGI. Ruddell earned his PhD from the School of Geographical Sciences and Urban Planning at Arizona State University. His teaching and research efforts use geospatial technologies to investigate and advance issues of human safety and security, climate change, and human-environment interactions.

Chapter 3

Introduction to fire intelligence—wildfires

| ***Tiffany Monicque Lee,*** *Center for Public Safety Management; Maxar Technologies*

Overview

This chapter is designed to provide a brief introduction to aspects of how GIST is used in wildfire disaster management and the role of fire intelligence in the field of HSGI.

This chapter and the tutorial workflow provide a basic overview and insight into some of the considerations used in fire intelligence. At its conclusion, you will have actionable insights into the strategic aspect of fire intelligence, which will lay the foundation for thinking spatially and multidimensionally to address the threat of wildland fire.

Learning objectives

- Perform exploratory geospatial data visualization by examining fire hazard severity zone (FHSZ) data to identify areas with high wildfire risk.
- Identify specific wildland urban interface (WUI) areas that are most susceptible by proximity or overlap with high-risk FHSZ areas.
- Visually analyze which areas within the potential danger areas are at highest vulnerability risk due to a census tract population demographic of more than 25 percent of people 65-plus years of age.
- Analyze which of the identified high-risk WUI areas within the study region experience the longest response times by local fire support.
- Learn how to identify complexities within wildfire disaster management and response in conflagration scenarios.

Technical components

- Software: ArcGIS Online (at least Publisher-level access)
- Datasets:
  - FHSZ data for California
  - WUI for California
  - LA County city and county boundaries
  - LA County Fire Stations layer
  - US Census Bureau American Community Survey (ACS) demographic data for target demographic of population with 25% > 65 years of age
  - LA County roads
  - Select fire station layer for western LA County

Prerequisites

- Basic knowledge of GIS and ArcGIS Online, Map Viewer.
- Technical skill level: Introductory to intermediate GIS.
- Time to complete chapter: One to two hours without intelligence brief, two-plus hours with intelligence brief.

Background and situational context

Fire intelligence is the science and study of fighting one of nature's most formidable foes—fire. In fire intelligence, the opposing force is not human, as is often the case in other areas of GEOINT; instead, fire is the "enemy" and requires specialized forces to combat the threat it poses. These forces encompass a broad range of teams in which GIS professionals play a crucial role:

- Strategic
  - Modeling worst-case scenarios for community preparedness and after-disaster crisis events
  - Evaluating wildfire fuels and developing wildfire management plans (for example, controlled burns)
  - Planning for localized fire and conflagration events
  - Selecting sites for potential evacuation routing and evacuation shelter setup
  - Identifying potential risk areas to help community managers take proactive measures
  - Aiding emergency management with actionable information to determine fire station efficiency and needs, such as new station placement and equipment and staffing acquisition and allocation
- Operational—conflagration management
  - Often multiagency with mobilization of resources specialized in wildfire events, usually coming from other regions, including geographic information systems specialist (GISS) teams

- Tactical
  - Multiagency emergency management with resources on the ground, in the air, and in the field, including fire line intel, critical for fire support and updating nearby residents and communities
  - Determining and declaring evacuation "Ready, Set, Go" scenarios

This chapter is not an all-encompassing guide to assessing risk and resources in a WUI. The WUI issue is often interagency, with regional and national response efforts, the complexities of which will vary according to the severity of the incident and from one county or municipality and fire department to another.

Understanding wildfire risk is a complex process requiring analysis of multiple factors:

- Current weather conditions (for example, wind speed, direction, and relative humidity)
- Vegetation type and cover (fuels)
- Geographic aspect
- Elevation
- Geographic slope
- Fire weather contributing factors, such as areas that experience more wind due to specific geography, which can drive flame heights and embers
- Response times from fire support once a wildfire start has been reported
- Proximity to added human threat potential (beyond that of the WUI neighborhoods) posed by trailheads or other types of areas that are used or traversed by higher numbers of people (for example, parks, preserves, recreation areas, roadways, railways, and so on)

Advanced GIS users can develop fire risk maps and models using data from these components to develop current, real-time condition risk assessment or fire behavior modeling, which are beyond the scope of this chapter.

Instead, the purpose of this chapter is to briefly introduce some of the methods and techniques GIS users can employ in assessing the potential threat to neighborhoods and communities within the FHSZ and WUI zone. The WUI has become increasingly important because of the encroachment of urban areas on natural wildland open space. Living near open spaces holds considerable appeal because it provides ease of access for recreation, exercise, and general communion with nature.

These urban fringe zones have evolved because of urban expansion and sprawl and are often not quickly accessible by established emergency fire services. Although local fire departments train in WUI situations, most local county or municipal fire departments are equipped to handle structural and small- to medium-scale outside fires. These departments do not usually have the equipment employed by crews enlisted in wildland fire support units. These county or municipal fire departments are, however, most often the first responders on site in a WUI situation.

It is important that community and emergency management have a means of understanding and estimating potential scenarios in wildfire emergencies within these WUI zones. Understanding vulnerabilities allows emergency managers and regional leaders to assess the potential threat to WUI zone neighborhoods and communities, as well as the efficiency of local municipality fire departments in responding to wildfire events in these high-risk areas.

Exercise workflow 3

Using the following datasets provided in the chapter data folder, you will follow a step-by-step guide to visually investigate spatial relationships between areas of potential threat due to high-risk wildland conditions, identify where these areas overlap with the WUI, filter ACS data to spatially identify specific census tracts with a population greater than 25 percent of 65-plus years of age, which presents a higher vulnerability factor and would require more resources in an evacuation scenario.

1 Add fire hazard layers to a map

1. In your browser, navigate to www.arcgis.com. Sign in to your ArcGIS Online account.
2. On the navigation bar, click the **Map** tab to open a new map in **Map Viewer**.
3. On the **Contents** toolbar (dark), click **Save and open** > **Save as**.
4. In the **Save map** window, apply the following settings:
    - **Title:** Wildfire Intelligence Ch. 3 <your initials>
    - **Folder:** SecurityFirst
    - **Tags:** Security First, Chapter 3
    - **Summary:** Wildfire intelligence case study in Los Angeles County.

Note: If you need to create the folder, click the down arrow, choose **Create new folder**, and type the folder name.

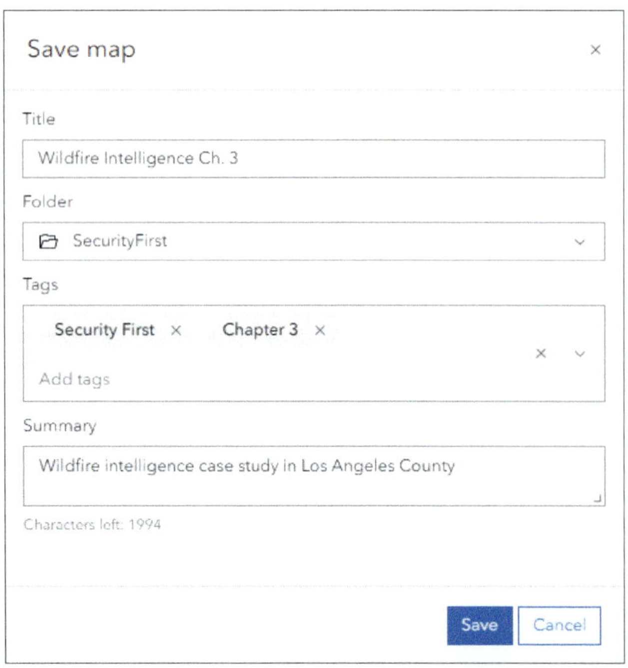

5. Click **Save**.
6. On the **Contents** toolbar, click **Add** > **Add layer from file**.
7. From the **SecurityFirst_ExerciseData** folder, in **Chapter03**, add **FHSZ LRA Very High Risk.zip**.
8. For **File type**, select **Shapefile**. Save your layer to the **SecurityFirst** folder and leave all other settings with the default values.
9. Click **Create and add to map**.
10. Repeat the previous steps to add the following zip files as new map layers:
    - **FHSZ SRA Very High Risk.zip**
    - **FHSZ SRA High Risk.zip**
    - **FHSZ SRA Moderate Risk.zip**

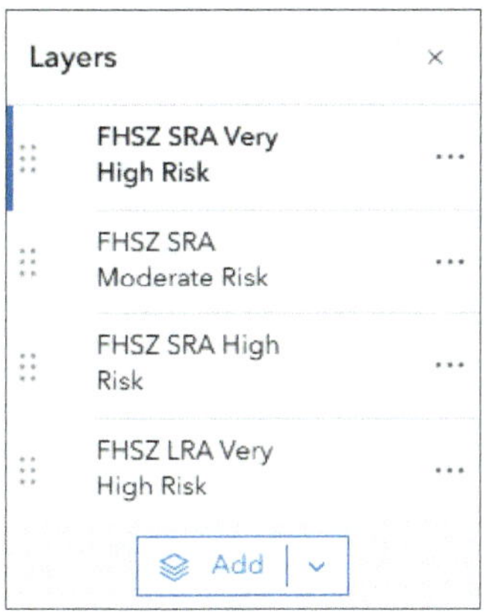

The FHSZ layer represents the California FHSZ. It uses various risk factors, such as slope, vegetation fuels, and weather patterns conducive to creating fire weather. It does not consider fuel modifications, such as prescribed burns or other wildfire fuel mitigation. The FHSZ layer reflects current conditions and estimates a burn probability with a 30-to-50-year window. The color codes are for local responsibility area (LRA) and state responsibility area (SRA). The zones determine the primary response and who is financially responsible for incidents in relevant boundaries.

Next, you'll color-code each layer.

11. In the **Layers** list, next to **FHSZ LRA Very High Risk**, click the **Options** button (three dots). Click **Show Properties**.
12. In the **Properties** pane, under **Symbology**, click **Edit layer style**.
13. Under **Pick a style**, for **Location (single symbol)**, click **Style options**.
14. In the **Styles** pane, under **Symbol style**, click the **Edit** button (pencil).
15. In the **Symbol style** window, edit the **Fill color**. In the prompt box next to #, type ff0000. Click **Done**.

16. For the **Outline color**, click the **No color** button.

    This will be done for all layers.

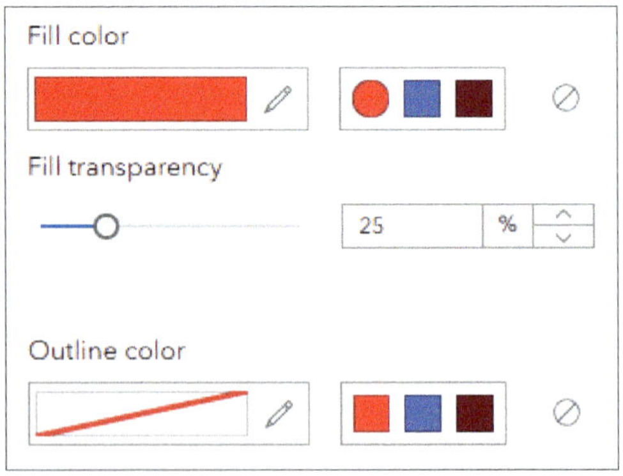

17. Repeat the previous steps using the following colors for each layer:
    - **FHSZ SRA Very High Risk:** #fac8e0
    - **FHSZ SRA High Risk:** #ffb400
    - **FHSZ SRA Moderate Risk:** #ff8519
18. Familiarize yourself with the layer's legend and contents. Make sure that the layers are arranged from top to bottom as follows:
    a. FHSZ LRA Very High Risk
    b. FHSZ SRA Very High Risk
    c. FHSZ SRA High Risk
    d. FHSZ SRA Moderate Risk
19. From **SecurityFirst_ExerciseData**, add the **WUI.zip** file to the map as a **Shapefile**. Set the **Title** of the layer to Wildlife Urban Interface. Save it to the **SecurityFirst** folder. Click **Create and add to map**.
20. In the **Layers** list, click **Wildland Urban Interface** to select it.
21. Open the **Styles** pane and edit the **Fill color** and **Outline color** using the following settings:
    - **Fill color:** #ffae0
    - **Outline color:** #fffb00
22. Click **Done** and click **Done** again.
23. On the map, zoom to the extent of Southern California, focusing on Los Angeles County.

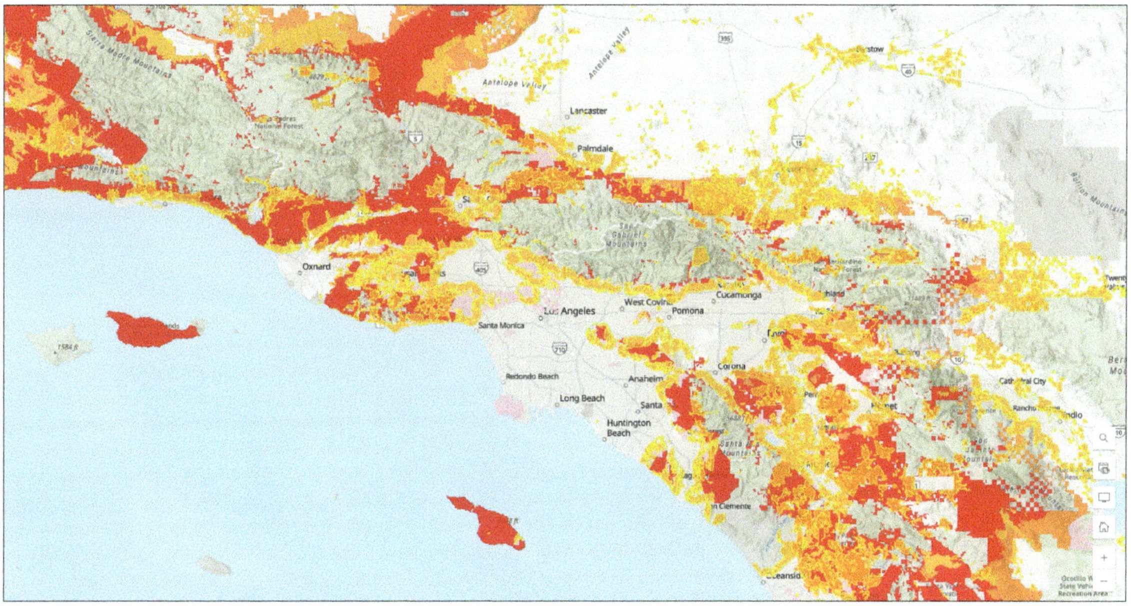

24. Add **LACounty_City_Boundaries.zip** to the map as a **Shapefile**. Set the **Title** of the layer to LA County and City Boundaries and save it to the **SecurityFirst** folder. Click **Create and add to map**.
25. In the **Layers** list, next to **LA County and City Boundaries**, click the **Options** button and click **Zoom to layer**.
26. In the **Styles** pane, change the **Outline color** to #153e96 and the **Fill color** to **No color**.
27. Click **Done** and click **Done** again.

2 Add population data

1. On the **Contents** toolbar, click **Add** > **Add layer from URL**. Add the ACS Census Tracts for a population of 65-plus years of age by typing or pasting the following URL: https://services.arcgis.com/P3ePLMYs2RVChkJx/arcgis/rest/services/ACS_Highlights_Senior_Well_Being_Boundaries/FeatureServer.
2. Click **Add to map**.
3. In the **Layers** list, expand **ACS Highlights Senior Well Being Boundaries** and click **Tract**.

   You will filter the layer attributes to show only the census tracts that have your target population: 25 percent or more of 65-plus years of age.
4. On the **Settings** (light) toolbar, click **Filter**. Click **Add new**.

   Note: Clear any existing filters.
5. Click the **Field** box and select **Percent of Population that is 65 Years and Over**. Set the expression to **is at least** and set the percentage value to 25. Click **Save**.

6. Click **Properties**, and under **Appearance**, change the **Transparency** to **50%**.
7. In the **Layers** list, click the **Options** button for **the ACS Highlights Senior Well Being Boundaries** layer. Click **Rename** and type 25% of population over 65 years.
8. On your map, zoom in until the **Tracts** layer is visible.

   Your map now shows where the highest-risk FHSZ, WUI, and highest-percent population for a vulnerable demographic overlap within the greater LA County area.

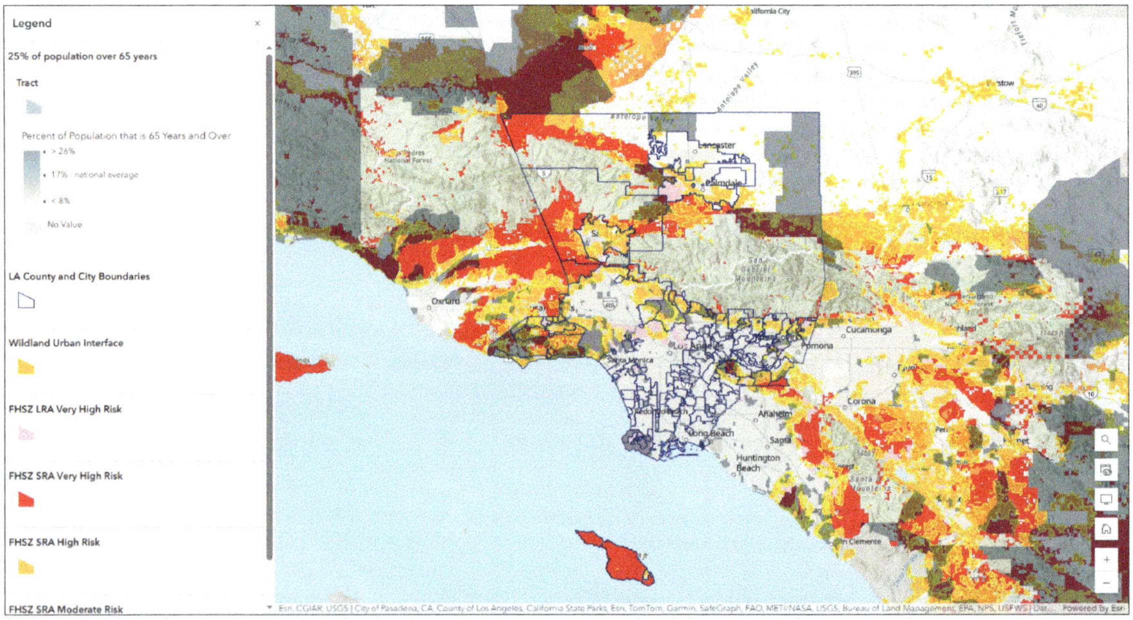

3 Add infrastructure data

1. Add **LACounty_Roads.zip** to the map as a **Shapefile**. Name the layer LA County Roads and save it to the **SecurityFirst** folder. Click **Create and add to map.**
2. Change the **Color** of the layer to #0BDEB2.
3. In the **Layers** list, hover over **LA County Roads** and click the **Visibility** button (eye) to turn the layer off.
4. From the **LACounty_FireStations** folder, add the **GEOJSON** file to your map. Click **Upload the file as an item and add it to the map.** Rename the layer LA County Fire Station Locations and save it to the **SecurityFirst** folder. Click **Create and add to map.**
5. On the map, examine the extent of the **LA County Fire Station Locations** points. Zoom to mainland Los Angeles.
6. Add the **Western_LACounty_Stations.zip** file to the map as a **Shapefile**. Rename the layer Western LA County Fire Stations and save it to the **SecurityFirst** folder. Click **Create and add to map.**

7. In the **Layers** list, confirm that this layer is at the top of the list.
8. In the layer's **Style** pane, under **Pick a style**, confirm that **Location (single symbol)** is checked. Click **Style options.**
9. Edit the **Symbol style.** Under **Current symbol**, click **Basic point**. Change the **Category** to **Public Safety**, scroll down, and select the fire truck icon. Click **Done.**

> **Tip:** You can switch to the List view and search for symbols by typing their name. To find the fire truck icon, search for Fire-Truck.

10. Change the **Size** to 20 px and confirm that the box for **Adjust size automatically** is unchecked.

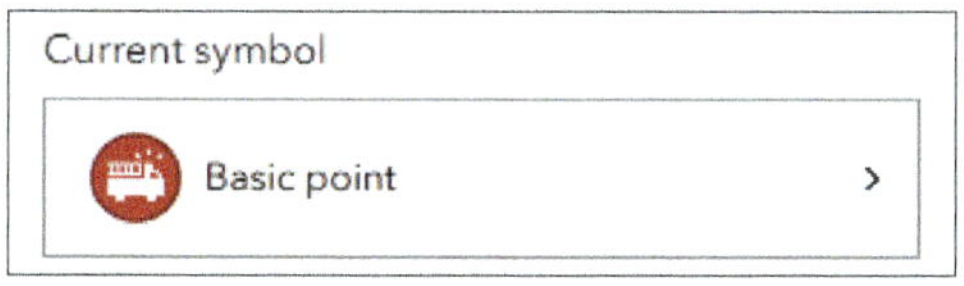

This will change the **Western LA County Fire Stations** points to red fire trucks.

11. Turn off the **Visibility** for this layer.

When you finish, the order of your layers list should be as follows:

a. Western LA County Fire Stations
b. LA County Fire Station Locations
c. LA County Roads
d. 25% of population over 65 years
e. LA County and City Boundaries
f. Wildland Urban Interface
g. FHSZ LRA Very High Risk
h. FHSZ SRA Very High Risk
i. FHSZ SRA High Risk
j. FHSZ SRA Moderate Risk

4 Analyze the data

1. Turn off the **Visibility** for the **LA County Fire Station Locations** layer. Zoom to the **Western LA County Fire Stations** layer.
2. Click a point or census tract to view its pop-up and examine the correlating information.

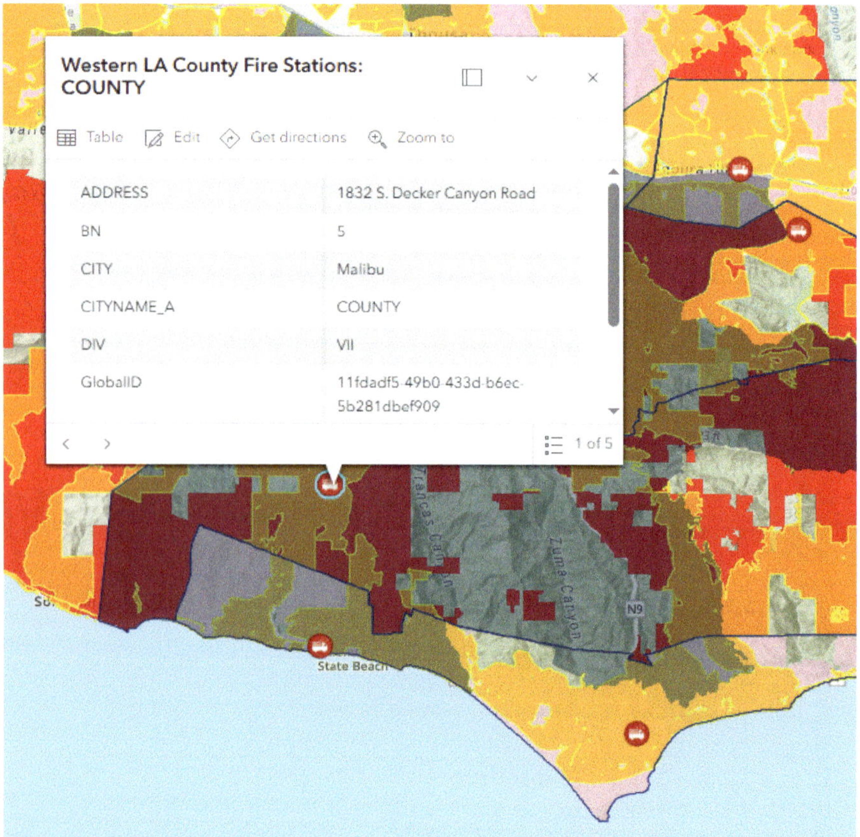

You will be analyzing the area of greater Malibu, where more than 30 percent of the population is over 65 years of age, creating potential challenges for timely, lifesaving evacuations in a "Ready, Set, Go" scenario. You will determine how far the current stations can reach in 610 seconds (10 minutes and 10 seconds).

3. Click the **Western LA County Fire Station** layer. On the **Settings** toolbar, click the **Analysis** tab.
4. Click **Tools** and expand the **Use Proximity** group. Open the **Generate Travel Areas** tool.
5. For **Input layer**, select **Western LA County Fire Station.**
6. Under **Analysis settings**, for **Cutoffs**, type 610 and click **Add.** For **Cutoff units**, select **Seconds.**

> **Tip:** Disregard the warning that appears below the Cutoffs value.

This is the National Fire Protection Association (NFPA) Standard 1710 for Initial Full Alarm High Hazard/High Rise response.

7. For **Departure time**, select **Now.**

8. Check the box for **Generate detailed polygons**.

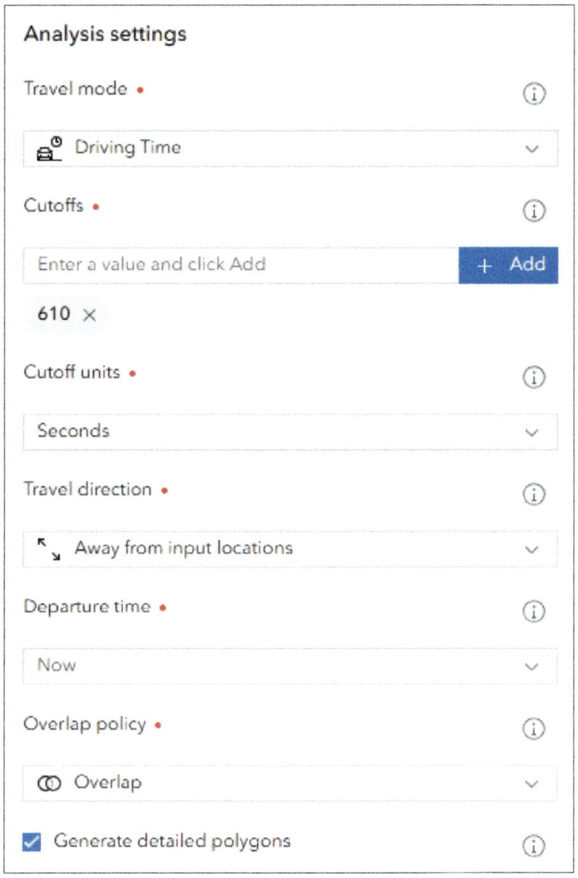

9. Under **Result layers**, set the **Output name** to 610 Second Response. Check the box for **Include reachable streets**. Save the results to the **SecurityFirst** layer.

10. Click **Run**.

    You can now see the spatial coverage of a 610-second response for each of the western LA County fire stations.

11. In the **Layers** list, reposition your layers so that the **Western LA County Fire Station** layer is on top and the newly created **610 Second Response** layer is below it.

12. Expand the **610 Second Response** layer and turn off the **Visibility** for the **Travelareas** layer to get a clear idea of the response coverage by accessible roads.

    The map now shows the 610-second response reachable streets and response areas, with the FHSZ layer, WUI, and high-vulnerability demographic. This is a high-risk area for all the elements under examination.

13. Turn on the **Visibility** of the **LA County Roads** layer that you added earlier.

    You can now see where the local fire support cannot reach within 610 seconds.

Interpretation of analysis

After completing this exercise, answer the following questions in detail. The answers to these questions, as well as the analysis results from the exercise workflow, will inform the intelligence brief.

1. What is the most important takeaway from the analysis workflow in this chapter?
2. What actionable information could local and regional emergency management use to protect residents existing in the WUI if a wildfire starts in proximity to the high-risk areas identified?
3. What communities in the county would benefit the most by implementing proactive measures, including control burns, or defensive space measures and community outreach and education on the importance of proactive measures and fire preparedness in a "Ready, Set, Go" scenario?

Sample response

The results of the preceding workflow reveal that [X number of] areas of LA County are at an elevated risk for wildfire and a possible conflagration event. This presents a human security challenge for those communities, as well as a national health and potential security threat because smoke exposure from conflagration events can significantly affect populations well beyond the extent of the local region. Smoke can potentially affect even military installations, and poor air quality can also affect military trainings. These effects would be akin to how the 2024 Canadian wildfires seriously affected air quality and population health on the US East Coast.

Intelligence brief

Based on the analysis from the exercise workflow, produce a three- to five-page intelligence brief by responding to the following question prompts. Incorporate your geospatial outputs from the exercise to enrich and illustrate your conclusions.

- **BLUF:** List three to five primary observations from your analysis in bulleted-list format that summarize findings of this work.
- **Top-level analysis:** Report the primary data points supporting your BLUF.
- **Secondary analysis:** Provide the data sources and deeper analytic products from this analysis.
- **Impact:** Describe the impacts of your findings on the department or organization.
- **Confidence:** Report the level of confidence of your assessment based on available data, accuracy, and so on.
- **Bias:** Explain any background or information you have that could have affected your analysis.
- **Conclusions:** Determine what type of analytic conclusions there may be and how the data requirements, analysis, and brief writing may change based on what analytic framework you are following (descriptive, predictive, prescriptive).

Applying and extending your skills

Wildfire data is often available in near real time from multiple sources.

- The Esri Wildfire Aware app is an excellent source for exploring active wildfires with information about the area in the vicinity of an active fire. The interface provides fire perimeter, area population, aspects of current fire weather, fuels, and the resources currently on incident. For more information, go to livingatlas.arcgis.com/wildfireaware.
- Another great source is Geographic Area Coordination Centers (GACC), which provides information about current and recent fires throughout the US. Visit gacc.nifc.gov.
- Visit the CAL FIRE Data Hub at hub-calfire-forestry.hub.arcgis.com.

Additional resources

Consider the following resources to learn more about fire intelligence.

- FEMA. Wildfire and WUI blog. www.usfa.fema.gov/wui/data.
- Nevada Fire and Aviation. https://nvfireintel-nifc.hub.arcgis.com.
- The Center for Climate & Security. "Wildfires in the US and Their Effects on Security" (Sept. 15, 2020). https://climateandsecurity.org/2020/09/wildfires-in-the-u-s-and-their-effects-on-security.
- USDA Forest Service. "Wildfire Risk in the Wildland Urban Interface." https://research.fs.usda.gov/treesearch/36420.
- Just Security. "Burning Threats: How Wildfires Undermine US National Security." www.justsecurity.org/87265/burning-threats-how-wildfires-undermine-u-s-national-security.

References

- ArcGIS Hub. Wildland Urban Interface [datasets]. n.d. https://hub.arcgis.com/search?q=Wildland%20Urban%20Interface.
- CAL FIRE. Fire Hazard Severity Zones. n.d. https://osfm.fire.ca.gov/what-we-do/community-wildfire-preparedness-and-mitigation/fire-hazard-severity-zones.
- County of Los Angeles. Open Data and Emergency Data. n.d. https://data.lacounty.gov.
- Esri. COVID-19 Overview. 2025. ACS Context for Senior Well-Being—Boundaries [authoritative dataset]. https://coronavirus-resources.esri.com/maps/e4b16658bc4749c58cb55ced3298d7d2/about.
- Hill, Alice C., and Tess Turner. July 2023. "Burning Threats: How Wildfires Undermine US National Security." Just Security. www.justsecurity.org/87265/burning-threats-how-wildfires-undermine-u-s-national-security.
- LA City GeoHub. April 2025. LACoFD Fire Station Locations (Feature Layer) 11/27/2023. https://geohub.lacity.org/datasets/lacounty::lacofd-fire-station-locations-feature-layer-11-27-2023/explore.
- National Fire Protection Agency. 2020. NFPA 1710: Standard for the Organization and Deployment of Fire Suppression Operations, Emergency Medical Operations, and Special Operations to the Public by Career Fire Departments. www.nfpa.org/codes-and-standards/nfpa-1710-standard-development/1710.
- Southwest Coordination Center. Accessed in 2025. gacc.nifc.gov/swcc.
- Sullivan, John P. 2019. "How Intelligence Can Map Wildfire Risk and Help Reduce Catastrophes." GTSC Homeland Security Today. www.hstoday.us/subject-matter-areas/intelligence/how-intelligence-can-map-wildfire-risk-and-help-reduce-catastrophes.
- TIGER/Line Shapefile. 2018. County, Los Angeles County, CA, All Roads County-Based Shapefile [US Census Bureau data]. https://catalog.data.gov/dataset/tiger-line-shapefile-2018-county-los-angeles-county-ca-all-roads-county-based-shapefile.

About the author

Tiffany Monicque Lee has worked in multiple capacities in the geospatial realm throughout her professional career. Her past work includes hydrographic and physical science roles with the USGS Grand Canyon Monitoring and Research Center (GCMRC) and the Department of Defense, US Army Corps of Engineers, New Orleans District, where she trained crews and conducted hydrographic surveys for the lower Mississippi River and Delta region of the Gulf of Mexico. She has since focused on environmental disaster response and community outreach. She works as a geospatial analyst and GIS specialist in GEOINT and emergency response, specifically HGIS and Fire and EMS. Monicque has an MS in GIST from the University of Southern California and a GISP certification from the GIS Certification Institute (GISCI.)

Chapter 4

Creating facility outline graphics and gridded reference graphics at multiple spatial scales in ArcGIS AllSource

Jacob Spear, *US Marine Corps*

Overview

In this chapter, you will learn how to use tools and methods in ArcGIS AllSource, intelligence analysis software, in the context of an HSGI problem by analyzing satellite imagery and creating graphic products to enable operations. The workflows introduced and the final deliverables produced will provide actionable information that can help coordinate military, emergency response, or humanitarian activities at various scales of analysis.

Learning objectives

- Analyze true-color imagery in ArcGIS AllSource.
- Identify critical infrastructure in facilities and extract features to feature classes within a geodatabase.
- Create a gridded reference graphic in ArcGIS AllSource at local and regional scales of analysis.
- Set definition queries.
- Export printable maps of various sizes to disseminate to customers.

Technical components

- Software: ArcGIS AllSource 1.2
- Dataset: OpenStreetMap (OSM)

Prerequisites

- Prior knowledge or experience with GIS software or spatial data is helpful but not required.
- Technical skill level: Introductory GIS.
- Time to complete chapter: Three to four hours.

Background and situational context

Commercial satellite imagery has given modern analysts unprecedented access to overhead imagery of critical locations around the globe. These images, many of which boast submeter spatial resolution with revisit rates within the same day, help us understand our world in ways that were not possible 10 years ago. We can prepare to conduct operations at locations we cannot access and gain almost immediate insights into damage caused by anthropogenic or natural disasters.

An essential product and a core skill of imagery analysts is to provide decision-makers and those on the ground with detailed facility overlay graphics to help them navigate locations and coordinate activities on the ground. To this end, facility outline graphics and gridded reference graphics help decision-makers understand the physical environment and ensure that those on the ground have detailed information to enable them to make timely decisions in high-stress situations. When disaster strikes, it can bring chaos, confusion, and uncertainty. Disaster events can damage critical infrastructure, cause unforeseen damage, or completely change a landscape. Despite these challenges, we must be able to mobilize emergency resources and respond to the impacted areas to help save lives and expedite the recovery process.

The following exercise workflow provides guidance for creating a facility overlay and gridded reference graphics over different locations and at two spatial scales of analysis. These products are designed to help first responders or other emergency personnel respond to the variety of situations they may face.

Exercise workflow 4

Before beginning this exercise, you must confirm that you have the necessary data and software. For this chapter, you will use the Esri Imagery basemap. The imagery and geodatabases for the workflow and intelligence brief are provided in the chapter resource folder. You can incorporate other data sources for future iterations to improve situational awareness. Examples can include underground pipes and electrical wiring locations, street information, and handheld pictures of key locations.

Gridded reference graphics at a local scale

In the first half of this exercise, you will work on a local scale. Later in the chapter, you will work on a regional scale.

Preparing your workspace

The first step of any project should be workspace preparation. You will begin by asking questions to understand the product's intent to help guide your analysis. Once you understand the project scope, you will need to collect the required data sources to facilitate your analysis. These sources may include Microsoft Excel workbooks, OpenStreetMap data, utility information, and so on. Once the data is gathered, you can create a project and add the data to a geodatabase in an ArcGIS AllSource project.

In ArcGIS, geodatabases are the native solution for housing, storing, and organizing data. You can think of geodatabases as repositories for geospatial information in various forms, including feature classes, feature layers, and tables. Using geodatabases is important within the ArcGIS product ecosystem because they are convenient ways to store, organize, and share geospatial data. Creating a new geodatabase is easy in ArcGIS, and each new project you create will automatically have one created for it. More instruction on creating geodatabases, feature classes, and additional information on these features in all ArcGIS products can be found using ArcGIS Help in ArcGIS AllSource or online (www.esri.com/en-us/arcgis/products/arcgis-allsource/overview).

For this example, you will create facility outlines for all major buildings in the University of Southern California Wrigley Institute to provide local emergency services so that the institute has accurate and current maps of the area to respond to an emergency at the facility. A project has been prepared for you with the necessary feature classes and geodatabases. It is located in the USC Wrigley folder in the provided resources.

1 Add data

1. In your files, locate the **SecurityFirst_ExerciseData** folder in the **Chapter04** folder. Open the **USC Wrigley** folder and double-click **USC Wrigley.asrx**.
2. On the top ribbon, click the **Project** tab and click **Save Project As**.
3. In the **Save Project As** window, navigate to your **C:/SecurityFirst** folder and name the project Chapter 4_USCWrigley. Click **Save**.
4. On the **Map** tab, in the **Navigate** group, click **Bookmarks** and click the **USC Wrigley** bookmark to center your view on the area of interest.
5. In the **Catalog** pane, on the right side of the screen, click the arrow next to **Databases** to expand the database contents.

   Note: There are two geodatabases in the Contents pane for this project. The one called USC Wrigley is for you to complete these steps. The one called USC Wrigley Example was created to use as a reference.

6. Expand **USC Wrigley.gdb** to view the geodatabase contents.
7. Press **Ctrl**, select the **Berthing**, **Buildings**, **Entrance**, and **Facility Outline** feature classes, and drag them to the **Map**.
8. On the ribbon, click **Save Project**.

   You will see the feature classes contained in the geodatabase populated in the **Contents** pane. You should see the layers **Entrance/Exit**, **Facility Outline**, **Buildings**, **Dock/Berthing**, and **World Imagery**. When the project is loaded and centered on the USC Wrigley Institute with the **USC Wrigley** geodatabase loaded, you can continue to the next step.

Note: The colors associated with the layers may differ. Different colors will not affect your progress.

2 Extract a feature

This step focuses on extracting features. Features may include building outlines, fences, gates, utility poles, road hazards, and so on. For this example, you will collect the building outlines, gates, and dock and berthing areas.

1. On the ribbon, click the **Map** tab. In the **Features** group, click **Edit**.

   When the map is in **Edit** mode, the **Edit Mode** toolbar is visible.
2. In the **Contents** pane, click the **Buildings** layer, and then click the **Polygon** tool on the **Edit Mode** toolbar.

This will allow you to click anywhere on the map to draw a free-form polygon.

3. Using the **Polygon** tool, create polygons for each building within USC Wrigley. Click to create vertices for each polygon while using your pointer to trace the building. When you are done outlining each building, double-click to finish the feature.
4. Click the **Finish** button on the bottom toolbar.

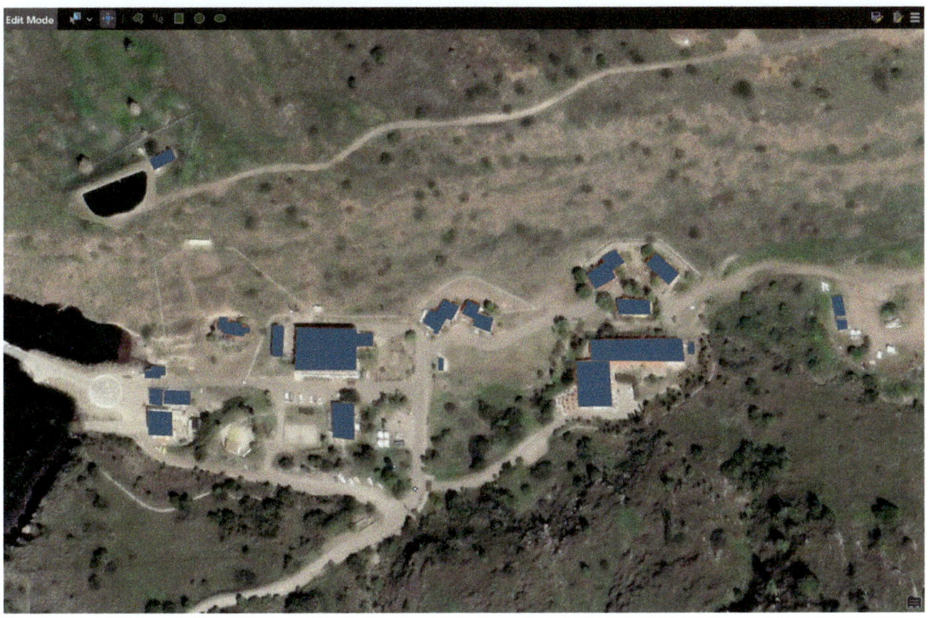

5. Repeat the preceding steps for the **Facility Outline** and **Dock/Berthing** layers.
6. For the **Entrance/Exit** layer, use the **Point** tool to add the feature.

> **Tip:** You can rearrange and hide the layers in the Contents pane to make the drawing process easier.

7. When you have completed all edits, click the **Edit** tab. In the **Manage Edits** group, click the **Save All Edits** button. Click the **Close Edit Mode** button.

Your map will show the representation of three features.

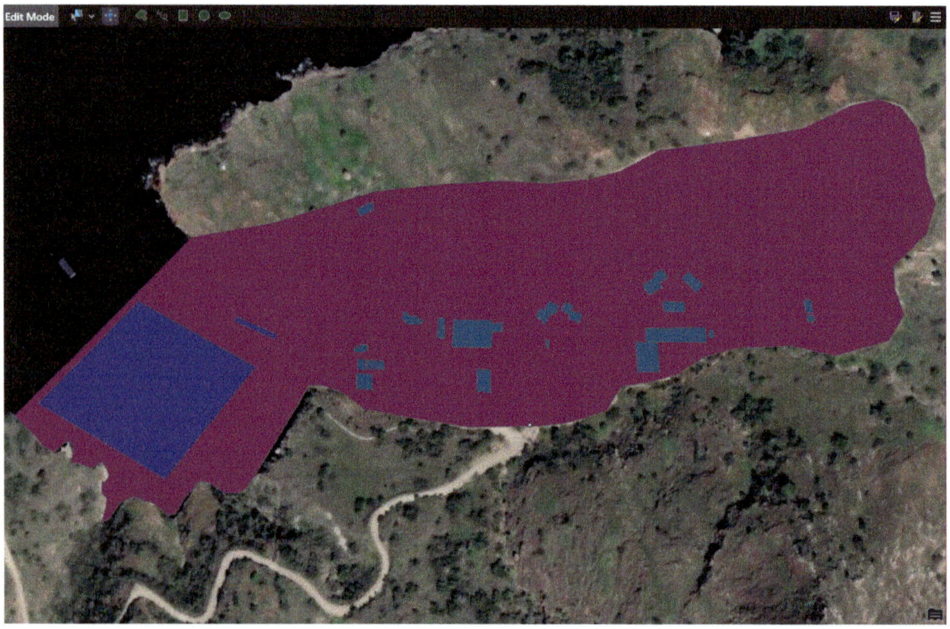

3 Create grids and labels

Now that you have completed the feature extraction, the next step is to create a gridded reference graphic, number the building polygons, and apply symbology to the feature classes.

1. On the top ribbon, click the **Imagery** tab. Click **Open Clearing Grids**.
2. In the **Draw Grids** group, check the box for **Use Existing Layer**. For **Grid Area**, expand the menu and select **Facility Outline**.
3. Set the **Grid Type** to **Reference** and the **Cell Size** to **100m × 100m Grid**.

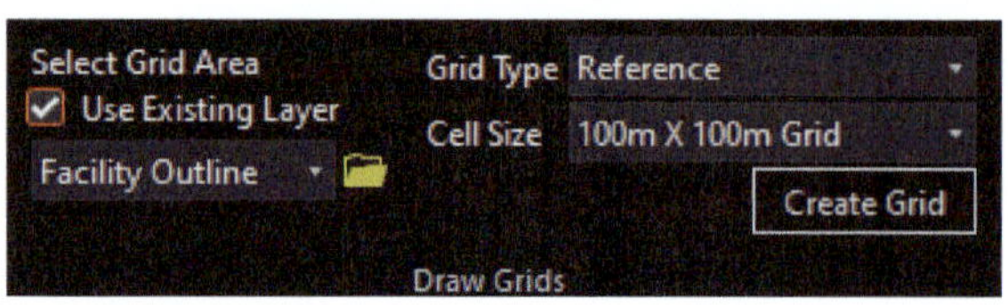

4. Click **Create Grid**.

   With the grid created over the area of interest, you now need to call out key buildings that may need to be referenced.

5. On the **Clearing Grids** tab, set the **Label Method** group to **Number**.
6. In the **Number/Letter Features** group, next to **Layer**, click the **Add New Layer** button. In the **Select Output** window, set the **Name** to KeyBuildings and click **Save**.

7. Next to **Field**, click the **Add New Layer** button. In the **Add Field** window, set the **Field Name** to BuildingNumber and click **OK.**

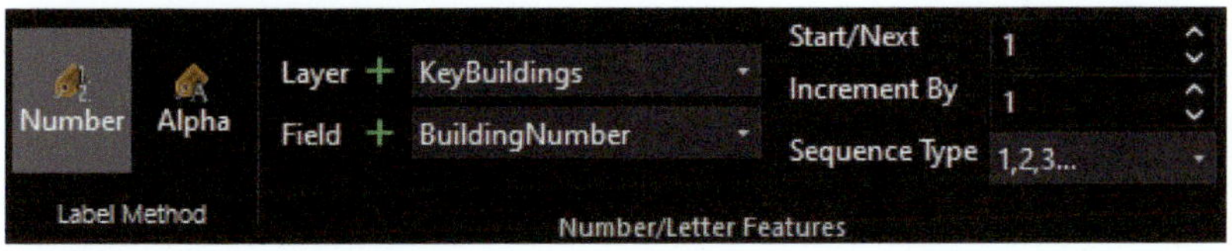

Note: Keep the default numbering options. Depending on the situation, you can customize these settings to meet specific requirements.

8. In the **Contents** pane, uncheck the boxes for the **Facility Outline** and **Dock/Berthing** layers to hide them.
9. On the **Clearing Grids** tab, open the **Add Number** tool. On the **Map**, apply a number to 10 buildings.

Tip: Refer to the following image if you want your map to match the example.

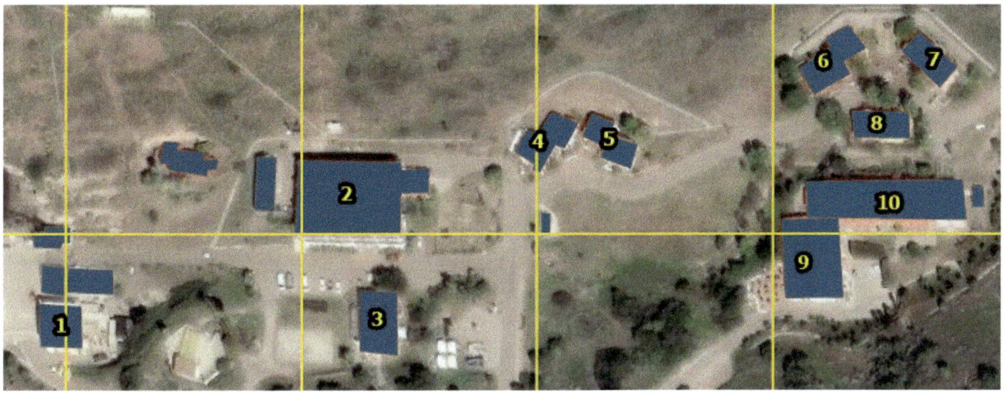

The following step is optional, but this practice will ensure that the building labels are recorded if you need to disseminate the feature classes you created separately from the grids and point labels.

10. Update the **Number** attribute in the **Buildings** feature class to reflect the number label that you just applied. Keep unmarked buildings **<Null>**.

4 Apply symbology

With the reference grid created and key buildings labeled, you will symbolize the feature classes to ensure that the final product is visually meaningful.

Note: For more information about the Clearing Grids toolset, see the additional resources in this chapter.

1. Check the box next to **Facility Outline** and **Dock/Berthing** to display the layers.
2. Right-click **Facility Outline** and click **Symbology**.
3. In the **Symbology** pane, click the colored square next to **Symbol**.
4. In the **Gallery**, select **Dashed Black Outline (1pt)**.
5. In the **Format Polygon Symbol** settings, click the **Properties** tab. Set the **Outline color** to **Tuscan Red** and the **Outline width** to 5 pt.

> **Tip:** Hover over the color to see its name.

6. Click **Apply**.

   You will now explore the gallery and customize the remaining layers to achieve your desired look. Symbolization ensures that the final product is legible and meaningful. Use colors that are unique and easily identifiable against the imagery basemap.

7. Apply appropriate symbology to the **Entrance/Exit**, **Buildings**, and **Dock/Berthing** layers.

> **Tip:** You can adjust the transparency of the color by going to Color > Color Properties > Transparency.

8. Click the **Save Project** button.

   After applying the symbology, compare your results with the following figure. You may see color variations, but the grids, key building labels, and overall color scheme should facilitate easy interpretation of the information.

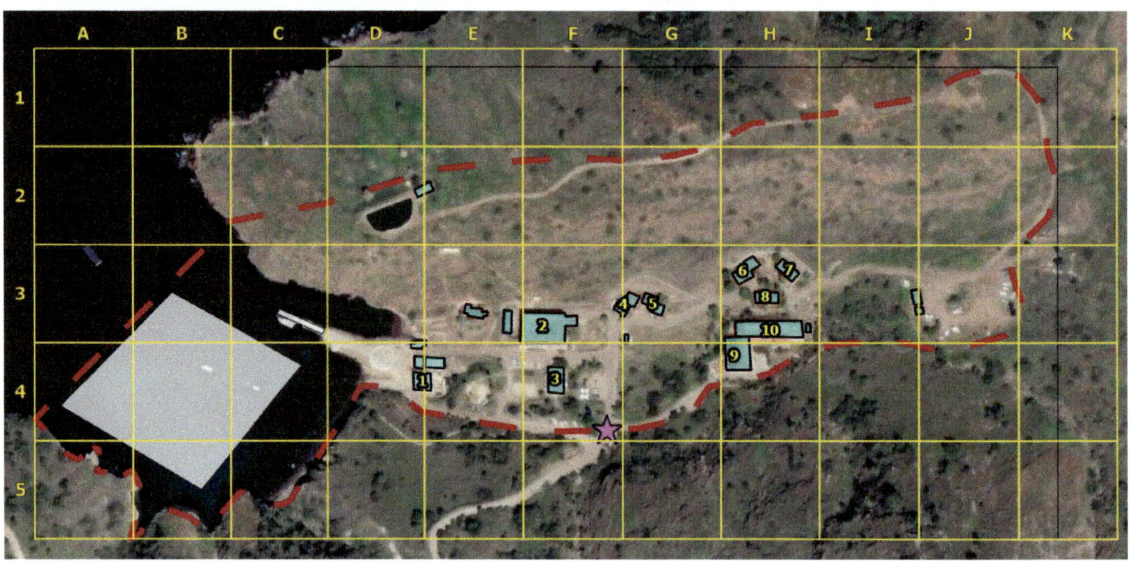

5 Disseminate the product

The next steps involve fine-tuning the product to create an 8.5-by-11-inch printable layout that can be easily disseminated and used. ArcGIS AllSource supports disseminating intelligence products through exporting data to ArcGIS Field Maps or publishing an online map, but you will produce a printable layout.

1. On the ribbon, click the **Disseminate** tab. In the **Insert** group, click **New Layout.**
2. On the **New Layout** menu, under the **ANSI–Landscape** section, select **Letter 8.5" x 11"**.

   Note: The layout appears in the center of your screen. Your original map is still accessible by clicking the Map tab.

3. On the ribbon, click the **Insert** tab. In the **Map Frames** group, click **Map Frame** > **USC Wrigley.**
4. In the **Layout**, drag your pointer to create the **map frame.**
5. Right-click the **Layout** and click **Activate.** Adjust your **map frame** so that the extent of the grid is fully within the **Layout** view. Leave space at the top for the title and legend.

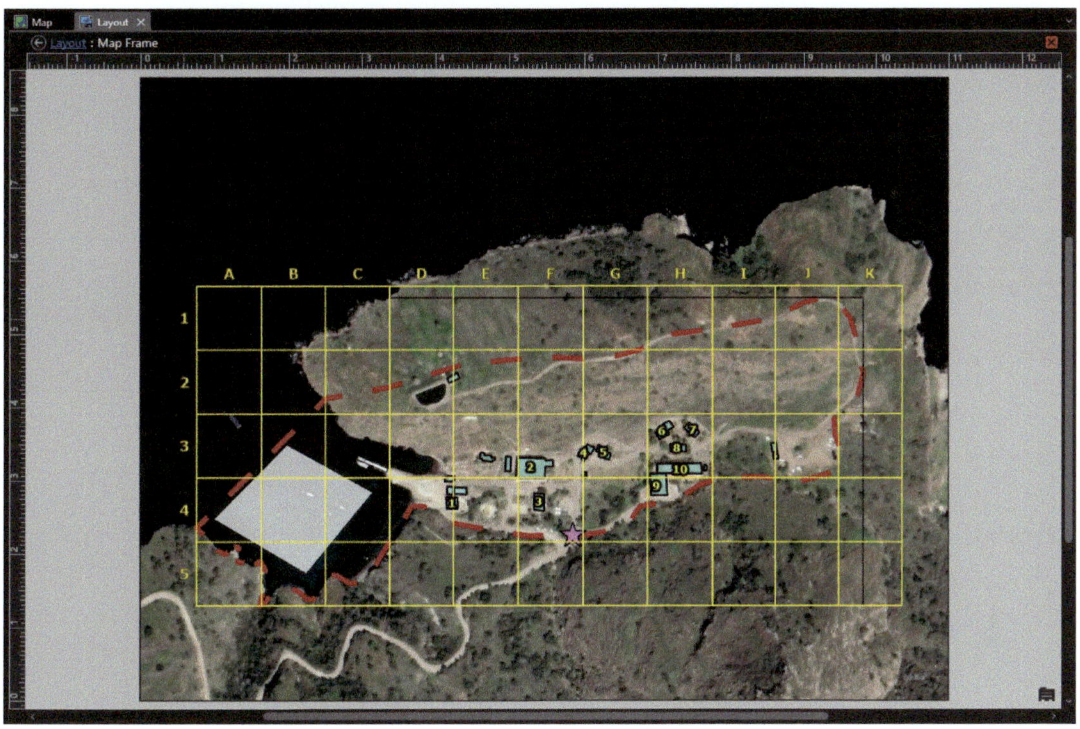

6. In the top-right corner of the **Layout**, click the **Close activated map frame** button.
7. On the **Insert** tab, in the **Graphics and Text** group, click the **Rectangle text** button.
8. In the top-right corner of the **Layout**, drag your pointer to create a **Text** box.
9. Type USC Wrigley Institute, Catalina Island, CA.

10. In the **Element** pane, on the right side of your screen, click the **Display** button (paintbrush). Apply the following settings:
    - **Border Symbol:** Black
    - **Border X gap:** 0.1 in
    - **Border Y gap:** 0.1 in
    - **Background Symbol:** White
    - **Background X gap:** 0.1 in
    - **Background Y Gap:** 0.1 in
11. Click the **Text Symbol** tab and set the **Size** to 20 pt. Adjust the size of the text box as needed to fit the **Text** in a single line.
12. On the **Insert** tab, in the **Map Surrounds** group, click **Legend**.
13. On the **Layout**, inside the **Text** box you just added, drag your pointer to create the **Legend**.
14. In the **Contents** pane, under **Legend**, ensure that only the boxes next to **Entrance/Exit**, **Buildings**, **Dock/Berthing**, and **Facility Outline** are checked.
15. Repeat the preceding steps to set the same **Legend** settings in the **Element** pane.

    Note: Instead of Text, you will see Legend in the Element pane.

16. In the **Map Surrounds** group, click **North Arrow**.
17. In the **Layout**, on the left of the **Legend**, drag your pointer to create the **North Arrow**.
18. In the **Map Surrounds** group, click **Scale Bar**.
19. In the **Layout**, under the **Legend**, drag your pointer to create the **Scale Bar**.

    Note: In the bottom left of the Layout pane, you will see a number format of 1:XXXX. This is the scale of the current extent. After the colon, you can type a number to achieve a precise scale that fits your needs. Generally, cartographers keep this as an even number, but certain maps will require a specific scale. For this example, 1:4,500 appears to fit well.

    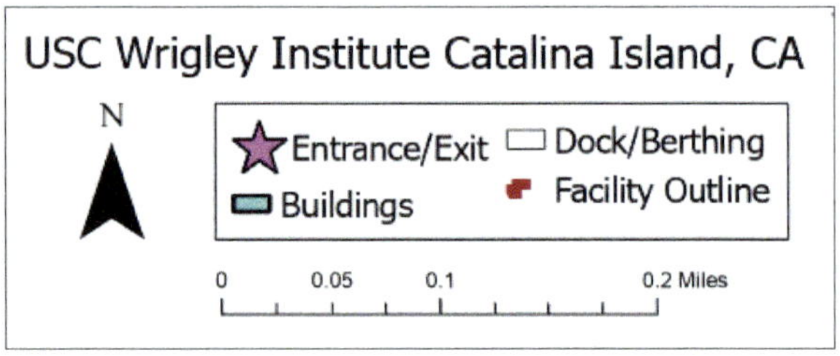

    At this point, you have all the basic elements required for a simple printable map.

20. Adjust the **Layout** elements to achieve a balance between visually appealing and practical.

21. In the **Contents** pane, uncheck the box next to **GridRectangle_Reference** to hide the layer.

    The following image is an example of a layout.

    Note: For more information about layouts, see the additional resources.

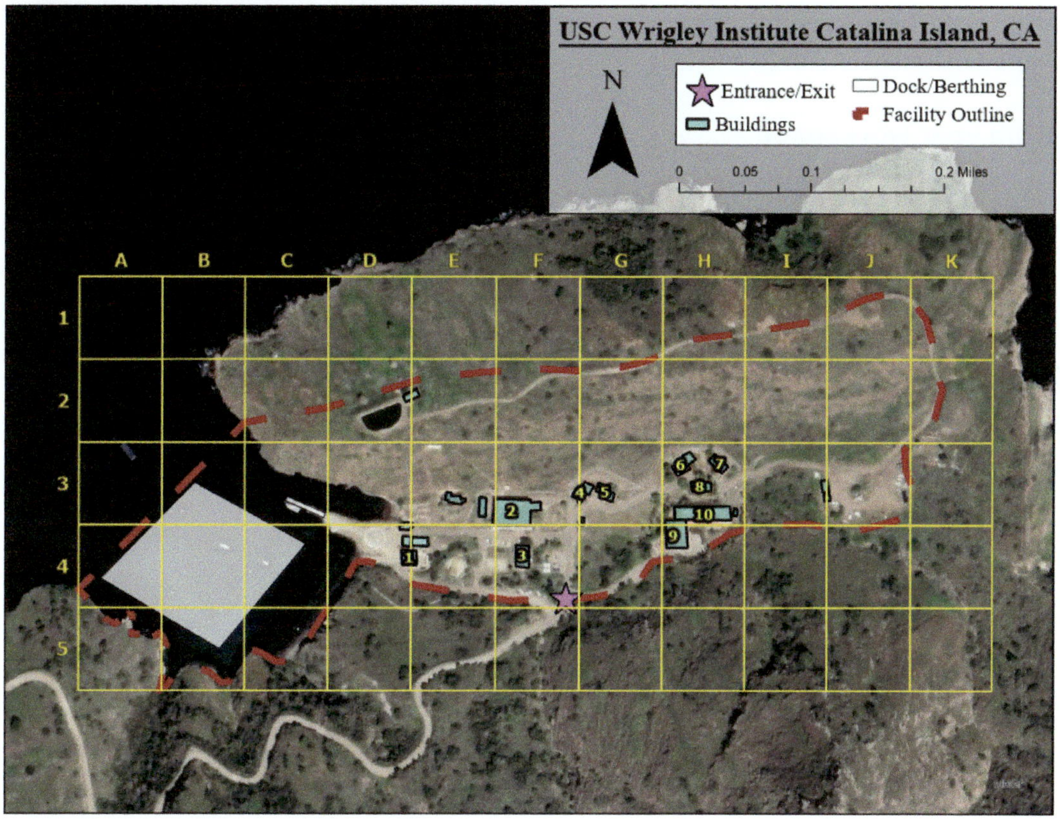

22. On the ribbon, click the **Disseminate** tab.
23. In the **Export** group, click **Export Layout** > **Flattened PDF**.

    Note: The other export formats may have useful information, but a flattened PDF file is appropriate for a simple printable map.

24. The **Export Layout** pane appears on the right side of your screen. Specify a location for saving and name the file USC Wrigley Overview.
25. Click **Export**.
26. When the export is completed, click **View exported file** to open your map or navigate to the file location to open it.
27. In ArcGIS AllSource, click **Save Project** and exit the program.

Gridded reference graphics at a regional scale

In this section, you will apply the local-scale workflow to a regional scale. You will build a gridded reference graphic over Catalina Island to help coordinate emergency services in some of the island's more remote areas. At the local scale, you highlighted USC Wrigley buildings and facility perimeters. Now, you will highlight larger infrastructures that are more impactful at the regional level, such as campsites, radio towers, helicopter landing pads, and major settlements.

Preparing your workspace

Before you begin analyzing data and producing the gridded reference graphic, you must prepare your workspace. As you did earlier, you should ensure that you understand the question being asked to collect the required data. This workflow focus is to create a graphic that can be used to coordinate emergency services across the island. At this scale, you do not need individual building or birthing polygons. The data you collect will consist mostly of lines and points to highlight roads and the general location of critical infrastructure. Authoritative data sources from government organizations would be preferred, but otherwise you can use open-source information, such as OpenStreetMap. This organization facilitates the mass collection of geographic information by using the input of hundreds of thousands of users. OpenStreetMap is not subject to the same quality control measures as an authoritative data source, so you must ensure that you validate the data.

Note: For more information about OpenStreetMap, see the additional resources in this chapter.

1 Create a project and zoom to a bookmark

A project file has already been prepared with the OpenStreetMap data in a geodatabase.

1. In the **Chapter04** folder, open the **Catalina Island** folder and double-click **Catalina Island.asrx**.
2. On the ribbon, click the **Project** tab and click **Save Project As**.
3. In the **Save Project As** window, navigate to your **C:/SecurityFirst** folder and name the project Chapter 4_CatalinaIsland. Click **Save**.
4. On the **Map** tab, in the **Navigate** group, click **Bookmarks** and click the **Catalina Island** bookmark to center your view on the area of interest.
5. In the **Catalog** pane, click the arrow next to **Databases** to expand the database contents.

   Note: There are three geodatabases in the Contents pane for this project. The one called Catalina_OSM contains the OpenStreetMap data that you will need for this section. The one called Catalina Island is for you to complete several steps. The one called Catalina Island Example is a reference.

6. Expand **Catalina_OSM.gdb** to view the geodatabase contents.

7. Select **gis_osm_buildings_a_free**, and while pressing **Shift**, select **gis_osm_waterways_free**. Drag the highlighted layers to the **Map**.

   All the feature classes in Catalina_OSM.gdb are added to your map. The feature classes in the geodatabase are displayed in the **Contents** pane; it may take a few seconds to load all the data to the map, and it will appear crowded. In the next section, you will explore these data layers to determine which ones are needed to inform the final product. When the project is loaded and centered on Catalina Island with the Catalina_OSM geodatabase loaded, you can continue.

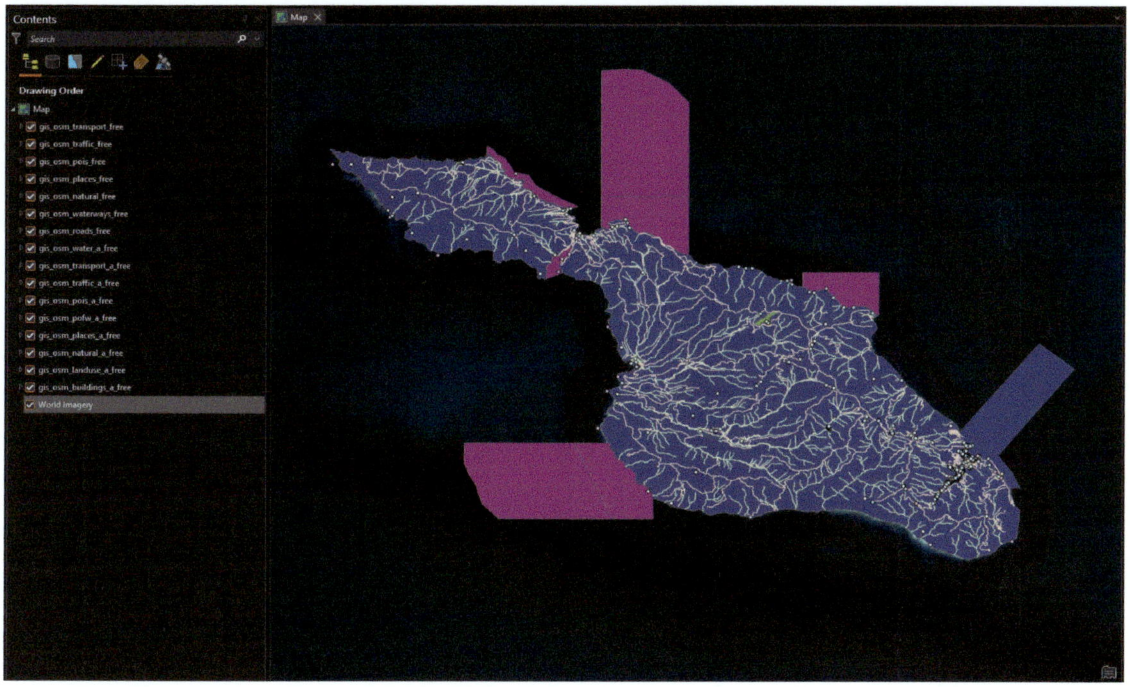

2 Extract a feature

Because you are using OpenStreetMap data, you will not need to extract the features you want to visualize manually. Instead, you will choose and prepare the feature classes required to inform the emergency personnel using the final map. You will not need every layer provided in Catalina_OSM.gdb, but you need to understand what information is shown with each one before removing the unnecessary feature classes. Examine each feature class's name and attribute table and write down what you think each one is showing. Table 4-1 lists what each feature class is displaying.

Table 4-1. Catalina OSM geodatabase feature class explanations

Layer Name	Description	Type
gis_osm_transport_free	Public transportation hubs or facilities	Point
gis_osm_traffic_free	Transportation network features	Point
gis_osm_pois_free	Places of interest	Point
gis_osm_places_free	Cities or settlements	Point
gis_osm_natural_free	Natural elements such as named mountains or urban trees	Point
gis_osm_waterways_free	Streams and rivers, both dry and running	Line
gis_osm_roads_free	Road networks	Line
gis_osm_water_a_free	Lakes or large bodies of water	Polygon
gis_osm_transport_a_free	Public transportation hubs or facilities	Polygon
gis_osm_traffic_a_free	Transportation network features	Polygon
gis_osm_pois_a_free	Places of interest	Polygon
gis_osm_pofw_a_free	Places of worship	Polygon
gis_osm_places_a_free	City area	Polygon
gis_osm_natural_a_free	Natural elements such as beaches or parks	Polygon
gis_osm_landuse_a_free	Designated land-use areas	Polygon
gis_osm_buildings_a_free	Building footprints	Polygon

Now, decide which feature class layers should be displayed on the map. Your selections may differ from those used in the final product, but working through what you want to visualize is a good exercise that will improve your ability to make logical decisions in the future.

1. In the **Contents** pane, press and hold **Ctrl** and select the following layers:
    - **gis_osm_traffic_free**
    - **gis_osm_waterways_free**
    - **gis_osm_water_a_free**
    - **gis_osm_transport_a_free**
    - **gis_osm_traffic_a_free**
    - **gis_osm_pois_a_free**
    - **gis_osm_pofw_a_free**
    - **gis_osm_places_a_free**
    - **gis_osm_natural_a_free**
    - **gis_osm_landuse_a_free**
    - **gis_osm_buildings_a_free**

2. Press and hold **Ctrl**, right-click any of the selected layers, and click **Remove**.

Note: You should have only the following six layers remaining in the Contents pane: gis_osm_transport_free, gis_osm_pois_free, gis_osm_places_free, gis_osm_natural_free, gis_osm_roads_free, and World Imagery.

Only the layers that you want to visualize remain in the Contents pane, but your exploration through their attribute tables revealed numerous features that you do not want to visualize because they are redundant or inconsequential. In the next steps, you will learn how to use definition queries to exclude data from visualization and analysis. Using definition queries is the preferred method of removing data from a feature class because this method preserves the original dataset. If you need the data that you removed with the query, you can remove the query, which would be impossible if you deleted the features from the feature class. More information about definition queries is available in the additional resources.

3. Open the **Attribute Table** for **gis_osm_natural_free** and look through the table.

   You can filter unnecessary points by hiding points with an **fclass** that does not indicate **peak.**

Note: Setting definition queries highlights the importance of ensuring consistent and accurate attribute information. If the attribute information is messy, with various capitalizations, abbreviations, or misspellings, using definition queries or other analyses based on attributes would be difficult or impossible.

4. In the **Contents** pane, right-click **gis_osm_natural_free** and click **Build Definition Query**.
5. In the **Layer Properties** window, click **New definition query**.
6. Use the drop-down menus to define **Query 1** as follows:

```
Where fclass is equal to peak
```

Note: The definition query follows basic logic formats. You can set multiple definition queries to filter based on multiple attribute fields.

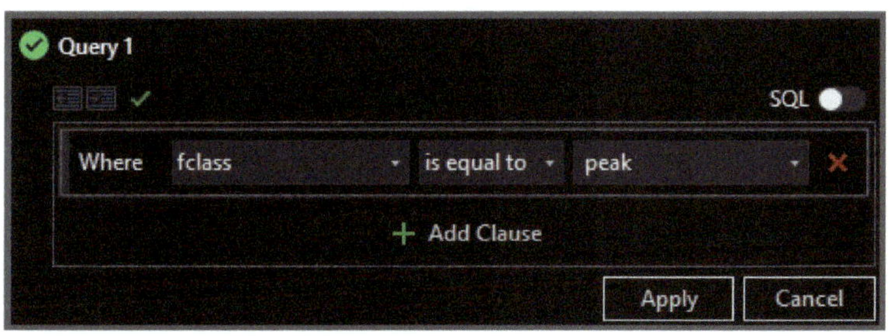

7. Click **Apply** and then click **OK.**

8. Repeat the preceding steps to apply **Definition Queries** to the **gis_osm_places_free**, **gis_osm_pois_free**, and **gis_osm_transport_free** feature classes. Use table 4-2 as a reference to set the definition query parameters.

   Note: You are not applying a definition query to the gis_osm_roads_free layer.

   Table 4-2. Feature class definition query parameters

Layer Name	Field	Logic	Value
gis_osm_places_free	fclass	includes the value(s)	hamlet, locality, town, village
gis_osm_pois_free	fclass	includes the value(s)	camp_site, comms_tower, fire_station, police, shelter, telephone, tower, viewpoint
gis_osm_transport_free	fclass	does not include the value(s)	bus_station, bus_stop

   When you have finished applying the definition queries to the feature classes, you can move on to the next section.

3 Create grids and labels

Now that you have finished preparing your data, you will create the grids and add labels to your feature classes.

Because this product is at a larger scale, you will use the Military Grid Reference System (MGRS) grid option. It is commonly used among military, first responders, and hikers because it facilitates navigation and provides an easy-to-read coordinate system. Additionally, MGRS is easier to plot on a paper map compared with DMS (degree, minute, second) coordinates. MGRS divides the globe into 60 sections with 100,000-meter grid zones that can be further divided to provide submeter accurate coordinates at 12 digits. More information on MGRS can be found in the additional resources.

1. On the ribbon, click the **Imagery** tab. Click **Open Clearing Grids**.
2. In the **Draw Grids** group, for **Select Grid Area**, click **Draw Grid Rectangle**.
3. On the **Map**, place the top-left corner of the **Grid Rectangle** northwest of Catalina Island and place the bottom-right corner southeast of Catalina Island.

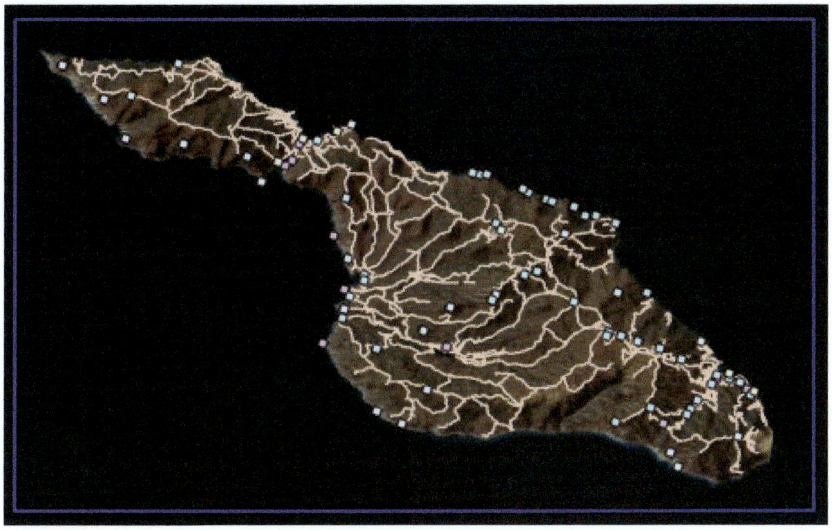

Note: Ensure that the blue box encompasses the entire island. This will be the area used to generate the grids.

4. In the **Draw Grids** group, set the **Grid Type** to **MGRS** and the **Cell Size** to **1000m X 1000m Grid**.
5. Click **Create Grid**.
6. In the **Contents** pane, expand the **MilitaryGridReferenceSystem** layer. Right-click **GridLines_ MGRSGrid** and click **Symbology**.
7. In the **Symbology** pane, click the **Symbol** option.
8. Click the **Properties** tab and set the **Line width** to 0.5 pt.
9. Click **Apply**.

   You reduced the size of the line to make the map more legible. With the grids created, you will now number the features.
10. On the **Clearing Grids** tab, set the **Label Method** to **Number**.
11. In the **Number/Letters Features** group, for **Layer**, open the **Layer** list and select **gis_osm_ pois_free**.
12. For **Field**, click the **Add Field** button. In the **Add Field** window, set the **Field Name** to Label. Click **OK**.

    You will also letter certain features.
13. Set **Label Method** to **Alpha**.
14. In the **Number/Letters Features** group, for **Layer**, open the **Layer** list and select **gis_osm_ transport_free**.

15. For **Field**, click the **Add Field** button. In the **Add Field** window, set the **Field Name** to Label. Click **OK**.

    Note: A new attribute field for the feature class is created and labels are enabled. But numbers and letters are not yet assigned.

16. Above the ribbon, locate the **Command Search** bar.
17. Type Number Features and click **Number Features (Defense Tools)**.

    The **Geoprocessing** pane appears on the right side of the screen with the **Number Features** tool open.

18. In the **Number Features** tool, apply the following parameters:
    - **Input Features:** gis_osm_pois_free
    - **Field to Number (Existing or New):** Label
19. Click **Run**.
20. In the **Command Search** bar, type Letter Features and click **Letter Features (Defense Tools)**.
21. In the **Letter Features** tool, apply the following parameters:
    - **Input Features:** gis_osm_transport_free
    - **Field to Letter (Existing or New):** Label
22. Click **Run**.

    When the tool finishes running, the labels automatically populate. To avoid cluttering the map with labels, you will use other labeling methods to call out the cities and peaks.

23. In the **Contents** pane, click **gis_osm_natural_free.** On the ribbon, click the **Labeling** tab.
24. In the **Layer** group, click **Label.**
25. In the **Label Class** group, for **Field**, confirm that **name** is selected.
26. In the **Text Symbol** group, click **Text Symbol Style.** From the menu, select **POI Black 25% Transparent.**
27. In the **Contents** pane, click **gis_osm_places_free** and confirm that you are still on the **Labeling** tab.
28. In the **Layer** group, click **Label.**
29. In the **Label Class** group, for **Field**, confirm that **name** is selected.
30. In the **Text Symbol** group, click **Text Symbol Style.** From the menu, select **Populated Place.**
31. In the **Label Placement** group, click **Label Placement Style** and select **Centered Point.**

    At this point, you have labeled all point features that will be visualized in the final product. Next, you will symbolize the road feature class to establish a simple road hierarchy.

4 Apply symbology

1. In the **Contents** pane, right-click **gis_osm_roads_free** and click **Symbology**.
2. In the **Symbology** pane, change the **Primary symbology** to **Unique Values**.
3. Set **Field 1** to **fclass**.
4. Under the **Classes** tab, hold **Ctrl** and select the following **fclass** values:
    - **footway**
    - **path**
    - **pedestrian**
    - **steps**
    - **unclassified**
5. Right-click the selected values and click **Group values**.
6. Double-click the grouped values **Label** and change the **Label value** to Secondary Roads.
7. Hold **Ctrl** and select the following **fclass** values:
    - **residential**
    - **service**
    - **tertiary**
    - **track**
    - **track_grade4**
    - **track_grade2**
8. Right-click the selected values and click **Group values**.
9. Double-click the grouped values **Label** and change the **Label value** to Primary Roads.
10. Right-click **<all other values>** and click **Remove**.
11. Click the **Symbol** for the **Secondary Roads** group. On the **Gallery** tab, select **Dashed 4:4**.
12. On the **Properties** tab, increase the **Line width** to 2 pt. Change the **Color** to one that is easily visible through the **Grid**. Click **Apply**.
13. Return to the **Primary symbology** pane. Drag the **Primary Roads** group to the top of the **Classes** list.
14. Click the **Symbol** button for the **Primary Roads** group. On the **Gallery** tab, select **2.0 Point**. On the **Properties** tab, change the **Color** to a lighter color than **Secondary Roads**.

    You grouped the **fclass** values to simplify the symbology for the roads. You do not need to separate each road value for this map scale. The following figure shows what the map should look like at this point if you chose the same feature classes as the text. Check your work and then continue.

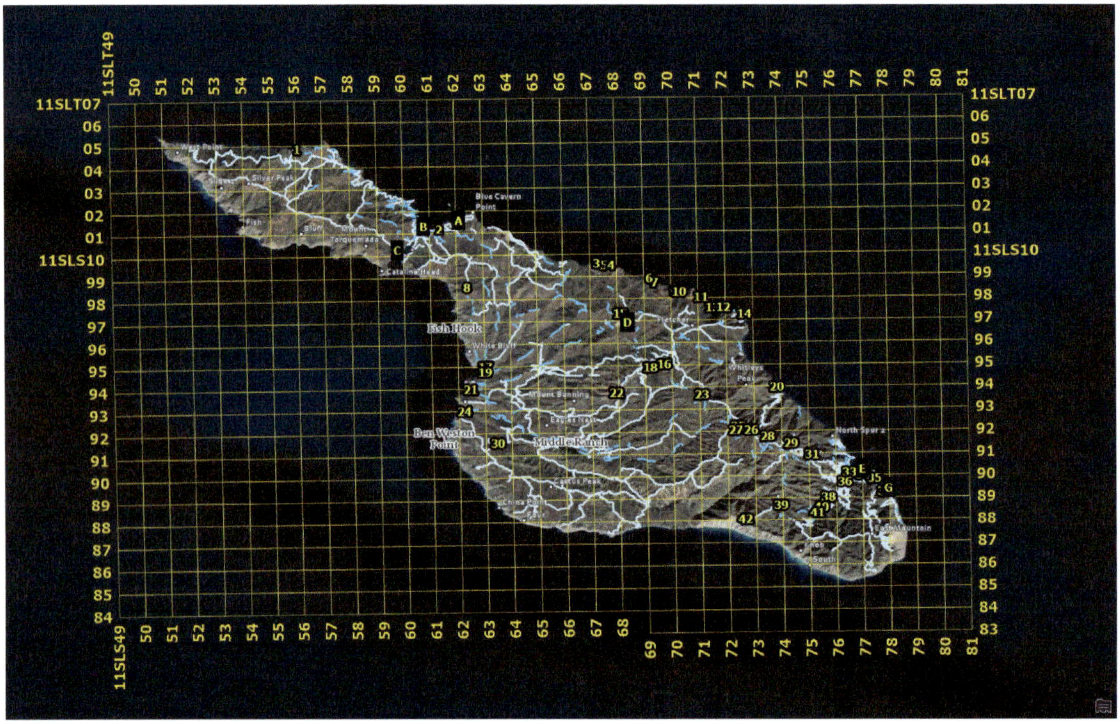

5 Disseminate the product

The final step in the workflow is to create a layout to disseminate the intelligence product. However, because of the map scale, an 8.5-by-11-inch format would be too small to be legible; instead, you will use a larger map. To address this challenge, some organizations have standardized layout templates that you may be required to use. For this example, you will import a prebuilt layout template. You will also create a table within the layout to explain each letter and number label that is used in the map.

1. On the ribbon, click the **Disseminate** tab. In the **Insert** group, click **New Layout**.
2. At the bottom of the **New Layout** menu, click **Import Layout File**.
3. Navigate to the **Catalina Island** folder and double-click **24x36 Map Template.pagx**.

   When the template opens, you need to set the map frame to reference the project map.
4. In the **Contents** pane, right-click the **Map Frame** element and click **Properties**.
5. In the **Element** pane, under **Map Frame**, change the **Map** to **Map**.

   When you set the map source, the map frame will populate the default global view. Additionally, the supporting dynamic marginalia, such as the legend, north arrow, and scale bars, will automatically reference the data within the map.
6. In the **Layout**, right-click the **Map Frame** and click **Activate**.

   Activating the map frame allows you to interact with its contents. On the ribbon, you can now access the **Map** tab as if you were in the map.

7. On the **Map** tab, in the **Navigate** group, click **Bookmarks** and select **Catalina Island.**
8. In the bottom left of the **Layout**, set the **Map scale** to 1:50,000 and press **Enter.**

Note: 1:50,000 is a preferred scale for this type of map because it is a standard scale that works with MGRS and facilitates coordinate plotting and land navigation.

9. On the top right of the **Layout**, click the **Close activated map frame** button.
10. In the **Layout**, double-click **Map Title** and rename it Catalina Island Infrastructure.
11. In the **Contents** pane, click the **Map Information** text box.

12. In the **Element** pane, after the last line of the **Text**, type - OpenStreetMap, Acquired: December 2023.

Tip: You should always credit the source and date of the data used in your maps.

13. In the **Contents** pane, expand the **Legend** element and ensure that **gis_osm_roads_free** is the only feature class checked. Right-click the layer and click **Properties**. In the **Element** pane, confirm that only the **Label (or layer name)** and **Description** boxes are checked.
14. In the **Contents** pane, expand the **Scale Bar** element and click **Scale Bar Metric**. In the **Element** pane, click the **Properties** button. Change the **Fitting Strategy** to **Adjust width** and set the **Division Value** to 5.15.
15. In the **Contents** pane, click **Scale Bar Imperial**. In the **Element** pane, click the **Properties** button. Change the **Fitting Strategy** to **Adjust width** and set the **Division Value** to 3.

Tip: You can change the scale bar division values to whole numbers whenever possible so that the scale bars are easier to use.

6 Add a table

You have now formatted the premade template to fit your map and data. The last element you need to add is a table to explain what each number and letter label represents on the map.

1. On the ribbon, click the **Insert** tab. In the **Map Surrounds** group, click **Table Frame**.
2. In the **Table Frame** menu, select **Blue Underline Heading with Alternating Rows**.
3. In the **Layout**, place the **Table Frame** under the **Map Information** element.
4. In the **Element** pane, under **Source**, change the **Table** to **gis_osm_transport_free**.
5. In the **Contents** pane, find the new **Table Frame** element. Click the **Table Frame** once, and then click it again to rename it. Change the **Name** to Transport.

6. Below **Transport**, confirm that only the boxes for **fclass**, **name**, and **Label** are checked.
7. In the **Element** pane, click the **Arrangement** button. Under **Sorting**, click the **Add** button and click **Label**. Keep the **Ascending** and **Case Sensitive** boxes checked.
8. Click the **Display** button.
9. Set the **Background Symbol Color** to white and the **Table Frame** column border to black.
10. In the **Contents** pane, expand the **Map Frame** element.
11. Right-click **gis_osm_transport_free** and click **Attribute Table**.
12. In the **Attribute Table**, on the right side, click the **menu** button and click **Fields View**.
13. Find the **fclass**, **name**, and **Label** fields and change their **Alias** to Type, Name, and Label, respectively.

    Note: You want to change the alias of the attribute fields so the table header information helps customers make better sense of the information in the table.

14. On the **Fields** tab, in the **Manage Edits** group, click **Save** to save your edits.

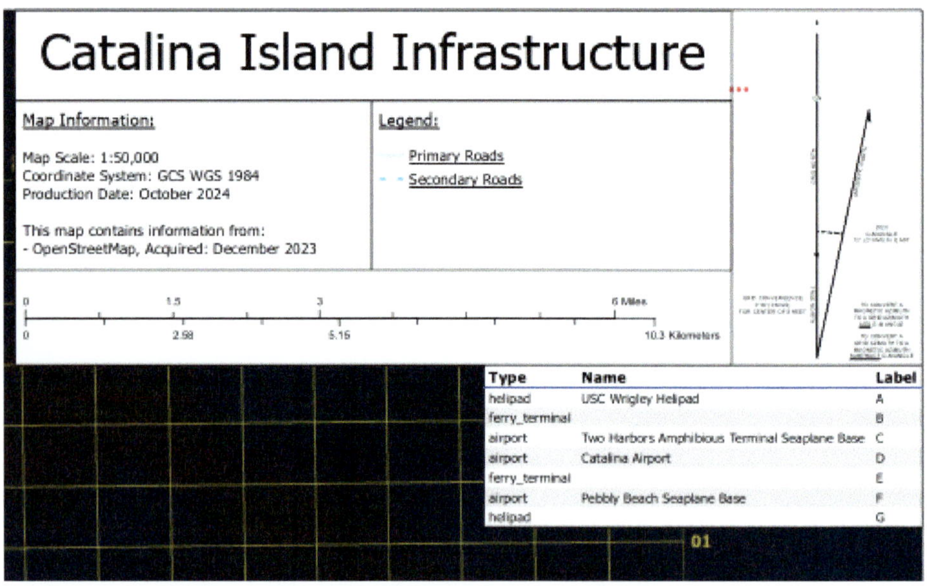

15. Create a new **Table Frame** and place it beneath the **Transport** table. This time, set the **Table source** to **gis_osm_pois_free** and change the name **Table Frame name** to POIs. Apply the same **Arrangement** and **Display** settings.

    Note: If you see the bracketed red ellipsis, you need to adjust the table frame size to display all the data. To adjust the table frame size after creating it, select the table frame and then adjust the box size by dragging it until the ellipsis disappears.

You have now imported a layout template, adjusted the template to be accurate for the current map, and added tables that display what each label element references on the map. When you view the layout on your computer, the labels and callouts may be too small to be legible. This is because you are not viewing the map at its true size. At the bottom left of the layout pane, next to the map scale, there is a percentage number. This tells you the percentage scale at which you are viewing the map. If you set this percentage to 100 percent, the map will look legible when it is printed in full size. You can use this feature to ensure that your map labels and information are legible at the desired printing size.

At this point, you can activate the map frame and adjust the position of Catalina Island so that the map information you added does not cover the island or the grids. However, be careful not to change the map scale by zooming in or out. When you have finished adjusting the layout, proceed to the next steps to export the final map.

16. On the ribbon, click the **Disseminate** tab. In the **Export** group, click **Export Layout** > **Flattened PDF**.

    The **Export Layout** pane appears on the right.

17. Specify your desired save location and set the **Name** of the file to Catalina Island Infrastructure.
18. Click **Export**.
19. When the export is complete, click **View exported file** to open your map or navigate to the file's location to open it.
20. In **ArcGIS AllSource**, click **Save Project** and exit.

    You can compare your final product with the PDF file Catalina Island Infrastructure Example in the Catalina Island folder.

Interpretation of analysis

After completing this exercise, answer the following questions in detail. The answers to these questions, as well as the analysis results from the exercise workflow, will inform the intelligence brief.

1. What are two to three valuable takeaways from the workflows you worked on in this chapter?
2. What information was produced at the local scale of analysis? Are there any critical vulnerabilities? How many entrances and exits are there? What is the most important thing a first responder would want to know if responding to an emergency at USC Wrigley?
3. What actionable information was produced at the regional scales of analysis? How might this information be used?
4. State and explain two advantages and two disadvantages of creating a gridded reference graphic at the local scale for a given human security threat.
5. State and explain two advantages and two disadvantages of creating a gridded reference graphic at the regional scale for a given human security threat.

This workflow has two significant outputs. One is a facility overlay with a gridded reference that identifies critical infrastructure, key locations, and other features. The second is a regional overview with MGRS grids that calls out major settlements, key terrain features, and critical infrastructure with labels and a reference chart. These outputs can be disseminated using hosted web maps through an ArcGIS portal or hard-copy printouts if you are in an environment with limited network connectivity. With these workflows, you can break down facilities into their key components to determine vulnerable areas, choke points, obstacles, and other places where hazards may exist that can affect operations at the location. The facility overlay and gridded reference graphics displayed in this chapter are small examples of the possible applications and use cases for this product type.

Intelligence brief

Produce an intelligence brief that describes how vulnerable the critical infrastructure of USC Wrigley and Catalina Island are to wildfires. Structure the brief with headings that match the following key points. Refer to Intelligence Community Directive 203–Analytic Standards to guide the writing of your intelligence brief.

- **BLUF:** List three to five primary observations from your analysis in bulleted-list format that summarize findings of this work.
- **Top-level analysis:** Report the primary data points supporting your BLUF.
- **Secondary analysis:** Provide the data sources and deeper analytic products from this analysis.
- **Impact:** Describe the impacts of your findings on the department or organization.
- **Confidence:** Report the level of confidence of your assessment based on available data, accuracy, and so on.
- **Bias:** Explain any background or information you have that could have affected your analysis.
- **Conclusions:** Determine what type of analytic conclusions there may be and how the data requirements, analysis, and brief writing may change based on what analytic framework you are following (descriptive, predictive, prescriptive).
- **Information gaps**: Discuss what additional data you would want to collect to improve your confidence rating or change your overall assessment.
- **Attachments**: Include the two exported gridded reference maps as attachments to the brief.

Applying and extending your skills

This chapter showcases two examples of how a facility overlay and gridded reference graphic can be used. However, it is not meant to be an authoritative explanation of this type of product format because it can be modified and combined with other workflows to create custom products that meet a user's needs. Facility overlays are essential for emergency response teams responding to natural disasters. Knowing how an environment has changed after a hurricane or earthquake will help protect emergency response personnel by ensuring that they avoid dangerous routes while still enabling them to reach people who need their help. The gridded reference graphic also helps personnel coordinate actions, especially in unfamiliar locations. Depending on the question you are answering and your customer's needs, you can modify the basic workflow to create customized products that meet those needs.

The two examples in this chapter show only two applications of this type of workflow. You can apply these steps to create overlay graphics in practically any area of interest and at any scale. Common examples include airports, port facilities, school campuses, and key areas of cities. Another example includes vertical obstructions for pilots and key terrain for operational planning. The primary difference between each application is the features you collect and the scale at which you display the grids.

Additionally, you can pair these workflows with more advanced analytic tools. One example is the ArcGIS AllSource Deep Learning extension, which can automatically extract building outlines over large areas using pretrained models. Other options include using spectral analysis to determine vegetation health or different types of imagery, such as synthetic aperture imagery, to perform change detection analysis. Those advanced imagery exploitation and geospatial analysis techniques provide greater depth for those using the products.

Data sources

- OpenStreetMap, November 2023. https://openstreetmap.org/copyright

Additional resources

- Information about OpenStreetMap can be found at www.openstreetmap.org/about.
- For more information about intelligence community directives (ICDs)—particularly ICD 203, which governs analytic standards for intelligence professionals—see www.dni.gov/index.php/what-we-do/ic-related-menus/ic-related-links/intelligence-community-directives.
- For more information about how the Universal Transverse Mercator (UTM) and MGRS coordinate systems are defined and read, see https://web.archive.org/web/20210224185834/earth-info.nga.mil/GandG/coordsys/mmr201.pdf.
- For documentation about specific tools and functions discussed in this chapter, go to Esri ArcGIS AllSource documentation (www.esri.com/en-us/arcgis/products/arcgis-allsource/overview) and search the Help for these entries:
  - **Clearing grids**: Resources > Data > Imagery > Clearing Grids > Get Started
  - **Command search**: Resources > Get Started > General > Around the Application
  - **Definition queries**: Resources > Visualization > Multidimensional Visualization > Interact with Maps and Scenes > Query Expressions
  - **Layouts**: Resources > Production > Production and Dissemination > Layouts
  - **Number features**: Resources > Analysis > Geoprocessing Tools > Defense Toolbox > Gridded Reference Graphic Toolset > Number and Letter Toolset

About the author

Jacob Spear is an imagery analyst in the US Marine Corps. In that capacity, he has supported national defense, humanitarian response, and civil works missions. He is also an instructor with the National Geospatial-Intelligence Agency College, where he teaches imagery analysis, sensor phenomenology, and GEOINT collections. He has a master's degree in human security and geospatial intelligence and holds the NGIA GEOINT Professional certifications for imagery analysis, geospatial analysis, cartography, and human geography.

Chapter 5

Mapping flooded urban areas from the Nova Kakhovka, Ukraine, dam destruction

Rúben Santiago, *Portuguese Navy, Geospatial, Meteorological, and Oceanographic Centre, Lisbon, Portugal*
Sofia Henriques, *Portuguese Navy, Geospatial, Meteorological, and Oceanographic Centre, Lisbon, Portugal*

Overview

In this chapter, you will learn how to use GIST tools and methods to determine the extent of flooding that resulted from the destruction of the Nova Kakhovka Dam in Kherson Oblast, Ukraine. The procedures presented here will provide valuable outputs for analysis and facilitating informed decision-making in HSGI.

Learning objectives

- Apply a series of preprocessing steps to Sentinel-1 satellite imagery, including updating orbit data, removing thermal noise, calibrating to retrieve a meaningful backscatter value, mitigating speckle noise, removing radiometric and geometric distortions, and scaling the image to a different unit.
- Generate RGB composite images from preprocessed satellite data to enhance visual interpretation, allowing for easy identification of features such as water bodies and flooded areas.
- Apply thresholding techniques to pre- and postflood images to create binary images of water or land and perform differential analysis to quantify the extent of the flooded areas, enabling accurate assessment of the impact of the flood event.
- Use geospatial data, such as RGB composite images, to remove noise from a binary image and refine flood-mapping results.

- Use ArcGIS software to combine processed satellite data with additional datasets to perform spatial analysis tasks, such as clipping the area of interest, converting raster to vector (polygons), and calculating the amount of affected adjacent urban areas.

Technical components

- Software:
  - ArcGIS Pro 3.4
  - ArcGIS Image Analyst extension
- Datasets:
  - Ukraine administrative boundaries
  - Sample adjacent urban/populated areas
  - Sentinel-1 imagery of study area
  - Digital elevation model (DEM) of study area

Table 5-1 provides full details about the imagery data.

Table 5-1. Imagery data details

Label	Product	Instrument	Date captured	Format	Data source
Preflood	Sentinel-1 IW GRDH 1SDV	C-SAR	June 2, 2023	ZIP	Copernicus Open Access Hub[1]
Postflood	Sentinel-1 IW GRDH 1SDV	C-SAR	June 9, 2023	ZIP	Copernicus Open Access Hub

Product	Spatial resolution	Date captured	Elevation unit	Data source
Merged DEM (SRTM1N46E032V3 & SRTM1N46E033V3)	1-ARC	February 11, 2000	Meters	Shuttle Radar Topography Mission (NASA)[2]

1. https://scihub.copernicus.eu/dhus/#/home
2. https://www.earthdata.nasa.gov/sensors/srtm

Prerequisites

- Intermediate knowledge or experience with GIS software or spatial data is required.
- Technical skill level: Intermediate GIS.
- Time to complete chapter: Two to three hours without intelligence brief, four to six hours with intelligence brief.

Background and situational context

During the escalating conflict between Russia and Ukraine, a series of strategic maneuvers and attacks unfolded with the objective of weakening the opposing forces' capabilities and diminishing their ability to retaliate effectively. One such pivotal event occurred on June 6, 2023, when the Nova Kakhovka Dam within Ukrainian territory was deliberately destroyed. The circumstances surrounding this incident suggest that it was a calculated action, potentially carried out by Russia (Glanz et al. 2023), which was positioned to gain considerable advantage from such an act. The timing of the dam's destruction coincided with Ukraine's preparations for a counteroffensive, intensifying suspicions that Russia aimed to disrupt and incapacitate Ukraine's strategic response.

During this period, Russia's control over the region where the dam was located further complicated the situation, hindering independent investigation (Wintour 2023). As a result, the precise details and motivations behind the destruction remained shrouded in uncertainty, amplifying the urgency to discern the extent of the aftermath, particularly the impact of the ensuing floods.

The repercussions of the dam's collapse reverberated through the landscape, inundating large swaths of land, including urban centers (AJLabs 2023). To comprehensively assess the aftermath and gauge the scale of the flooding's impact, a multidisciplinary approach should be adopted. This approach merges advanced satellite imagery analysis techniques with geospatial data resources, aiming to map and quantify the flooded regions. The objective is to provide actionable intelligence by effectively capturing and visualizing the extent of flooding, contributing to a comprehensive understanding of the human security threat caused by this orchestrated event.

Considering these circumstances, this study embarks on a focused endeavor to employ Sentinel-1 satellite imagery in tandem with OpenStreetMap data. The integration of these diverse datasets offers a unique opportunity to shed light on the consequences of the dam's destruction and contribute valuable insights into the intricate puzzle of a conflict-driven environmental disaster. Through the meticulous analysis of the flooded areas, including urban zones, this research seeks to delineate the scope of the affected regions and quantify the impact on populated areas. The amalgamation of advanced remote sensing technologies and geospatial intelligence aims to address a critical information gap in the wake of this strategically significant incident.

Exercise workflow 5

This exercise provides an in-depth exploration of the methodologies and findings arising from this interdisciplinary approach, underscoring the importance of harnessing geospatial analysis to inform responses to complex and dynamic security challenges.

1 Detect flooded areas

1. Start ArcGIS Pro. Under **New Project**, click **Map.**
2. In the **New Project** window, apply the following settings:
    - **Name**: Chapter_05
    - **Location:** C:\SecurityFirst

    Note: This exercise uses the C:\SecurityFirst folder.

3. Click **OK.**
4. On the **Map** tab, in the **Layer** group, click **Add Data.**
5. In the **Chapter05** folder, navigate to the **imagery** folder where the Sentinel-1 images are stored. Open the folder named **S1A_IW_GRDH_1SDV_20230602T154509.**
6. In the folder, click **manifest.safe.** Click **OK** to load the image data in the map.
7. If the option to **Build Pyramids and Calculate Statistics** appears, click **OK.**

    The **manifest.safe** file contains the image data and serves as the image itself within the software's processing context.

8. Add data from the **S1A_IW_GRDH_1SDV_20230609T153657** folder. Load **manifest.safe** in the map.

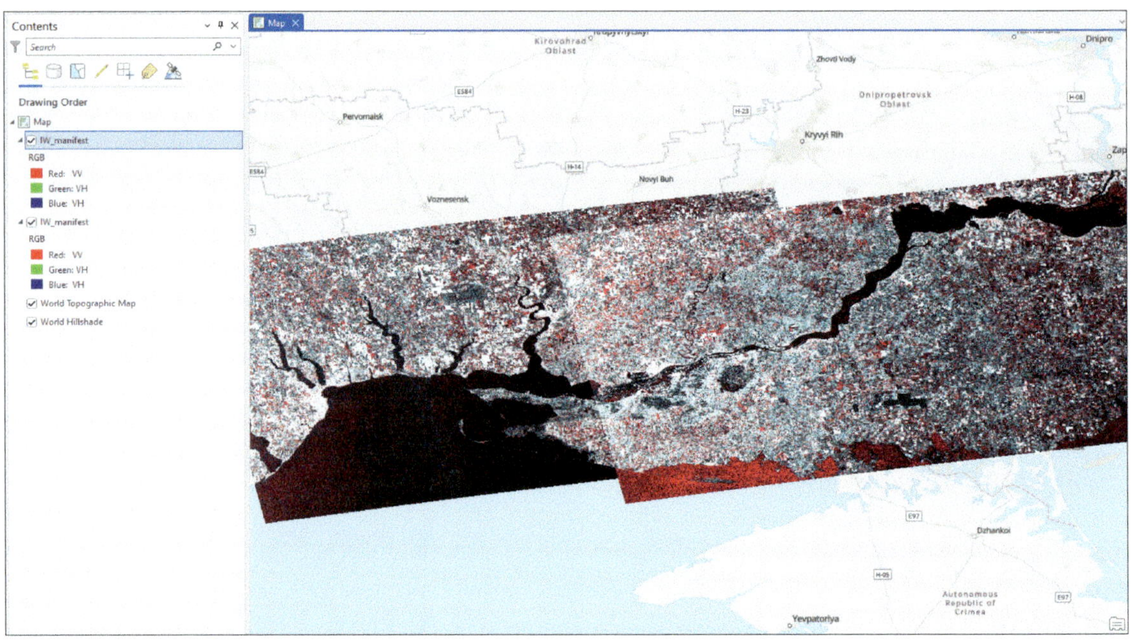

It is important to confirm that you have both VV (vertical-vertical) and VH (vertical-horizontal) polarizations available for your analysis. These polarizations contain valuable information that can aid in accurately identifying flooded areas.

9. In the **Contents** pane, first uncheck and then check the **IW_manifest** layer at the top. Determine which raster layer represents the image on the left and which represents the image on the right in the map.
10. Double-click the raster layer that represents the image on the left.
11. In the **Layer Properties** window, click **General.** Change the **Name** of the raster to Pre_flood_02062023. Rename the other raster layer Post_flood_09062023.
12. Right-click **Pre_flood_02062023** and click **Symbology.** Apply the following settings:
    - **Primary Symbology**: Stretch
    - **Band:** VH (55000000-55000000 nm)
    - **Number of standard deviations:** 2
    - **Gamma:** 2.0

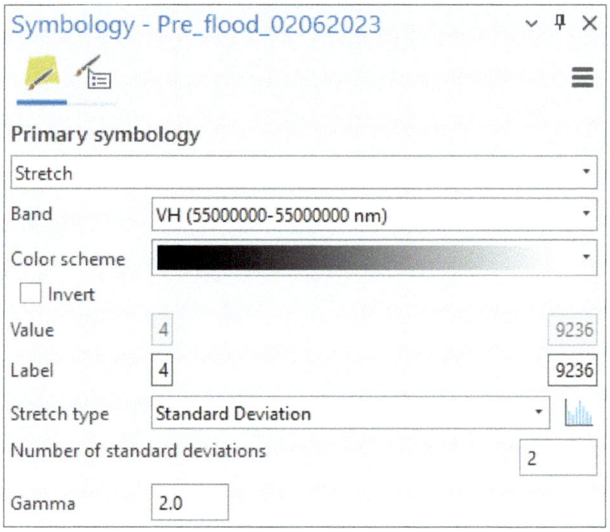

13. Apply the same **Symbology** settings to **Post_flood_09062023.**

    Before proceeding, it is essential to add the administrative divisions of Ukraine and the DEM to the project. Additionally, you need to delineate an area of interest covering the region from the Nova Kakhovka Dam to the mouth of the Dnieper River, where both the preflood and postflood images intersect. This area of interest will serve as the basis for clipping the provided DEM to focus the processing. When the area of interest is defined, you will clip it from the northern border of Kherson Oblast. You will add these layers in the following steps.

14. From the chapter data folder, in the **vector_data** subfolder, add **ukr_admin_boundaries_level1.shp** to the map. Next, add **dem.tif** and **area_of_interest.shp**.

You will now clip the DEM to this extent.

15. On the **Analysis** tab, in the **Geoprocessing** group, click **Tools**.
16. In the **Geoprocessing** pane, search for Clip Raster. Open the tool and apply the following settings:
    - **Input Raster**: dem.tif
    - **Output Extent**: area_of_interest
    - **Output Raster Dataset**: dem_Clip
    - **Use Input Features for Clipping Geometry**: checked
17. Click **Run**.

    Clipping the DEM ensures that the final products of the radiometric terrain flattening and geometric terrain correction are limited to the spatial boundaries of the area of interest. This enhances processing efficiency by reducing unnecessary computations outside the specified area.

18. Ensure that the resulting clip raster is the first layer in the **Contents** pane, above the **area_of_interest** layer.

    Next, you will continue processing the synthetic aperture radar (SAR) images using tools in the SAR toolset of ArcGIS Image Analyst. You will repeat these steps for the pre- and postflood SAR datasets.

    When preparing SAR images for analysis, an important step is to apply orbit correction. No matter how fine-tuned a satellite's orbit may be, the location of the satellite will drift owing to various factors such as atmospheric drag and gravitational forces from other celestial bodies. Thus, having up-to-date orbital files is crucial for precise georeferencing.

2 Apply orbit correction and remove thermal noise

1. In the **Geoprocessing** pane, search for the Download Orbit File tool. Select it, add the following settings, and click **Run**:
    - **Input Radar Data**: Pre_Flood_02062023
    - **Orbit Type**: Sentinel Precise
    - **Folder**: Chapter_05

    Note: Chapter_05 is the project data folder.

2. Rerun the tool for the post-flood dataset, adding the following settings and then clicking **Run**:
    - **Input Radar Data**: Post_Flood_09062023
    - **Orbit Type**: Sentinel Precise
    - **Folder**: Chapter_05

    You will now use the downloaded orbit file to correct geometric distortions by using the **Apply Orbit Correction** tool.

3. In the **Geoprocessing** pane, search for the Apply Orbit Correction tool. Add the following settings and click **Run**:
    - **Input Radar Data**: Pre_Flood_02062023
    - **Input Orbit File**: Chapter_05\ S1A_OPER_AUX_POEORB_OPOD_20230622T080837_V20230601T225942_20230603T005942.EOF
    - **Folder**: Chapter_05

4. Rerun the tool for the postflood dataset, adding the following settings, and click **Run**:
    - **Input Radar Data**: Post_Flood_09062023
    - **Input Orbit File**: Chapter_05\ S1A_OPER_AUX_POEORB_OPOD_20230629T080846_V20230608T225942_20230610T005942.EOF
    - **Folder**: Chapter_05

    You will now correct backscatter disturbances caused by thermal noise in the SAR data by using the **Remove Thermal Noise** tool. Thermal noise is most noticeable in areas of low backscatter, such as water bodies, and can affect image quality.

5. Search for and open the Remove Thermal Noise tool. Add the following settings and click **Run**:

   - **Input Radar Data**: Pre_Flood_02062023
   - **Output Radar Data**: Chapter_05\Pre_Flood_02062023_TNR.crf
   - **VV** and **VH** boxes: checked

6. Rerun the tool for the postflood dataset, adding the following parameters and clicking **Run**:

   - **Input Radar Data:** Post_Flood_09062023
   - **Output Radar Data**: Chapter_05\Post_Flood_09062023_TNR.cr
   - **VV** and **VH** boxes: checked

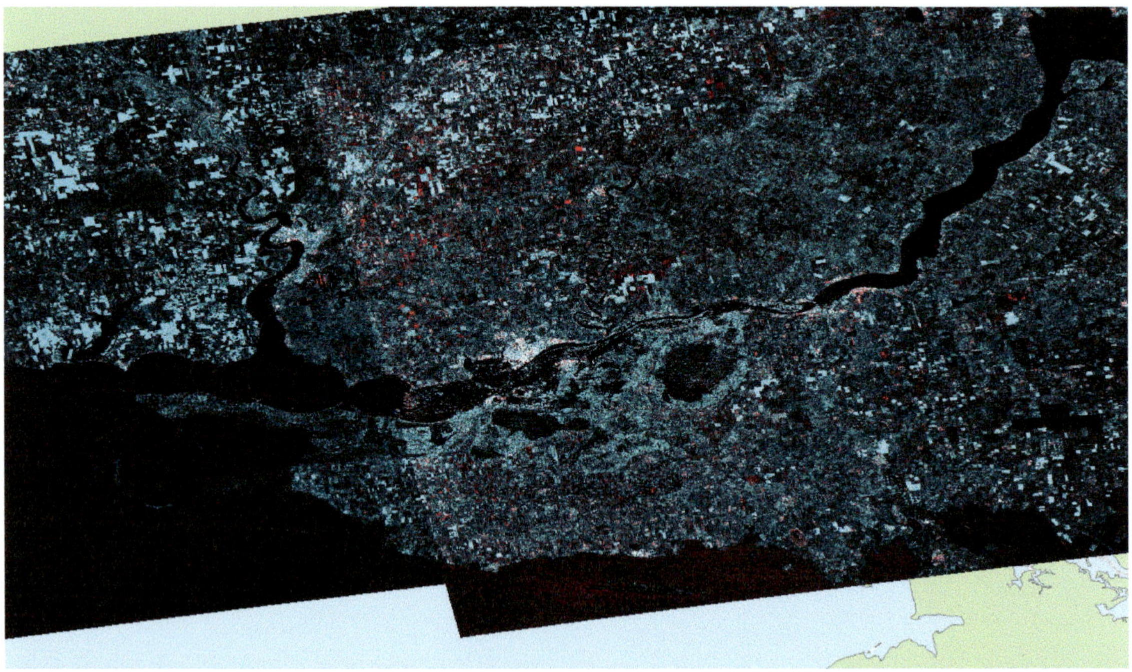

Tip: Both VV and VH polarizations should be active to allow for a comparative analysis of their effectiveness in detecting flooded areas. Whereas VV polarization is generally more responsive to smooth water surfaces, VH polarization enhances contrast in flooded vegetation and urban zones. Keeping both active ensures a comprehensive assessment, allowing you to determine which polarization provides the most accurate results for this specific flood event.

You will now standardize intensity values across the SAR images by using the **Apply Radiometric Calibration** tool. In this process, the tool will use the clipped DEM previously prepared, which accounts for variations in terrain elevation, ensuring accurate calibration of intensity values across the image.

7. Search for and open the Apply Radiometric Calibration tool. Add these settings and click **Run**:
   - **Input Radar Data:** Pre_Flood_02062023_TNR.crf
   - **Output Radar Data:** Pre_Flood_02062023_TNR_CalB0.crf
   - **VV** and **VH** boxes: checked
   - **Calibration Type:** Beta nought
8. Rerun the tool for the postflood dataset with the following parameters:
   - **Input Radar Data:** Post_Flood_09062023_TNR.crf
   - **Output Radar Data:** Post_Flood_09062023_TNR_CalB0.crf
   - **VV** and **VH** boxes: checked
   - **Calibration Type:** Beta nought

You will now correct radiometric distortions due to topography by using the **Apply Radiometric Terrain Flattening** tool. This process normalizes backscatter values to eliminate distortions caused by terrain variations. The tool uses the clipped DEM to compensate for variations in terrain elevation, allowing for precise adjustment of backscatter values and ensuring uniformity in the image.

9. Search for and run the Apply Radiometric Terrain Flattening tool with the following settings:
   - **Input Radar Data:** Pre_Flood_02062023_TNR_CalB0.crf
   - **Output Radar Data:** Pre_Flood_02062023_ TNR_CalB0_RTFS0.crf
   - **DEM raster:** dem_Clip
   - **Apply geoid correction** box: checked
   - **VV** and **VH** boxes: checked
   - **Calibration Type:** Sigma nought
10. Rerun the tool for the postflood dataset with the following settings:
   - **Input Radar Data:** Post_Flood_02062023_TNR_CalB0.crf
   - **Output Radar Data:** Post_Flood_02062023_ TNR_CalB0_RTFS0.crf
   - **DEM raster:** dem_Clip
   - **Apply Geoid Correction** box: checked
   - **Calibration Type:** Sigma nought

You will now correct speckle noise using the **Despeckle** tool. Speckle noise is a major challenge when analyzing SAR images. It appears as a grainy or salt-and-pepper effect, making it difficult to clearly distinguish between land and water areas.

11. Search for and run the Despeckle tool with the following settings:
    - **Input Radar Data**: Pre_Flood_02062023_ TNR_CalB0_RTFS0.crf
    - **Output Radar Data**: Pre_Flood_02062023_ TNR_CalB0_RTFS0_Dspk.crf
    - **VV and VH** boxes: checked
    - **Filter Type**: Lee
    - **Filter Size**: 7 × 7 (check results and adjust as needed)
12. Rerun the tool for the postflood dataset with the following parameters:
    - **Input Radar Data**: Post_Flood_02062023_ TNR_CalB0_RTFS0.crf
    - **Output Radar Data**: Post_Flood_09062023_ TNR_CalB0_RTFS0_Dspk.crf
    - **VV and VH** boxes: checked
    - **Filter Type**: Lee
    - **Window Size**: 7 × 7 (check results and adjust as needed)

Tip: It is best to use the Despeckle tool with a Lee filter type and a window size of 7 × 7 for optimal results. Although the default parameters for the remaining settings are acceptable, you can adjust them as needed during processing. The objective is to generate two images with minimal noise and high resolution while ensuring clear differentiation between water and land.

You will now correct geometric distortions and orthorectify the SAR data by using the **Geometric Terrain Correction** tool. This process ensures accurate positioning of ground objects by accounting for sensor view angles and ground terrain. The DEM is also used. It provides crucial terrain elevation data used to adjust the SAR image's perspective, compensating for terrain variations and ensuring accurate positioning of ground features.

13. Search for and run the Apply Geometric Terrain Correction tool with the following settings:
    - **Input Radar Data**: Pre_Flood_02062023_ TNR_CalB0_RTFS0_Dspk.crf
    - **Output Radar Data**: Pre_Flood_02062023_TNR_CalB0_RTFS0_Dspk_GTC.crf
    - **VV and VH** boxes: checked
    - **DEM Raster**: dem_Clip
    - **Apply geoid correction** box: checked
14. Rerun the tool for the postflood dataset with the following settings:
    - **Input Radar Data**: Post_Flood_09062023_ TNR_CalB0_RTFS0_Dspk.crf
    - **Output Radar Data**: Post_Flood_09062023_TNR_CalB0_RTFS0_Dspk_GTC.crf
    - **VV and VH** boxes: checked
    - **DEM Raster**: dem_Clip
    - **Apply geoid correction** box: checked

You will now convert the scaling of the SAR data from linear to decibels using the **Convert SAR Units** tool.

15. Search for and run the Convert SAR Units tool with the following settings:
    - **Input Radar Data:** Pre_Flood_02062023_TNR_CalB0_RTFS0_Dspk_GTC.crf
    - **Output Radar Data:** Pre_Flood_02062023_TNR_CalB0_RTFS0_Dspk_GTC_dB.crf
    - **Conversion Type:** Linear to dB
16. Rerun the tool for the postflood dataset with the following settings:
    - **Input Radar Data:** Post_Flood_09062023_TNR_CalB0_RTFS0_Dspk_GTC.crf
    - **Output Radar Data:** Post_Flood_09062023_TNR_CalB0_RTFS0_Dspk_GTC_dB.crf
    - **Conversion Type:** Linear to dB

Tip: Converting to dB simplifies the interpretation and display of SAR images by reducing the range of amplitude values, with positive values representing backscatter toward the sensor and negative values representing backscatter away from the sensor. This compression of wide signal intensity ranges enhances contrast, making it easier to perceive differences in radar returns from different surfaces, such as land and water. The dB scale is more effective in highlighting variations and is better suited for human visual perception.

17. In the **Contents** pane, right-click the **Post_Flood_09062023_TNR_CalB0_RTFS0_Dspk_GTC_dB.crf** layer and click **Symbology**.
18. In the **Symbology** pane, change the **Primary symbology** to **Stretch**. In the **Band** field, select **VV** or **VH** to switch between the two polarizations.

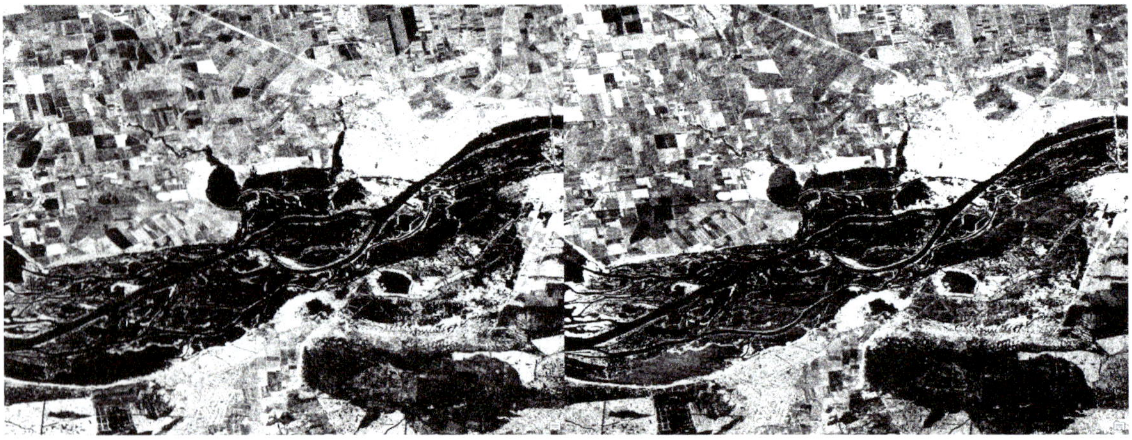

Based on the final products generated, VH polarization radar images reveal clearer distinctions between water and land surfaces as compared with VV polarization. This clarity can be attributed to the way that land surfaces interact with radar waves, including factors such as surface roughness, structure, and vegetation. VH polarization might accentuate these differences, making land features more discernible from water bodies in radar images. The choice of polarization should be based on a careful assessment of the specific characteristics and objectives of your study area.

> **Tip:** It is always recommended to experiment with both polarizations and compare their results to determine the most suitable choice for a particular analysis.

Several methods can be used to identify flooded areas. Two of the most popular ones are using an RGB composite or creating a binary image based on a threshold. In visual terms, the RGB composite is superior; however, applying a threshold method is more practical for quantifying flooded areas. Therefore, you will use the RGB composite to eliminate the noise resulting from the threshold application.

3 Create a composite raster layer

Before applying the RGB composite, you will integrate the bands of both pre-flood and postflood images into a single raster layer by using the **Composite Bands** tool.

1. Search for and run the Composite Bands tool with the following settings:
   - **Input Rasters**: Pre_Flood_02062023_TNR_CalB0_RTFS0_Dspk_GTC_dB.crf and Post_Flood_09062023_TNR_CalB0_RTFS0_Dspk_GTC_dB.crf
   - **Output Raster**: PrePost_Flood_CompositeBand

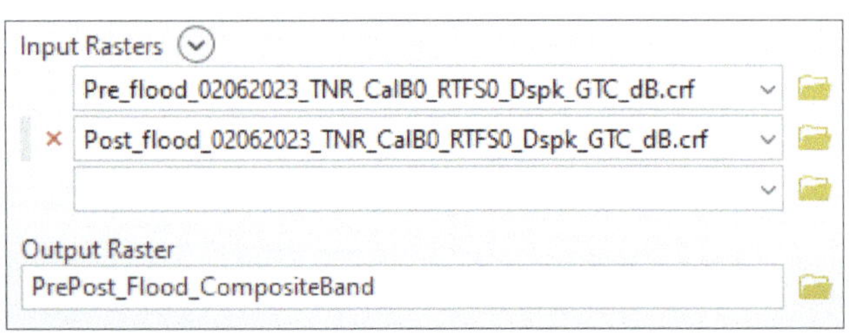

The order in which you select input rasters in the **Composite Bands** tool is crucial because the generated composite will not maintain the band names. Instead of having **VV** or **VH** for each of the images, as you had initially, you will have **Band_1**, **Band_2**, **Band_3**, and **Band_4**. The bands of interest in this composite will be **Band_2** and **Band_4**, associated with preflood **VH** and postflood **VH**, because those polarizations best differentiate land and water areas. If you inserted the preflood image first, **Band_2** will be associated with that image, and **Band_4** will be associated with the postflood image. If you inserted the postflood image first, the association will be the opposite.

Next, you will create the RGB composite image using the **Create Color Composite** tool.

2. Search for and run the Create Color Composite (Image Analyst Tools) tool with the following settings:

    - **Input Raster**: PrePost_flood_CompositeBand
    - **Output Raster**: PrePost_flood_CompositeBand_RGB.crf
    - **Method**: Band names
    - **Red Expression**: Band_2
    - **Green Expression**: Band_4
    - **Blue Expression**: Band_4

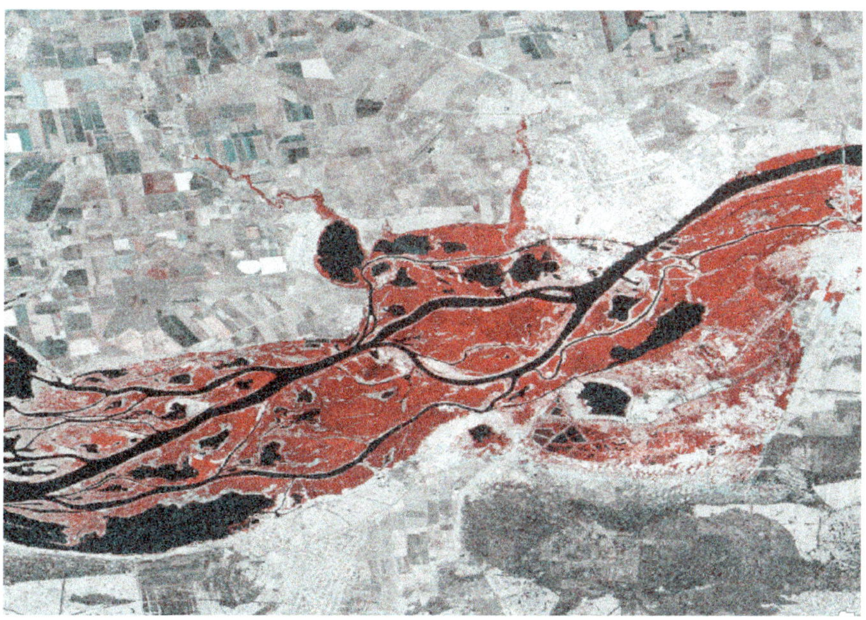

The RGB composite is crucial for reducing noise in binary images derived from thresholding. One method to establish a threshold involves creating polygon layers within the postflood image, particularly at the border zones between land and water, using a new layer. Subsequently, you will use the **Extract Values to Points** tool to extract raster values at the polygon points. You can then generate a histogram, focusing solely on these polygon zones, to provide statistics exclusively for those areas. This process facilitates the identification of two distinct value peaks, with the "valley" between them establishing the threshold value.

You will now create a new shapefile (feature class) for the digitized polygons.

3. In the **Catalog** pane, expand the **Folders** folder. Right-click the **Chapter_05** project folder and click **New** > **Shapefile**.
4. Run the **Create Feature Class** tool with the following settings:
   - **Feature Class Name:** Polygons
   - **Geometry Type:** Polygon
   - **Coordinate System:** GCS_WGS_1984
5. To begin drawing the polygons, click the **Edit** tab on the ribbon. In the **Features** group, click **Create**.
6. In the **Create Features** pane, click the **Polygons** layer that you created earlier to activate it. Beneath the layer name, select the **Rectangle** tool.
7. On the map, draw at least 15 rectangles in areas where land meets water as described earlier when establishing the threshold using the RGB composite.
8. When you have finished, click the **Save** button in the **Manage Edits** group.

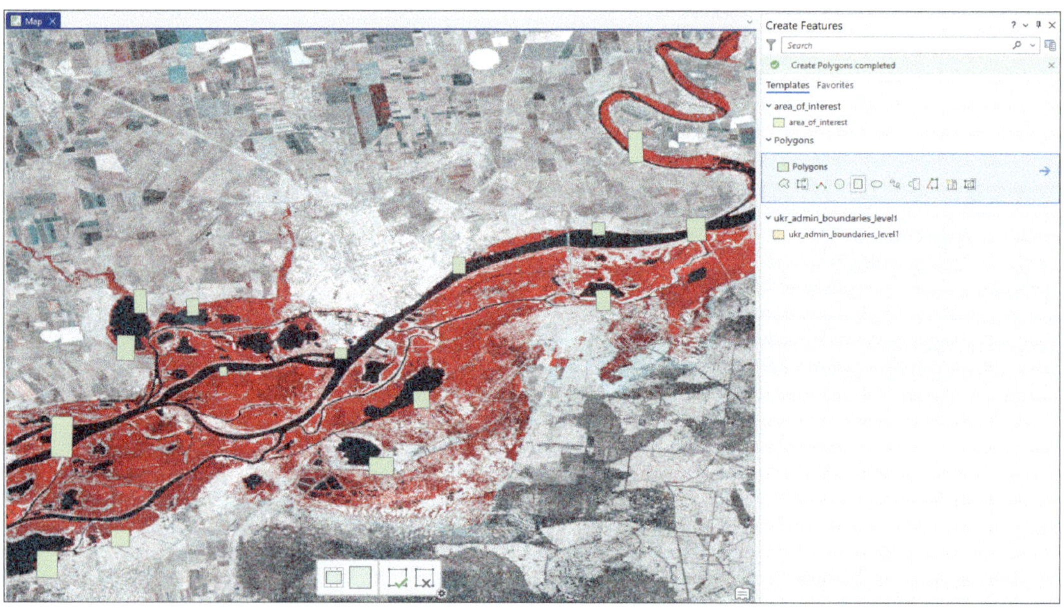

You will now extract the **VH** band from the postflood image by using the **Extract Bands** tool.

9. On the ribbon, click the **Imagery** tab. In the **Analysis** group, click **Raster Functions**.
10. In the **Raster Functions** pane, search for and open the Extract Bands tool. Enter the following settings and then click **Create new layer**.
    - **Raster:** Post_Flood_09062023_TNR_CalB0_RTFS0_Dspk_GTC_dB.crf
    - **Method:** Band IDs
    - **Band:** 2
    - **Combination:** 2

A new layer containing Band 2 (VH polarization) and named **Extract Bands_Post_Flood_09062023_TNR_CalB0_RTFS0_Dspk_GTC_dB** is created.

You will now use the **Clip Raster** tool to clip the extracted VH band using the previously created polygons as the clip layer.

11. In the **Geoprocessing** pane, search for and run the Clip Raster tool with the following settings:
    - **Input Raster:** Extract Bands_Post_Flood_09062023_TNR_CalB0_RTFS0_Dspk_GTC_dB
    - **Output Extent:** Polygons
    - **Use Input Features for Clipping Geometry** box: checked
    - **Output Raster:** ExtractBands_Post_Flood_Clip

    You will now generate a more detailed histogram, which will offer higher resolution than the 64-bin histogram available through the standard method. To do this, you will use the **Reclassify** tool on the newly clipped raster and use this reclassified dataset to determine the threshold.

12. Search for and run the Reclassify tool with the following settings:
    - **Input Raster:** ExtractBands_Post_Flood_Clip
    - **Reclass Field:** VALUE

13. For **Reclassification**, click **Classify**. In the window, select **Histogram** and make the following selections:
    - **Method:** Natural Breaks (Jenks)
    - **Number of Classes:** 2

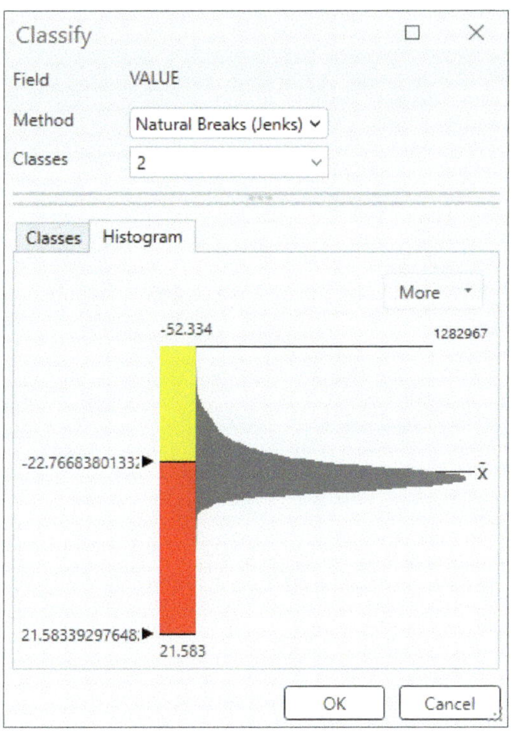

14. Click **OK** to apply the settings.

15. In the **Reclassification** table, under the **New** column, change **2** to 0.

    According to the histogram, the optimal threshold value is approximately −22.769 dB, which means that you can proceed with reclassifying the extracted (**VH**) band to generate the binary image with land-water values. Of course, this value may be adjusted to avoid misclassification of pixel noise. Pixels with values below −22.769 will be assigned the value 1 (water), whereas those above −22.769 will be assigned the value 0 (land).

16. Save the **Output raster** as Post_Flood_Binary. Click **Run**.

17. Repeat the steps for the preflood image, running the **Extract Bands** raster function on the preflood dataset, clipping the raster to the **Polygons** layer and reclassifying the data. Name the output raster Pre_Flood_Binary.

18. For both pre- and postflood reclassed rasters, change the symbology of the **Value** colors so that 1 is represented with white and 0 is represented with dark gray.

    The preflood dataset is on the left and the postflood dataset is on the right. Comparing the two reclassified images, you can see that there are areas that were improperly classified. The problem is that many areas in the terrain have identical backscatter values because of their properties. Backscatter values in radar images are influenced by factors such as surface roughness, dielectric properties, composition, vegetation, structures, incidence angle, and radar frequency sensitivity. Similar backscatter values between some land areas and water can result from shared characteristics. This is why you are going to rely on the RGB composite that you created earlier to determine that certain areas were not flooded.

19. In the **Contents** pane, remove or uncheck all layers except the following:

    - Pre_Flood_Binary
    - Post_Flood_Binary
    - PrePost_Flood_CompositeBandRGB.crf
    - ukr_admin_boundaries_level1
    - area_of_interest

20. From the chapter data folder, add the **adjacent_urban_populated_areas.shp** file to the map.

4 Apply fusion and analyze the data

You will now use the **Extract by Mask** tool to subtract the preflood binary raster from the postflood binary raster to find the newly flooded areas.

1. In the **Geoprocessing** pane, search for and run the Extract by Mask tool with the following settings:
    - **Input Raster**: Post_Flood_Binary
    - **Input raster or feature mask data**: Pre_Flood_Binary
    - **Output raster**: difference_flood_raster
2. In the **Contents** pane, right-click the **difference_flood_raster** layer and select **Attribute Table**.
3. On the left of the table, click the second row of values, which represent raster cells that have a value of 1, to select and highlight the cells on the map.

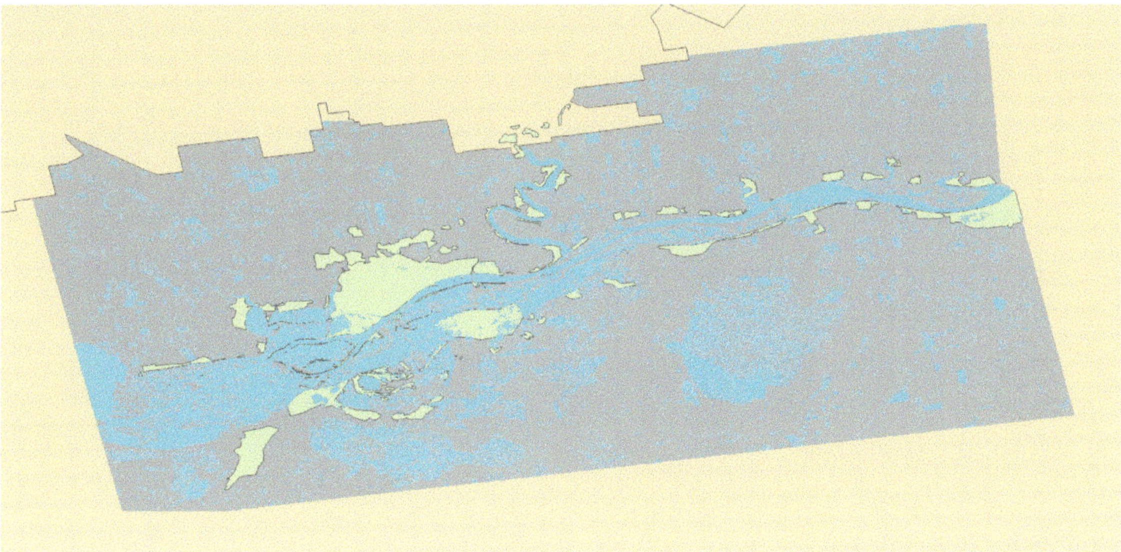

You will now transform the selection from the binary image into a vector dataset using the **Raster to Polygon** tool.

4. Open and run the Raster to Polygon tool with the following settings:
    - **Input raster**: difference_flood_raster
    - **Field**: Value
    - **Output polygon features**: difference_flood
    - **Simplify polygons** box: unchecked
5. In the **Contents** pane, rearrange the layers so that **difference_flood** is at the top and **adjacent_urban_populated_areas** is below it.

    Along with the **ukr_admin_boundaries_level1** layer, these will be the only layers that are turned on.

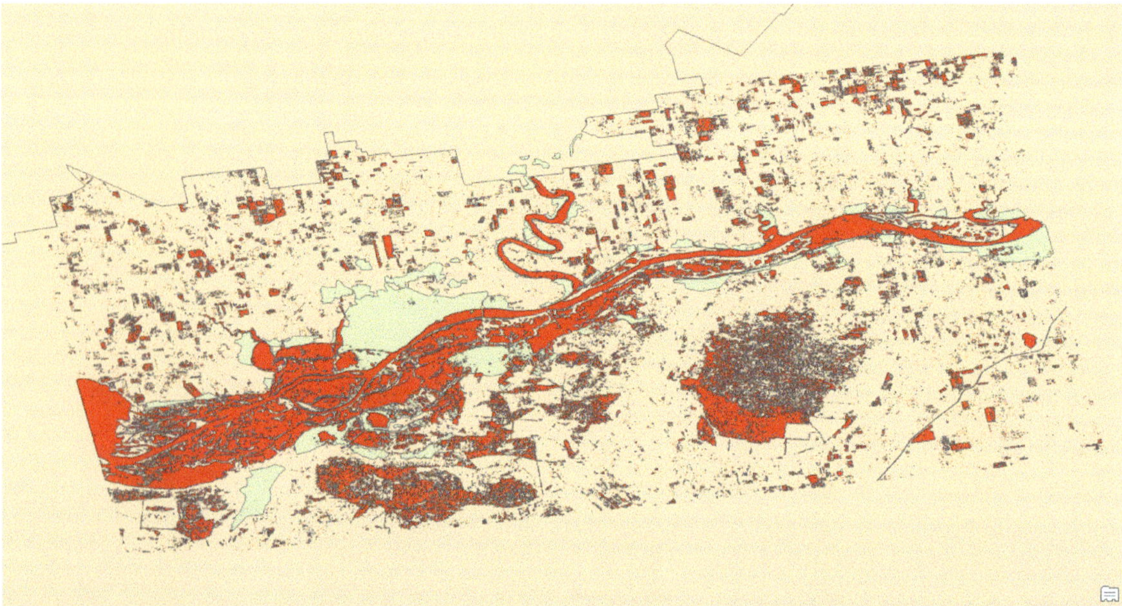

Next, you will quantify the size of the flooded area using the attribute table of the newly generated **difference_flood** dataset.

6. Open the attribute table for **difference_flood**.

   Because there is no existing field for area, you will use this newly created field to store the calculated area values.

7. On the table ribbon, click the **Add Field** button and create a new field using the following parameters:
   - **Field Name**: Area
   - **Data Type**: Double or Float
8. On the ribbon, on the **Fields** tab, click the **Save** button.
9. In the table, right-click the header of the newly created **Area** field and select **Calculate Geometry**.
10. In the **Calculate Geometry** window, set the **Property** for the **Area** field to **Area (geodesic)**. For **Area Unit**, select **Square Meters**. Click **OK**.

    Now that you have calculated the areas for each polygon, you can examine their statistics to determine the total area of the newly flooded areas.

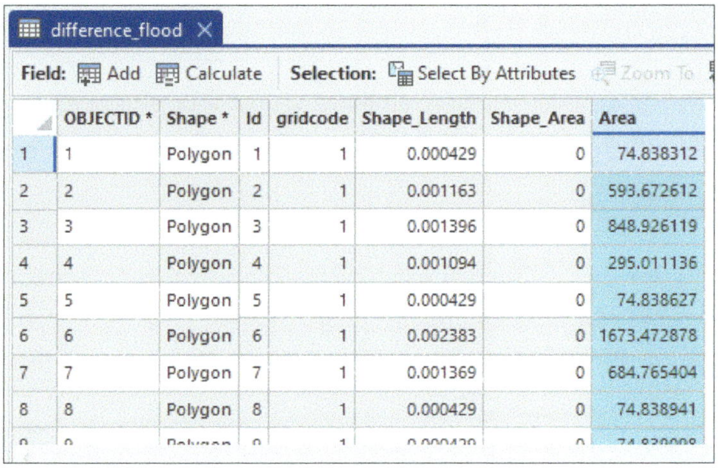

	OBJECTID *	Shape *	Id	gridcode	Shape_Length	Shape_Area	Area
1	1	Polygon	1	1	0.000429	0	74.838312
2	2	Polygon	2	1	0.001163	0	593.672612
3	3	Polygon	3	1	0.001396	0	848.926119
4	4	Polygon	4	1	0.001094	0	295.011136
5	5	Polygon	5	1	0.000429	0	74.838627
6	6	Polygon	6	1	0.002383	0	1673.472878
7	7	Polygon	7	1	0.001369	0	684.765404
8	8	Polygon	8	1	0.000429	0	74.838941

Interpretation of analysis

1. What important things did the analysis show? How can local authorities use this information to better handle disasters and plan for flooded urban areas caused by the dam break?
2. Looking at the flooded areas, did you see any patterns or trends that can help decide which areas to focus on during recovery? How can these patterns help decide where to use resources effectively?
3. How does using different types of data, such as OpenStreetMap and administrative information, help us understand the flood impact better? How can these types of data help authorities make smart choices about what to do after the flood, especially because it was not a natural event that caused it?

Intelligence brief

Based on the analysis from the exercise workflow, produce a three- to five-page intelligence brief by responding to the following question prompts. Incorporate your geospatial outputs from the exercise to enrich and illustrate your conclusions.

- **BLUF:** List three to five primary observations from your analysis in bulleted-list format that summarize findings of this work.
- **Top-level analysis:** Report the primary data points supporting your BLUF.
- **Secondary analysis:** Provide the data sources and deeper analytic products from this analysis.
- **Impact:** Describe the impacts of your findings on the department or organization.
- **Confidence:** Report the level of confidence of your assessment based on available data, accuracy, and so on.
- **Bias:** Explain any background or information you have that could have affected your analysis.
- **Conclusions:** Determine what type of analytic conclusions there may be and how the data requirements, analysis, and brief writing may change based on what analytic framework you are following (descriptive, predictive, prescriptive).

Applying and extending your skills

Expand your expertise in urban flood analysis and geospatial intelligence. The following resources can help you tackle complex challenges and contribute to comprehensive disaster management strategies.

Exploring advanced flood analysis techniques

Delve deeper into advanced flood analysis methods, such as machine learning algorithms and time-series analysis, to enhance the accuracy of flood extent mapping and prediction. These techniques can provide insights into changing flood patterns and facilitate early warning systems.

Integrating multisensor data

Broaden your skills by integrating data from multiple satellite sensors, such as optical and thermal sensors, with Sentinel-1 radar imagery. This approach can offer comprehensive insights into flood dynamics, urban resilience, and impact assessment, enhancing the effectiveness of your analysis.

Real-time monitoring and alerts

Extend your capabilities to incorporate real-time or near-real-time monitoring of flood events using continuous data feeds. Learn to set up automated alert systems that trigger notifications based on predefined flood thresholds, enabling rapid response and mitigation strategies.

Spatial analytics for urban planning

Apply geospatial analysis techniques to urban planning scenarios beyond flood assessment. Explore how GIS tools can aid in optimizing city infrastructure, identifying vulnerable zones, and designing sustainable urban development strategies that enhance disaster resilience.

Additional resources

For further information about urban flood analysis and geospatial intelligence, consider the following resources, which offer valuable insights, tools, and perspectives to enhance your expertise.

- United Nations Office for Disaster Risk Reduction (UNDRR): Access resources on disaster risk reduction strategies and urban resilience planning at www.undrr.org.
- NASA Earth Science Data Systems (ESDS): Learn about satellite data sources, remote sensing techniques, and geospatial tools for environmental monitoring at https://earthdata.nasa.gov.
- World Bank Global Facility for Disaster Reduction and Recovery (GFDRR): Explore resources for disaster risk management, including geospatial data and tools, at www.gfdrr.org.
- United Nations Office for the Coordination of Humanitarian Affairs (OCHA) Humanitarian Data Exchange (HDX): Access global humanitarian data and information, including geospatial data, from OCHA at https://data.humdata.org.

- Sentinel Hub: Explore Sentinel Hub for access to Sentinel satellite data and advanced remote sensing tools at www.sentinel-hub.com.
- National Geospatial-Intelligence Agency (NGA): Explore resources, tools, and geospatial data for intelligence purposes, including mapping, geospatial analysis, and geospatial intelligence, at www.nga.mil.
- Armed Conflict Location & Event Data Project (ACLED): Access the ACLED database to explore geospatial data and information on political violence and armed conflicts worldwide at https://acleddata.com.

References

- AJLabs. 2023. "Satellite Images Show Kakhovka Dam Before and After Destruction." *Al Jazeera*, June 7. www.aljazeera.com/news/longform/2023/6/7/before-and-after-satellite-images-show-kakhovka-dam-destruction.
- Glanz, James, Marc Santora, Pablo Robles, et al. 2023. "Why Evidence Suggests Russia Blew Up the Kakhovka Dam." *New York Times*, June 16. links.esri.com/KakhovkaDam.
- Wintour, Patrick. 2023. "Russia Floundering in 'Mud of Lies' over Kakhovka Dam Destruction, Ukraine Tells UN." *The Guardian*, June 7. links.esri.com/RussiaUkraineDam.

About the authors

Rúben Santiago is a geospatial information analyst at the Portuguese Navy Geospatial, Meteorological, and Oceanographic Centre, with six years of military experience. He has a bachelor's degree in geographic engineering and a postgraduate certificate in geographic information systems and science. Rúben is also certified in geospatial intelligence by the US Geospatial Intelligence Foundation.

Sofia Henriques, also a geospatial information analyst at the Portuguese Navy Geospatial, Meteorological, and Oceanographic Centre, has nine years of military expertise. She has a bachelor's degree in geographic engineering and has further qualifications in navigation and geomatics through her postgraduate studies.

Part 2

Climate challenges

Chapter 6

Accelerating electric vehicle uptake: Locating charging stations in an urban context

Scott Kelley, *University of Nevada, Reno*

Overview

In this chapter, you will learn how to use GIST tools and methods in the context of locating electric vehicle (EV) charging stations in an urban area, which is a critical HSGI issue. The workflows you follow and the final deliverable you produce will provide information that could help practitioners and policymakers effectively deploy new EV charging stations in a fast-growing urban area while considering traditional and emerging location criteria.

Learning objectives

- Perform exploratory data visualization of the current distribution of EV charging stations in the greater Las Vegas, Nevada, metropolitan area.
- Visualize and explore geospatial data that represents important criteria that is relevant to the process of EV charging infrastructure siting.
- Apply several geoprocessing tools that help narrow where new stations can be built.
- Produce a recommendation that can inform the locations of new EV charging stations in the greater Las Vegas, Nevada, metropolitan area using timely data.

Technical components

- Software: ArcGIS Pro 3.4
- Data: Consists of five distinct datasets
  - Street Centerlines, Nevada Department of Transportation
  - Electric Vehicle (EV) Charging Station Locations, US Department of Energy, Alternative Fuels Data Center
  - Disadvantaged Communities (DACs), US Department of Energy, US Department of Transportation
  - Multifamily Housing Unit Prevalence, US Census, American Community Survey
  - Zoning, Clark County, Nevada

Prerequisites

- Introductory knowledge or experience of GIS software or spatial data is required.
- Technical skill level: Introductory GIS.
- Time to complete chapter: one to two hours without intelligence brief, three to five hours with intelligence brief.

Background and situational context

Accelerating the adoption of EVs in the US vehicle fleet is essential to mitigating rising emissions from the transportation sector, which now exceeds total emissions from all other economic sectors in the nation. Each year, weather-related emergencies attributable to a warming climate increase in number and severity, threatening human populations and critical infrastructure throughout the world. On a local scale, exhaust and emissions from conventional vehicles operating with liquid petroleum fuels contribute to health-related issues, particularly in vulnerable and historically disadvantaged communities. In coordination with other transportation planning strategies, greater EV adoption can help address these urgent issues.

For several decades, efforts have been devoted to promoting the adoption of EVs, but uptake has remained modest until recently. Although there is optimism regarding recent EV adoption statistics, several barriers continue to challenge more widespread EV adoption, including high cost, lack of familiarity with the technology and recharging infrastructure, and especially range anxiety, or fear of running out of charge before reaching a destination. The lack of a convenient public charging station network has long been acknowledged as the primary concern of prospective adopters, and this is particularly acute in areas where people do not have access to reliable home charging. In recent years, federal and state policymakers have determined that for EV adoption targets to be met, vehicle and charging station network rollout must better account for equity in infrastructure investment to ensure that EVs can be more readily adopted by the general population beyond traditional early adopters.

Public charging stations offer a means to do this, because an adopter need not necessarily have access to charging at home to make their trips.

Recent web-mapping tools and geospatial data such as those you will use in this activity have helped provide decision-makers with better access to information to locate new charging infrastructure, although several factors must be accounted for to provide effective new station location recommendations. In this chapter, you will use a variety of data recommended for consideration in current charging station location methodology. Using this geospatial data and GIS analyses, you will help provide a recommendation for locations within one of the nation's fastest-growing urban areas, Las Vegas, Nevada, to prioritize for the next wave of public charging stations.

Exercise workflow 6

The primary task is to recommend locations for new EV charging stations in the greater Las Vegas, Nevada, area. You will narrow the search area and ensure that any new station recommended meets the following conditions:

1. Is at least one-half mile away from any existing charging station. This helps mitigate "charging deserts," or areas that lack access to nearby charging stations. This also helps prioritize accessibility for those who currently lack it.
2. Is within an area where at least one-third (33.3 percent) of the housing units are multifamily housing units. Locating a public charging station in such an area will ensure that those without reliable access to home charging can still feasibly recharge.
3. Is in a DAC identified in a National Electric Vehicle Infrastructure (NEVI) dataset, helping to meet federal infrastructure deployment targets in these areas.
4. Is within 250 feet of an existing road centerline, ensuring reasonable access to the street network for drivers.
5. Is in one of two commercial zoning designations for Clark County, Nevada (C-1 or C-2). For the purpose of this activity, you will co-locate your stations in areas that are conveniently close to nearby commercial establishments.

Your task will be to narrow the search area to delineate suitable places to construct stations within the metropolitan area, helping station developers to select effective locations for the next phase of EV charging infrastructure.

1 Identify areas that already have access to EV charging infrastructure

In this exercise, you will use several geoprocessing tools that generate output data. You will create a folder to store and organize the output as you work toward narrowing the search area for station locations. Having such a folder helps with data organization and storage while you conduct geospatial analysis.

1. From the **SecurityFirst_ExerciseData** folder, navigate to **Chapter06.** Open the ArcGIS Pro project called **EV_Charging_Chapter.aprx**.
2. In the **Catalog** pane, expand the **Folders** group.
3. Right-click the active folder in which you are working (it should be called **Chapter06**).
4. Select **New** > **Folder** and name the new folder Intermediary_Data.

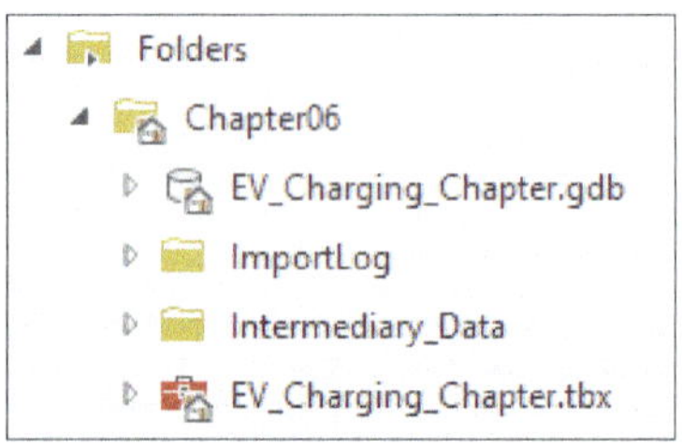

Make a note of this folder's location; you will save geoprocessing output here in each of the following steps.

You want to explore the map that shows all the input data with which you will be working.

5. In the Contents pane, find and explore the **EVCharging_Stations** layer.

This data reflects all available level 2 public charging stations in Clark County, Nevada, as of August 2024 that do not have any access restrictions. Note the different charging station networks and types of facilities and their locations and distribution in the greater Las Vegas area and the context in the areas immediately surrounding them.

6. Review the areas that currently lack stations.

What patterns can you identify in terms of current coverage?

Next, you will generate 0.5-mile buffers around existing charging station locations. This represents an upper limit of what many people would be willing to walk from a charger to reach another destination while their EV recharges.

7. On the **Analysis** tab, in the **Geoprocessing** group, click **Tools.**
8. In the **Geoprocessing** pane, search for Pairwise Buffer. Open the tool and enter the following settings:
   - **Input Features:** EVCharging_Stations
   - **Output Feature Class:** Intermediary_Data\EVCharging_Buffer
   - **Distance:** 0.5 US Survey Miles

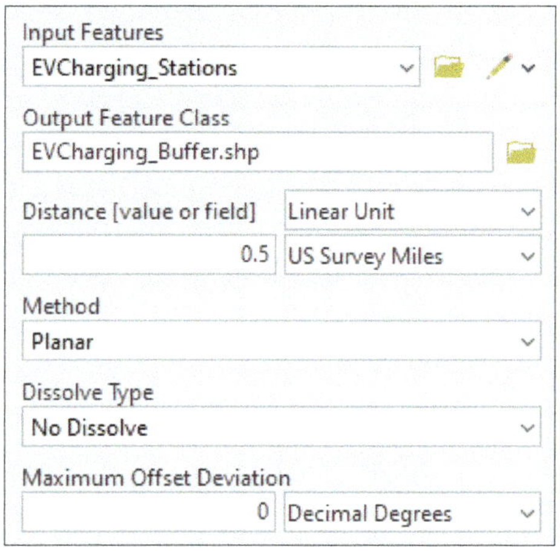

> **Tip:** Be sure to save the output to the Intermediary_Data folder that you created earlier. Also, confirm that the file type is a shapefile (.shp). Click the folder icon to redirect the result's location, if necessary.

The **Dissolve Type** for this tool is kept as **No Dissolve**. You are keeping the default in case you want to review which stations are at the center of a 0.5-mile buffer, as shown on the map.

9. Click **Run**.
10. Examine the results and, as before, note which parts of the city have access to nearby stations and which do not.

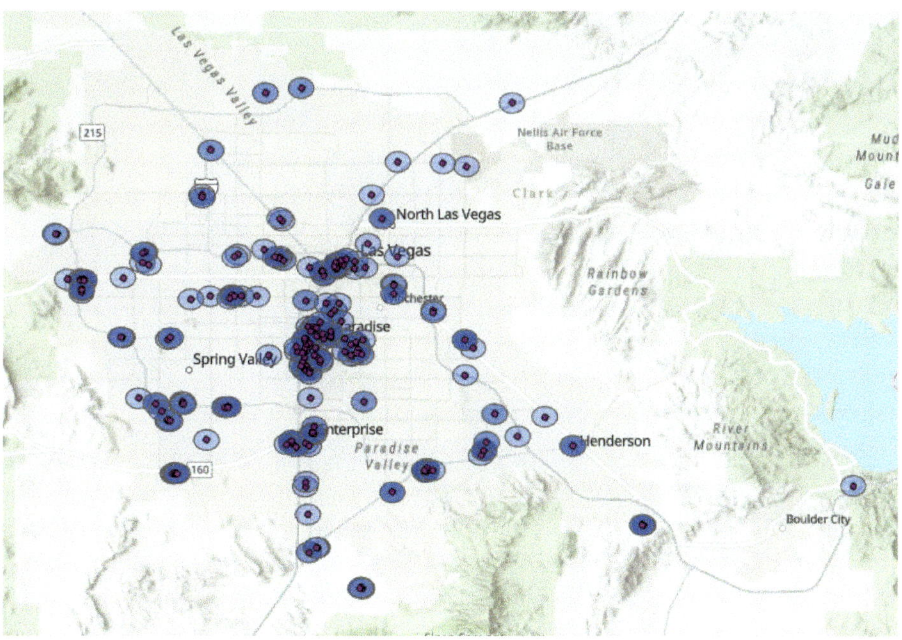

The geospatial output needed for the station proximity criteria is now ready for further analysis.

2 Identify areas within 250 feet of an existing road centerline

You will again generate a buffer output in this task. This time, though, you will create 250-foot buffers away from road centerlines.

1. In the **Pairwise Buffer** tool, enter the following settings:

   - **Input Features**: Roads
   - **Output Feature Class**: Intermediary_Data > RoadsBuffer.shp
   - **Distance**: 250 US Survey Feet
   - **Dissolve Type**: Dissolve all output features into a single feature

   At this time, you will dissolve the outputs because you do not necessarily need to know which road name is nearest to each buffer segment.

2. Click **Run** and review the output.

   In the denser parts of the city, the area within 250 feet of a road centerline is fairly continuous, but this is not the case throughout the entire metropolitan area. As with the EV charging station buffers generated earlier, you will use this file later to narrow your search area.

   The buffer generation portion of the analysis is complete, so you will now turn to other geoprocessing tasks. If your results are similar to the following image, you are ready to proceed to the next section.

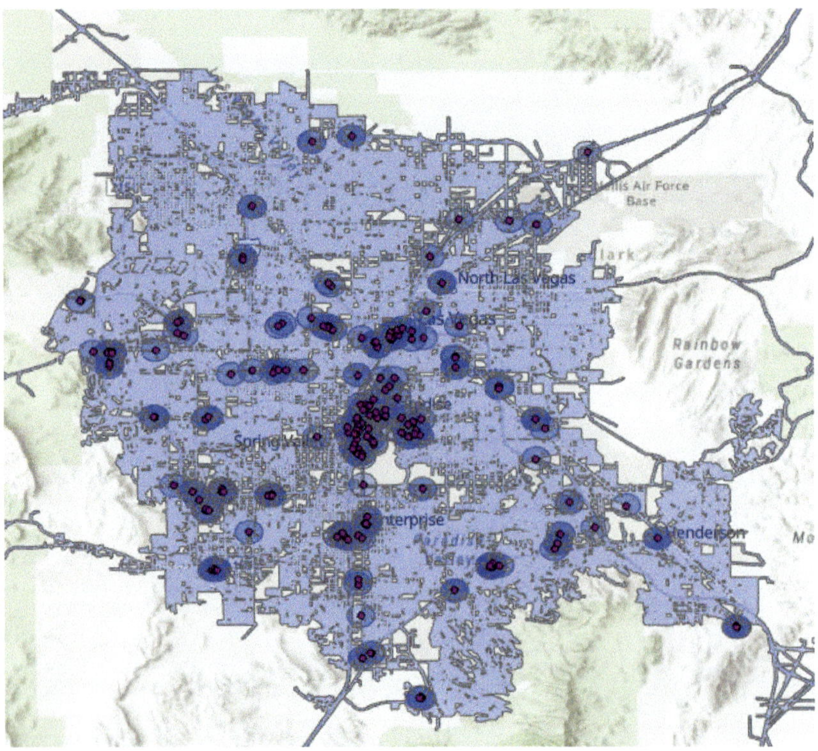

3 Identify areas where at least one-third of the housing stock includes multifamily housing units

1. In the **Contents** pane, right-click the **MultiFamily_Pct** layer dataset and click **Attribute Table.**

   This dataset's attribute table has only one field (column) of interest to you: the percentage of housing units in the block group that have multiple units in the structure. That field is called **MFH.**

   You will select all polygons (rows) using an attribute selection query that meets the requirement of having at least one-third of their housing units being multifamily (MFH) units.

2. On the table ribbon, click **Select By Attributes.**

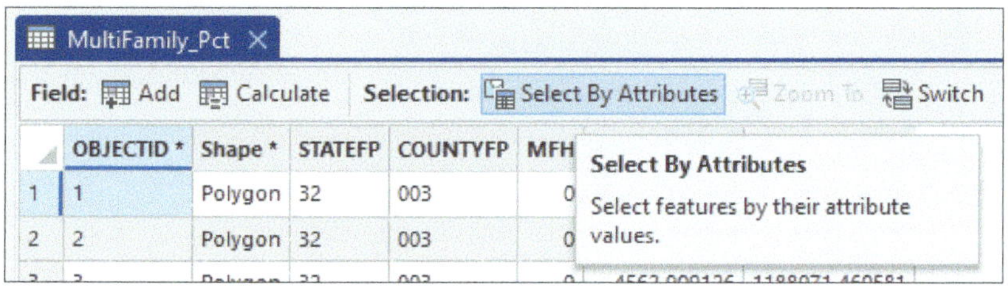

3. In the **Select By Attributes** window, confirm that **Input Rows** is set to **MultiFamily_Pct** and the **Selection type** is set to **New selection.**

4. For **Expression**, create the following query:
   - **Field:** MFH
   - **Operator:** is greater than
   - **Value:** 33.3

5. **Click OK.**

   You should have 459 selected records. If 459 out of the 1,325 records are selected, proceed.

   You will now save a copy of the currently selected block groups that meet this requirement to help narrow your search area.

6. With the records still selected, right-click the **MultiFamily_Pct** layer in the **Contents** pane. Click **Data > Export Features.**

7. In the **Export Features** window, set the **Output Feature Class** to HighMFH as a shapefile in the **Intermediary_Data** folder. Click **OK.**

8. Review the output, making a note of how these polygons align with the charging station buffers and road buffers that you created earlier.

   You will use this file later to narrow the search area.

9. Remove the **MulitFamily_Pct** layer from the **Contents** pane (you no longer need the full dataset). Keep the **HighMFH** layer visible.

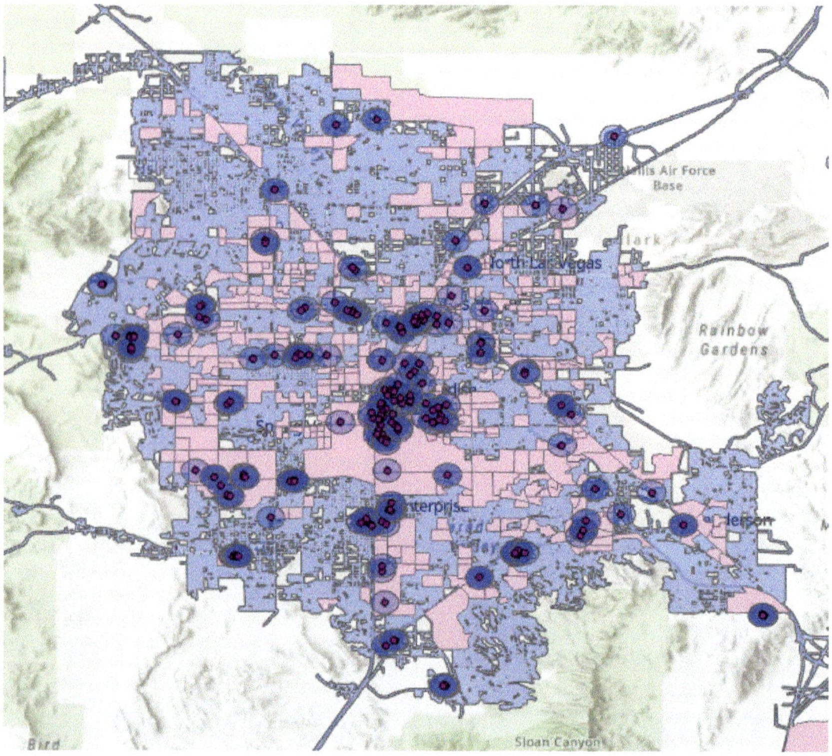

4 Identify disadvantaged communities using the NEVI data

1. In the **Contents** pane, open the **NEVI_LasVegas_Project** layer's table.

   In contrast to the **MFH** data, this is an extensive attribute table.

2. Find the field (column) called **NEVI_DAC.**

   This binary variable determines whether a US census tract is characterized as disadvantaged using the NEVI framework. You will create an attribute selection query to identify all tracts identified by the NEVI analysis as disadvantaged (**NEVI_DAC** value of 1).

3. Use the **Select By Attributes** tool to create a query using these settings:

   - **Field**: NEVI_DAC
   - **Operator**: is equal to
   - **Value**: 1

4. Click **OK.**

   There should be 219 such tracts in the Las Vegas area.

   As you did with the selected **MFH** polygons, you will export these selected polygons for further analysis.

5. With the selections still active, use the **Export Features** tool. Name the output DACs and save it as a shapefile in the **Intermediary_Data** folder.

6. Review the output, making a note of how these polygons align with the **HighMFH** polygons, charging station buffers, and road buffers you created earlier.
7. Remove the **NEVI_LasVegas_Project** file (you only need the **DACs** layer from this point forward).

   If your working map resembles the following screenshot (your symbology may vary), proceed to the next steps.

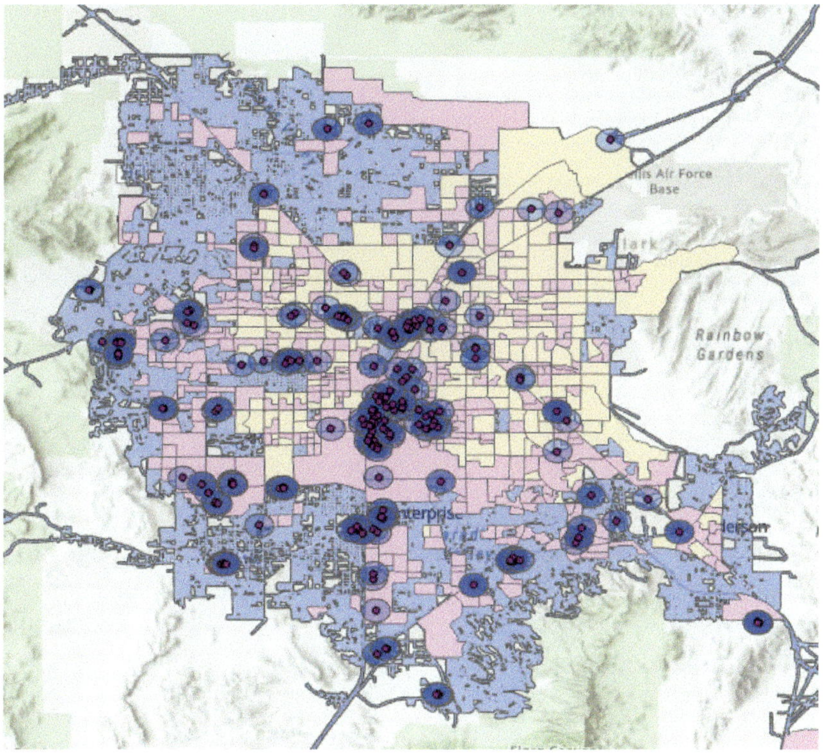

5 Intersect the common areas where criteria are met

You now have the spatial data representations of all criteria necessary to narrow the search area. The only layer you have yet to use is the **Commercial_Zoning** shapefile. This layer has been prepared for you in that only C-1 and C-2 zoned areas include this layer. Using this and the other layers you created, you can proceed with identifying suitable areas for new EV charging station construction.

Specifically, you will use the **Intersect** tool to identify areas where common criteria spatially overlap. This will be useful for overlaying each of your inputs, with the exception of the EV charging station buffers because you do not want stations within these buffered areas. You will use a different tool for that. You do want the common areas shared by the **HighMFH**, **DACs**, **RoadsBuffer**, and **Commercial_Zoning** areas, though, because if locations are located within each of these shared geographic extents, they are suitable for your analysis.

1. In the **Geoprocessing** pane, search for and populate the Intersect tool with the following settings:
   - **Input Features**: DACs, HighMFH, RoadsBuffer, Commercial_Zoning
   - **Output Feature Class**: Intermediary_Data > Intersect_Criteria.shp
   - **Attributes To Join**: Only feature IDs

   Note: Do not use the Pairwise Intersect tool.

   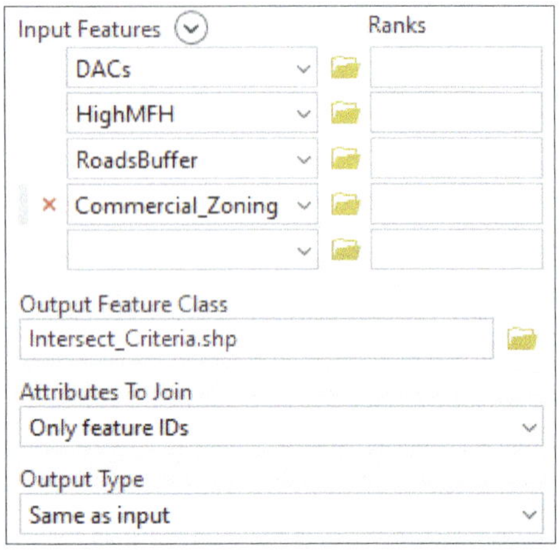

   Note: For the Attributes To Join field, you use Only feature IDs because you do not necessarily need each attribute from each input to narrow your search area. In other applications, that is often something you would want to do, but not in this case.

2. Click **Run**.
3. In the **Contents** pane, uncheck all layers except for the **Intersect_Criteria** layer.

   One part of each of your layers seemed to cover most of the city, but notice how much narrower your search area has become.

4. Zoom in to the intersected areas output to get a sense of where within the city these suitable locations are.

   The polygon you just produced should represent areas that satisfy each of the criteria: in a **DAC**, in high **MFH** areas, within 250 feet of a road, and in a commercially zoned area. Remember, though, that this is not your final search area—you still need to remove from consideration all areas within 0.5 miles of an existing charger. You will do so next. You can turn on the EV charging station buffer layer to get a sense of what areas will be further excluded in that step if you want.

6 Remove from consideration areas near existing stations

To identify your final suitable areas for new EV charging stations, you will use the **Erase** tool to remove areas produced previously that are within the buffers you produced at the start of the exercise.

1. In the **Geoprocessing** pane, search for and populate Erase with the following settings:
    - **Input Features:** Intersect_Criteria
    - **Erase Features:** EVCharging_Buffer
    - **Output Feature Class:** EV_Charging_Chapter.gdb > FinalAreas

    Note: Do not use the Pairwise Erase tool.

2. Click **Run**.
3. In the **Contents** pane, turn off all layers except **FinalAreas.**
4. Zoom to and pan these areas to get a sense of landmarks and features that are in these areas or nearby.

    You have successfully narrowed the search area to help station developers quickly and readily identify promising new sites for stations. Further geospatial analysis could be used to ultimately select final station locations. For example, parcel data may be used to identify particular properties and contact information of property owners that might be helpful. DMV data on EV adoption at the neighborhood scale might be cross-referenced against these sites, along with traffic volumes on roads. Specific datasets and considerations may vary by geographic region, but this workflow assures that you are able to provide EV station coverage to communities of interest to meet recent EV station location priorities and strategies.

    Optionally, depending on your level of access to ArcGIS Online, you may share the **FinalAreas** layer as a web layer (or the full map as a web map). This information can be readily used by a stakeholder or decision-maker in the ArcGIS Online environment.

Interpretation of analysis

After completing this exercise, answer the following questions in detail. The answers to these questions, as well as the analysis results from the exercise workflow, will inform the intelligence brief.

1. Think about the questions that you first considered: What sorts of patterns can you initially identify in terms of where there is current coverage or not? From your visual assessment, did you notice any readily apparent spatial patterns in terms of where coverage was good or not? What were those patterns? How did they compare to areas that lacked coverage?

    To the best of your ability, briefly characterize which areas of the city had good EV station coverage before you performed the exercise and which areas did not.

2. What is the final square mileage of suitable areas within which you can build an EV charging station in the greater Las Vegas area based on the workflow?

> **Tip:** If you need assistance, open the attribute table for the FinalAreas feature class and use the Statistics function on the Shape_Area field. The coordinate system of the file is in UTM, so the units will be in square meters, which you can convert to square miles.

3. What is the most important takeaway from the analysis workflow? What actionable information can station developers in the area use to decide where they can best place five new stations in the greater Las Vegas, Nevada, metropolitan area?
4. What other sorts of geospatial data or local information might station developers or local stakeholders want to identify or collect before finalizing five new sites for construction?

Intelligence brief

Based on the analysis from the exercise workflow, produce a three- to five-page intelligence brief by responding to the following question prompts. Incorporate your geospatial outputs from the exercise to enrich and illustrate your conclusions.

- **BLUF:** List three to five primary observations from your analysis in bulleted-list format that summarize findings of this work.
- **Top-level analysis:** Report the primary data points supporting your BLUF.
- **Secondary analysis:** Provide the data sources and deeper analytic products from this analysis.
- **Impact:** Describe the impacts of your findings on the department or organization.
- **Confidence:** Report the level of confidence of your assessment based on available data, accuracy, and so on.
- **Bias:** Explain any background or information you have that could have affected your analysis.
- **Conclusions:** Determine what type of analytic conclusions there may be and how the data requirements, analysis, and brief writing may change based on what analytic framework you are following (descriptive, predictive, prescriptive).

Applying and extending your skills

For more information about the geoprocessing tools that you used in this chapter, see ArcGIS Pro online resources at esri.com/en-us/arcgis/products/arcgis-pro and use the page search engine to find the following topics:

- Resources > Tool Reference > Geoprocessing Tools > Analysis Toolbox > Proximity Toolset > Buffer (Analysis)
- Resources > Tool Reference > Geoprocessing Tools > Data Management Toolbox > Layers and Table Views Toolset > Select Layer By Attribute (Data Management)
- Resources > Tool Reference > Geoprocessing Tools > Analysis Toolbox > Overlay Toolset > Intersect (Analysis)
- Resources > Tool Reference > Geoprocessing Tools > Analysis Toolbox > Overlay Toolset > Erase (Analysis)

Buffer vs. network distances

You used the **Buffer** tool in this analysis to assess proximity to existing EV charging stations, the output of which is a polygon with a Euclidean distance radius equal to the distance that you entered (0.5 miles). In many transportation-related geospatial analyses, it is common to use network-based distances. However, this requires you to use the Network Analyst extension and the accompanying license with your ArcGIS Pro subscription and a network dataset in ArcGIS Pro. If you have access to such a license, see the following links:

- Details about a network-based equivalent of buffers (service areas) can be found at esri.com/en-us/arcgis/products/arcgis-pro: Resources > Tool Reference > Geoprocessing Tools > Ready to Use Toolbox.
- Details about network datasets, including how to build them in ArcGIS Pro, can be found at esri.com/en-us/arcgis/products/arcgis-pro: Resources > Help > Analysis and Geoprocessing > Network Analyst > Network Datasets.

Equity in EV station planning

This exercise included some elements related to the emerging topic of equity in EV station planning. You can learn more about this effort, the considerations accounted for, and what kinds of geospatial data are used to represent this effort at www.transportation.gov/rural/ev/toolkit/ev-infrastructure-planning.

Additional resources

- EV types: www.transportation.gov/rural/ev/toolkit/ev-basics/vehicle-types
- EV charging basics: www.transportation.gov/rural/ev/toolkit/ev-basics/charging-speeds
- EV charging station planning: www.transportation.gov/rural/ev/toolkit/ev-infrastructure-planning/project-planning-checklist
- Alternative fuel corridors: https://afdc.energy.gov/corridors
- Joint Office of Energy and Transportation: https://driveelectric.gov/resources

About the author

Scott Kelley is an associate professor in the Department of Geography at the University of Nevada, Reno. He has more than 10 years' experience in studying how early adopters of alternative fuel vehicles evaluate and use refueling or recharging infrastructure to support their travel. He has a bachelor's degree in geography from the University of Wyoming and received his master's and PhD in geography at Arizona State University. He teaches introductory, intermediate, and advanced GIS classes at the undergraduate and graduate levels.

Chapter 7

Object detection and segmentation of trees using Text SAM in ArcGIS Online

Yifan Yang, *University of Southern California*
Dominic Borrelli, *University of Southern California*

Overview

In this chapter, you will learn how to use GIST tools and methodologies to analyze urban greenery by integrating aerial imagery and deep learning techniques. You will focus on applying the Segment Anything Model (SAM) for tree and building segmentation and explore how these technologies address the time and cost limitations of traditional manual monitoring and planning, particularly in large-scale and complex urban environments. Specifically, you will learn a nonprogramming approach to identifying and segmenting trees and buildings from aerial imagery using Text SAM and ArcGIS Online without relying on Python scripting. This process-oriented analytic method provides decision-makers and urban planners with precise, efficient greening data and strategic insights, enhancing their ability to implement scientifically optimized strategies. By offering data faster and more cost-effectively than conventional methods, this approach has the potential to transform urban planning, enabling more informed and data-driven decisions for sustainable city development.

Note: Text SAM is an open-source sample model that can be prompted using free-form text prompts to extract features of various kinds. This is achieved by using the Grounding DINO model and SAM.

Learning objectives

- Convert aerial TIF images into image layers using ArcGIS Online.
- Apply deep learning tools for accurate identification and segmentation of trees without coding, using the Text SAM model and Map Viewer in ArcGIS Online.

- Use aerial imagery for exploratory data visualization to assess the current state of urban greenery.
- Evaluate discrepancies between current urban conditions and planning objectives to propose data-driven optimization strategies for sustainable city development.

Technical components

- Software:
  - ArcGIS Image for ArcGIS Online
  - ArcGIS Online
  - ArcGIS Pro 3.4
- Datasets:
  - Aerial map of Estella Public School in New South Wales, Australia, available for download at links.esri.com/OpenAerialMap
  - Text SAM available for download at links.esri.com/TextSam

Prerequisites

- Beginner to intermediate proficiency with ArcGIS Online.
- Basic knowledge of or experience with ArcGIS Online Map Viewer.
- No prior knowledge or experience with the Python programming language, machine learning, or deep learning required.
- Technical skill level: Introductory GIS, introductory deep learning.
- Time to complete chapter: one hour without intelligence brief, two to three hours with intelligence brief.

Background and situational context

With the rapid acceleration of global urbanization, urban greenery has become a central issue in environmental governance and sustainable community development. Well-planned urban greenery plays a crucial role in regulating microclimates, improving quality of life, mitigating the urban heat island effect, enhancing air quality, and strengthening the stability of urban ecosystems. However, owing to the complexity of urban structures and the diversity of green spaces, traditional manual monitoring and planning methods often face challenges related to efficiency, cost, and precision management.

In recent years, advancements in remote sensing technology and deep learning algorithms have introduced groundbreaking opportunities for urban green space planning, monitoring, and optimization. The use of aerial imagery allows real-time capture of urban greenery, and AI-powered object detection and segmentation techniques enable precise analysis. Compared with traditional methods, deep learning models can automatically identify and segment objects, such as trees, buildings, and roads, with high accuracy. This not only provides comprehensive data on green space types, spatial distribution, and coverage but also helps identify areas requiring improved greenery, offering data-driven recommendations for urban planners.

To make these advanced AI techniques accessible to nonprogrammers, we have chosen to use ArcGIS Online, enabling users to perform object detection, recognition, and segmentation without the need for programming. Additionally, this approach supports the integration of ArcGIS Notebooks and ArcGIS API for Python, using the pretrained SAM model for tree identification and segmentation. Without using complex coding, users can complete data processing and analysis through an intuitive visual interface, allowing urban planners and city managers to harness the power of AI and GIS for more precise and efficient urban greenery management.

In this chapter, we provide a detailed guide on how to use ArcGIS Online, Text SAM, and other tools to perform automated tree and building detection and segmentation from aerial imagery in a no-code environment, ultimately supporting sustainable urban development and intelligent planning.

Exercise workflow 7

1 Load the imagery and add metadata

> **Note:** This step requires the license for ArcGIS Image for ArcGIS Online. You can confirm that you have this license by clicking your profile icon in the top right > My settings > Licenses, and search for ArcGIS Image for ArcGIS Online.

1. In a browser, navigate to links.esri.com/OpenAerialMap and download the image of **Estella Public School** in New South Wales, Australia, to your computer. Rename the file Estella_Public_School.tif.
2. In a browser, navigate to www.arcgis.com and sign in using your ArcGIS organization account.
3. On the ribbon, click the **Content** tab. On the top left, click **New Item.**
4. In the **New item** window, select **Imagery layer.**

   The **Create imagery layers window** opens.
5. Check the box for **Dynamic Imagery Layer.** Uncheck the box for **Tiled Imagery Layer.** Click **Next.**

   To prepare for your subsequent Text SAM tasks, it is best to select **Dynamic Imagery Layer.** Tiled imagery layers are suitable for fast display and access to static images. Dynamic imagery layers provide more powerful processing and are suitable for scenarios that require flexible analysis.
6. Select **One Image** and click **Next.**

   Choosing **One Image** simplifies data management, improves computing efficiency, and is compatible with deep learning and GIS analysis. When performing aerial image segmentation in ArcGIS Online, using a single dynamic image layer is a reasonable choice.
7. Select the **Estella_Public_School.tif** image from your computer and upload it.

8. Under **Set Item Details**, name the imagery Estella_Public_School_[your initials], add aerial imagery, deep learning, and SAM as tags, and type A high-resolution aerial image of the Estella Public School in New South Wales, Australia as the summary. Click **Create**.

> **Tip:** This step may take several minutes to complete.

If someone else in your ArcGIS Online organization has already completed this step, it is likely that there will be a dataset with that name (**Estella_Public_School**) already. To save a distinct copy of it, it is recommended that you include your initials in the name.

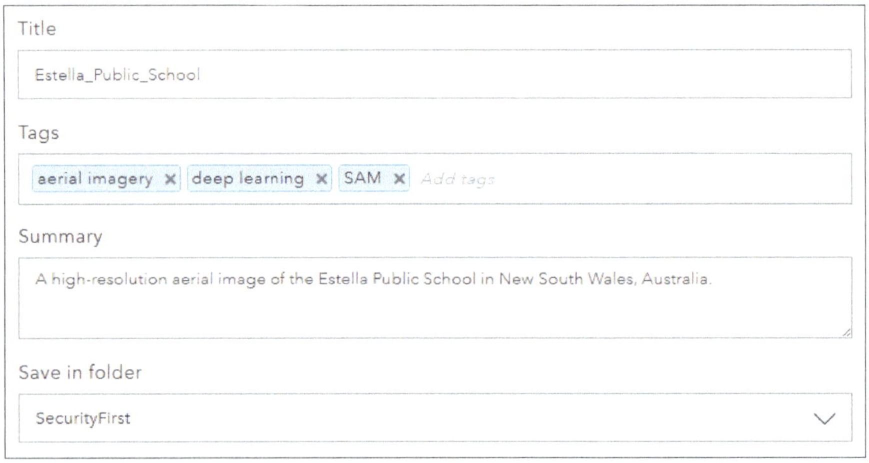

9. When the **Dynamic Imagery Layer** is created, open it in **Map Viewer** to confirm that it rendered correctly. After confirming, close the map without saving it.

2 Download and run the Text SAM model

Next, you will download the Text SAM model.

1. In a browser, go to ArcGIS Living Atlas of the World (livingatlas.com), search for TextSAM, and download the model.
2. Return to your ArcGIS Online home page and confirm that you are still signed in. On the **Content** page, click **New Item** and upload the Text SAM model that you just downloaded.

   Note: The size of this package exceeds 1 GB.

3. On the ribbon, click **Map** to access Map Viewer. Click **Add layer**. From **My Content**, click the dynamic imagery layer with your initials.
4. In the **Settings** pane, click **Analysis**. In the search bar, type detect objects using deep learning to find this deep learning tool. Alternatively, you can navigate to the **Use Deep Learning** toolbox to locate the **Detect Objects Using Deep Learning** tool.
5. Open the **Detect Objects Using Deep Learning** tool and add the following settings:
   - **Input imagery or feature layer:** Estella_Public_School_[your initials]
   - **Processing mode:** Process as mosaicked image
   - **Model settings:** Select **Text SAM** model from **My Content**
6. When the **Text SAM** model loads, add the following model arguments, and click **Run**.
   - **Text Prompt:** Tree
   - **Padding:** 256
   - **Batch Size:** 4
   - **Box threshold:** 0.2
   - **Text Threshold:** 0.2
   - **Perform test time augmentation while predicting using different scales:** 1
   - **NMS Overlap:** 0.1
   - **Output Class Label Field Name:** ClassLabel
   - **Output Name:** Estella_Segmented_Trees_[your initials]

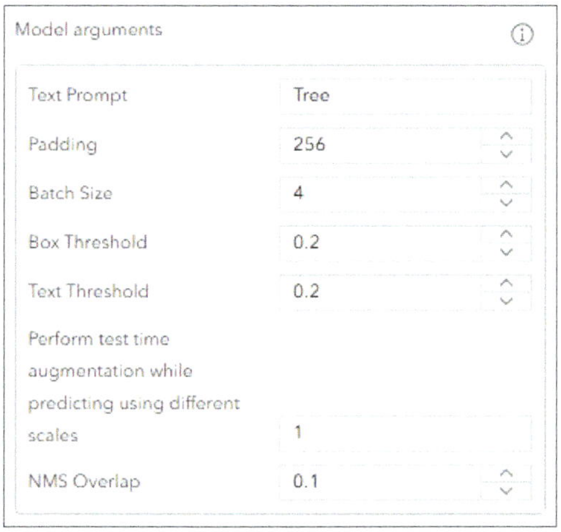

Table 7-1 provides descriptions of each of these model arguments.

Table 7-1. Model arguments

Argument	Value	Description
Text Prompt	Tree	The text input to the Text SAM model, instructing the model to detect objects related to "tree."
Padding	256	When processing images, add 256 pixels of padding to the image to confirm that the target object is not lost owing to boundary cropping, which improves the detection effect.
Batch Size	4	The batch size of images processed at a time, set to "4." This means that the model will process four images at the same time in one iteration to improve computational efficiency.
Box Threshold	0.2	Target box (bounding box) confidence threshold; objects below 0.2 will be filtered to avoid detecting less-reliable objects.
Text Threshold	0.2	Text-matching confidence threshold, used to control the matching degree of text prompts (such as "tree") and detected objects; detection results below 0.2 will be ignored.
Perform test time augmentation while predicting using different scales	1	Enable test time augmentation (TTA) to use different scales when predicting to improve the success of detection and reduce the impact of scale changes on the results.
NMS Overlap	0.1	Non-maximum suppression (NMS) overlap threshold, used to remove overlapping detection boxes. Setting it to 0.1 means that if the overlap between two target boxes exceeds 10 percent, only the one with the highest confidence will be retained to reduce redundant detection results.
Output Class Label Field Name	ClassLabel	
Output Name	Estella_Segmented_Trees_[your initials]	

7. When the model has run, add the new dataset, **Estella_Segmented_Trees_[your initials]**, to your map and review it. Take note of the segmented trees.

Interpretation of analysis

After completing this exercise, answer the following questions in detail. The answers to these questions, as well as the analysis results from the exercise workflow, will inform the intelligence brief.

1. What factors might affect the accuracy of Text SAM's object detection in ArcGIS Online, and how can you validate the results?
2. How do the quality and type of imagery affect the effectiveness of Text SAM, and what imagery preprocessing steps might improve performance?
3. In what ways might biases in training data or model limitations affect decision-making in geospatial analysis using Text SAM?
4. How might you optimize the workflow for large-scale analyses while maintaining efficiency in ArcGIS Online?
5. How might Text SAM be integrated with other geospatial tools or datasets to enhance real-world applications, such as disaster response or recovery, and other scenarios?

Intelligence brief

Based on the analysis from the exercise workflow, produce a three- to five-page intelligence brief by responding to the following question prompts. Incorporate your geospatial outputs from the exercise to enrich and illustrate your conclusions.

- **BLUF:** List three to five primary observations from your analysis in bulleted-list format that summarize findings of this work.
- **Top-level analysis:** Report the primary data points supporting your BLUF.
- **Secondary analysis:** Provide the data sources and deeper analytic products from this analysis.
- **Impact:** Describe the impacts of your findings on the department or organization.
- **Confidence:** Report the level of confidence of your assessment based on available data, accuracy, and so on.
- **Bias:** Explain any background or information you have that could have affected your analysis.
- **Conclusions:** Determine what type of analytic conclusions there may be and how the data requirements, analysis, and brief writing may change based on what analytic framework you are following (descriptive, predictive, prescriptive).

Applying and extending your skills

SAM is a conceptual representation of a model that combines vision (image segmentation) with language (text cues) to segment images based on given cues. The idea is compelling because it allows the use of natural language descriptions to segment objects or features in images without extensive training on specific datasets. The actual accuracy of a SAM model (or any model with similar capabilities) depends on several factors:

- **Training data**: The variety and quality of the data on which the model is trained plays a crucial role. If a model has seen a wide variety of scenes, objects, and contexts during training, it is more likely to perform well on different test data.
- **Prompt specificity**: The specificity or vagueness of the text prompt affects the performance of the model. For example, "trees with leaves in autumn" might produce more accurate results than "trees" in general if the model understands this distinction.
- **Image complexity**: Images with clear, unique features are easier to segment than cluttered images with overlapping objects and different patterns.
- **Postprocessing and thresholding**: The way that the results are postprocessed (for example, setting confidence level thresholds) can affect the accuracy of perception. Sometimes, adjusting these parameters, depending on the context, can lead to better or worse results.

SAM is poised for significant advancement as a useful visual model in remote sensing, with extensive research focused on its enhancement. A notable development is Efficient SAM, a segmentation model that uses SAM's mask image pretraining along a cross-attention decoder. This innovation enables Efficient SAM to deliver high performance and speed, even when deployed on smaller models.

Understanding and applying SAM technologies is increasingly crucial, given its potential to transform the landscape of remote sensing and geospatial analysis.

Additional resources

Esri Academy (www.esri.com/training) offers a variety of instructor-led and self-paced training courses about ArcGIS and its applications in different fields, including mineral exploration and environmental monitoring.

In addition, you can search *ArcGIS Blog* (esri.com/arcgis-blog/overview) for blog articles on Text SAM and other topics, including the following:

- "Multiresolution Object Detection with Text SAM" (May 16, 2024)
- "Text SAM: Extracting GIS Features Using Text Prompts" (May 14, 2024)
- "Advancing Image Segmentation with SAM: Segment Anything Model" (May 14, 2024)

References

- Carraro, Alberto, Marco Sozzi, and Francesco Marinello. 2023. "The Segment Anything Model (SAM) for Accelerating the Smart Farming Revolution." *Smart Agricultural Technology* 6: 100367.
- Degerickx, Jeroen, Martin Hermy, and Ben Somers. 2020. "Mapping Functional Urban Green Types Using High-Resolution Remote Sensing Data." *Sustainability* 12, no. 5: 2144.
- Ji, Wei, Jingjing Li, Qi Bi, Tingwei Liu, Wenbo Li, and Li Cheng. 2023. "Segment Anything Is Not Always Perfect: An Investigation of SAM on Different Real-World Applications." *arXiv* preprint: 2304.05750.
- Kirillov, Alexander, Eric Mintun, Nikhila Ravi, et al. 2023. "Segment Anything." *Proceedings of the IEEE/CVF International Conference on Computer Vision*: 4,015–26.
- Osco, Lucas Prado, Qiusheng Wu, Eduardo Lopes de Lemos, et al. 2023. "The Segment Anything Model (SAM) for Remote Sensing Applications: From Zero to One Shot." *International Journal of Applied Earth Observation and Geoinformation* 124: 103540.
- Ren, Yougui, Xiaomei Yang, Zhihua Wang, et al. 2023. "Segment Anything Model (SAM) Assisted Remote Sensing Supervision for Mariculture—Using Liaoning Province, China, as an Example." *Remote Sensing* 15, no. 24: 5781.
- Shahtahmassebi, Amir Reza, Chenlu Li, Yifan Fan, et al. 2021. "Remote Sensing of Urban Green Spaces: A Review." *Urban Forestry & Urban Greening* (57): 126946.
- Xiong, Yunyang, Bala Varadarajan, Lemeng Wu, et al. 2023. "Efficient SAM: Leveraged Masked Image Pretraining for Efficient Segment Anything." *arXiv* preprint: 2312.00863.
- Xu, Zhiyu, Yi Zhou, Shixin Wang, et al. 2020. "A Novel Intelligent Classification Method for Urban Green Space Based on High-Resolution Remote Sensing Images." *Remote Sensing* 12, no. 22: 3845.

About the authors

Yifan Yang is currently pursuing a PhD in geography at Texas A&M University. He graduated from the University of Southern California with a master of science degree in spatial data science in May 2024. His research interests include spatial data science and geographic artificial intelligence. He was recently awarded a lifetime membership in the Gamma Theta Upsilon Nu Theta Chapter of the International Geographic Honor Society. He also won the ArcGIS StoryMaps℠ Competition Award at the 2024 Los Angeles Geospatial Summit for "Best Analytical Methods Using Maps."

Dominic Borrelli currently works as an ArcGIS Enterprise consultant at Esri Australia, where he provides technical support, consulting, and training services to Esri clients based in Australia. He graduated from the University of Southern California in 2024, completing both a bachelor of science in environmental studies and a master of science in spatial data science in May and August, respectively. Dominic has several years of experience in the geospatial industry, with a wide breadth of knowledge including ArcGIS Enterprise, Python programming for GIS, and deploying deep learning for aerial imagery analysis.

Chapter 8

Estimating populations vulnerable to flood risk in Florida between 2001 and 2019

Jinwen Xu, *University of California, Berkeley, Institute of Transportation Studies*
Levente Juhász, *Florida International University, GIS Center*

Overview

In this chapter, you will learn how to use GIST tools and methods to identify urban development in flood zones, analyze its change during the past 20 years, and assess the population living in flood zones in Florida in the context of HSGI. The workflows and the final deliverable will provide information to enhance public awareness of flood risks and improve future disaster preparation.

Learning objectives

- Work with publicly available land-use data and conduct spatial and temporal analysis on raster datasets using Python.
- Compare the past (2001) and near-present (2019) urban flood exposure in Florida.
- Perform exploratory data visualization using interactive maps to identify changes in urban flood exposure in Florida using ArcGIS Notebooks and ArcGIS Online.
- Estimate populations living in flood-prone areas using dasymetric mapping and visualize the estimation and its ratio to the total population.

Technical components

- Software:
  - ArcGIS Notebooks in ArcGIS Online (Advanced level access)
- Datasets:
  - FL_counties.zip (Florida county shapefile)
  - FL_flood_zone.img (100-year flood zone layer)
  - LULC_Florida_2001.tif
  - LULC_Florida_2019.tif (2001 and 2019 Florida land-use data)
  - FEMA National Flood Hazard Layer
  - Multi-Resolution Land Characteristics (MRLC) Land Use Land Cover (LULC) dataset
  - Census block group boundaries
  - 2020 American Community Survey (ACS) data

Prerequisites

- Basic knowledge of or experience using GIS software or computer scripting (Python) required.
- Technical skill level: Intermediate GIS.
- Time to complete chapter: One to two hours without intelligence brief, three to five hours with intelligence brief.

Background and situational context

Flooding is one of the most frequent and expensive natural hazards; globally, more than 40 percent of all natural disasters in the past 50 years have been associated with floods (World Meteorological Organization 2021). Flood exposure measures the risk of valued societal elements located in floodplains, including people, critical infrastructure, and properties (Koks et al. 2015; Tate et al. 2021). Currently, more than 40 percent of the population in the United States resides in coastal areas, with approximately 13 percent living within 100-year flood zones, signifying high flood exposure (Hauer et al. 2022; Wing et al. 2018). The percentage of the population and urban lands susceptible to flooding varies by location and is influenced by numerous factors. These factors include local floodplain management (Committee [US] and Task Force 1994), public awareness of flood risk (Burningham et al. 2008), waterborne transportation facilities (Mitchell 2003), and agricultural irrigation (Schultz 2001). Collectively, these elements can affect urban development and the population spread in floodplains (Aerts et al. 2018; Qiang et al. 2017; Xu and Qiang 2024), resulting in spatial variations in flood exposure. The current assessment for flood exposure considers the potential inundation on a smaller scale but falls short of providing accurate estimates of the at-risk population on a larger scale. Besides, owing to data limitation, only a few studies focus on the long-term change of flood exposure.

This chapter aims to assess flood exposure with two approaches—that is, by assessing urban lands in flood zones and population living in flood zones. First, you will analyze the urban lands (indicating urban development and built environment) in Florida at the county level, including changes in urban lands between 2001 and 2019. The first approach includes raster analysis tools and zonal statistics tools. Second, you will estimate the number of people living in flood exposure areas using zonal statistics and dasymetric mapping (Maantay and Maroko 2009; Mennis 2009). The datasets used in this chapter include land use, community survey population data, and the 100-year flood map. The anticipated outputs of this workflow are a series of graduated color maps, which will highlight counties in Florida with high and low flood exposure, as well as trends in flood exposure.

Exercise workflow 8

The exercise is organized into three parts to present learners with a step-by-step guide to investigate urban flood exposure and estimate vulnerable population under flood risk.

The first part uses land-use data from the MRLC Consortium National Land Cover Database (NLCD) 2019, with a time series from 2001 to 2019 at two- to three-year intervals. The 100-year flood map is collected from the Federal Emergency Management Agency (FEMA) National Flood Hazard Layer and used as floodplains. Specifically, land-use data in 2001 and 2019 is selected and overlaid with a flood zone layer to analyze urban flood exposure and its temporal change over 20 years. The urban areas in flood zones will be calculated in each county in 2001 and compared with the total urban areas in 2001 to obtain the ratio of urban areas in flood zones. This ratio will also be calculated for 2019. Two choropleth maps will be made to indicate the flood exposure in Florida in 2001 and 2019. The exercise then compares the ratio difference between 2001 and 2019; the flood exposure change can be calculated, and the spatial distribution of the difference will be analyzed. Similarly, a graduated colored map will be used to indicate this flood exposure change.

The second section of the exercise retrieves 2020 US Census demographic data from ArcGIS Living Atlas of the World to assess population living in flood-prone areas. The dasymetric mapping method will be used to aggregate populations from a fine scale (census block groups) to a larger scale (county level) to accurately estimate people living in flood zones. Some data processing tools in Python will be introduced to organize the census dataset.

1 Load the data

1. In your browser, navigate to www.arcgis.com and sign in to your ArcGIS Online account.
2. On the top navigation bar, click **Notebooks**.
3. Click **New notebook** and click the **Advanced** notebook.

   Note: The advanced Python notebook uses three credits per hour. You can also use the advanced with GPU support notebook, but that version will consume 30 credits per hour.

4. In the notebook, click **Files** and click the **home** folder.

   Before uploading any files, it is always best to organize the work files.

5. In the **home** folder, click **New folder** and name it Flood_exposure_analysis.
6. In the **Flood_exposure_analysis** folder, create a new folder named Results to store output from the exercise.
7. At the bottom of the **Flood_exposure_analysis** folder, click **Choose file to upload.** From the **Chapter08** folder, select the following datasets:
    - **FL_counties.zip** (Florida county shapefile)
    - **FL_flood_zone.img** (100-year flood zone layer)
    - **LULC_Florida_2001.tif**
    - **LULC_Florida_2019.tif** (Florida land-use data in 2001 and 2019)

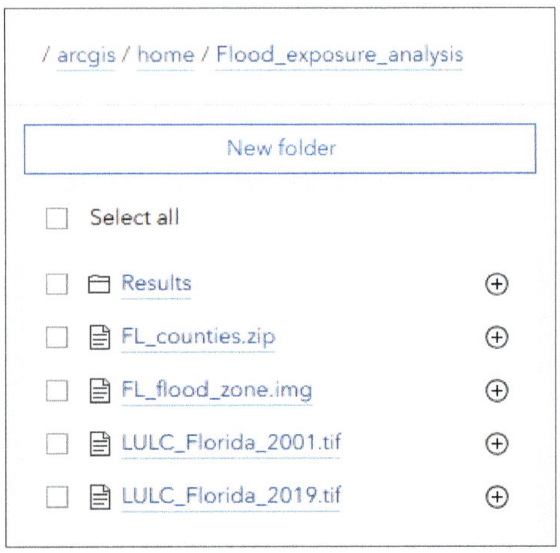

8. On the top right of the notebook, click **Save** > **Save As.**
9. In the **Save notebook as** window, apply the following parameters:
    - **Title:** Ch08_FloodExposureAnalysis
    - **Folder:** SecurityFirst

    Note: If you do not already have a folder called SecurityFirst, create a folder with that name.

10. Click **Save notebook.**

2 Use ArcGIS Notebooks to estimate urban flood exposure

1. In the notebook, click the first cell and click **Run** to connect the notebook to your GIS.

    Tip: You will know whether your cell ran properly if a number appears next to the cell.

    The following code will run the first few cells to load necessary packages (for example, `arcgis Python API` and `ArcPy`). ArcGIS Notebooks in ArcGIS Online uses a cloud-based environment, so you do not need to worry about installing Python packages on your own computer. After running the following cells, you will be able to install Python packages and load and test datasets.

2. In the next cell, type or copy the following code:

```
import sys
!conda install --yes --prefix {sys.prefix} geopandas==1.0.1 folium==0.14.0
↪ mapclassify==2.5.0 rasterstats
```

Note: The arrow in the code block indicates a continuation of the code line. When you type the code, it will all be on one line.

After the code runs, your output will resemble the following:

```
/bin/bash: /opt/conda/lib/libtinfo.so.6: no version information available
(required by /bin/bash)
Retrieving notices: ...working... done
Retrying (Retry(total=2, connect=None, read=None, redirect=None, status=None))
after connection broken by 'NewConnectionError('<urllib3.connection.HTTPS-
Connection object at 0x7f2fc52556d0>: Failed to establish a new connection:
[Errno -2] Name or service not known')': /builds/11.4.0/conda-bld/noarch
/repodata.json

Channels:
 - esri
 - defaults
Platform: linux-64
Collecting package metadata (repodata.json): done
Solving environment: done
```

Compatibility issues may occur in ArcGIS Notebooks owing to conflicts with older versions of GeoPandas. The current version of the `geopandas` package (1.0.1) works with the Python environment in ArcGIS Notebooks. However, in future releases of ArcGIS Notebooks, users may need to test and update the package version of GeoPandas to resolve potential compatibility issues.

3. In the next cell, type or copy the following code:

```
import arcpy
from arcpy.sa import *
import numpy as np
import geopandas as gpd
import pandas as pd
import requests
import json
import rasterio
import matplotlib.pyplot as plt
from rasterstats import zonal_stats
from pyproj import Proj, Transformer
import geopandas as gpd
from shapely.ops import transform

arcpy.env.overwriteOutput = True
```

4. Click **Run** or use these shortcuts:

   - Run and Move to Next Cell: press **Shift + Enter**
   - Run and Stay in Current Cell: press **Ctrl + Enter**
   - Run and Insert a New Cell Below: press **Alt + Enter**

   The Anderson Land Cover Classification System uses numbers to signify classes of land use. Urban land uses are coded as 21, 22, 23 or 24. You will use the **Reclassify** tool in ArcPy to reclassify the dataset to select only urban lands in a flood zone and exclude all other land types.

5. In the next cell, type or copy the following code and click **Run** (or use the shortcuts):

```
%%time
for i in ["2001","2019"]:
out_raster = arcpy.sa.Reclassify("/arcgis/home/Flood_exposure_analysis
↪ /LULC_Florida_"+i+".tif", "Value", RemapRange([[0,20.5,'NODATA'],
↪ [21,24.5,1],[24.5, 255,'NODATA']]), "NODATA")
    out_raster.save("/arcgis/home/Flood_exposure_analysis/Results/urban_"+i+
    ↪ ".tif")
print(i)
del out_raster
```

   After you run the cell, you will see the following response:

```
2001
2019
CPU times: user 53 s, sys: 3.54 s, total: 56.5 s
Wall time: 55.4 s
```

   Your times may be different, based on your processing resources.

   Tip: Adding `%%time` in the first line of the code cell helps track the computing time of the cell. It is especially useful when handling large datasets.

   You will now visualize the dataset.

6. In the next cell, type or copy the following code and click **Run**.

```
Raster("/arcgis/home/Flood_exposure_analysis/Results/urban_2019.tif")
```

   The result returns a visualization of points representing the urban lands of Florida in 2019.

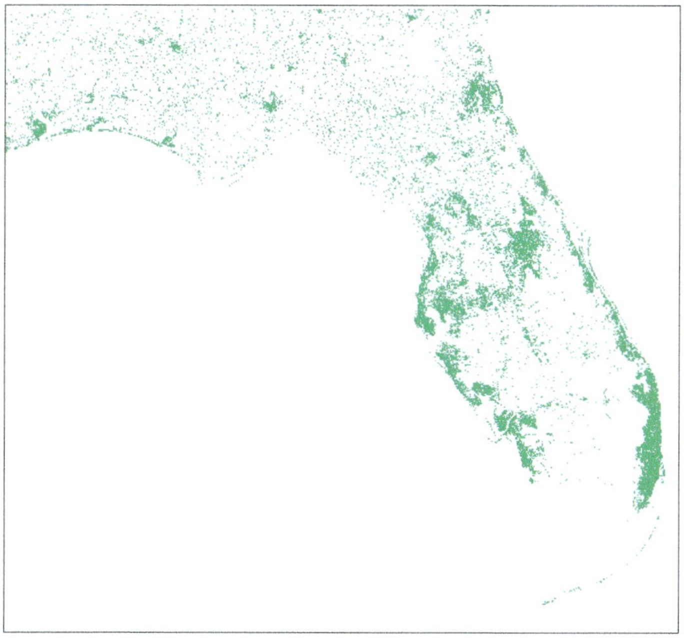

Next, you will find locations where urban lands and flood zones overlap. You will use the **Raster Calculator** tool for that task. You want to keep only urban lands in flood zones, so you can use the `multiply` function. Lands outside floodplains or in nonurban areas will use the **Nodata** value.

7. In the next cell, type or copy the following code and click **Run**.

```
for i in ["2001","2019"]:
    in_raster1 = "/arcgis/home/Flood_exposure_analysis/Results/urban_"+i+".tif"
    in_raster2 = "/arcgis/home/Flood_exposure_analysis/FL_flood_zone.img"
    out_raster = arcpy.sa.RasterCalculator([in_raster1, in_raster2],["x","y"],
    ↪ 'x * y')
    out_raster.save("/arcgis/home/Flood_exposure_analysis/Results/fz_urban_" +
    ↪ i + ".tif")
    print(i)

Raster("/arcgis/home/Flood_exposure_analysis/Results/fz_urban_2001.tif")
```

The outcome should resemble the following figure (the color on your screen may be different):

Different measures can be used to represent flood exposure. In this workflow, the following equation is used:

Flood exposure = (urban lands in flood zones)/(total urban lands)

You have already extracted both raster layers in previous steps. You will now use those datasets to calculate how many pixels each county contains. You will use **Zonal Statistics** (in the **rasterstats** package and the `sum` function) to perform this task, using Florida counties and zones to calculate the total number of urban lands in flood zones (the value of each pixel is 1, so the total number will represent the total number of pixels).

8. In the next cell, type or copy the following code and click **Run**:

```
FL_shp = gpd.GeoDataFrame.from_file('/arcgis/home/Flood_exposure_analysis
↪ /FL_counties.zip')
FL_shp.loc[FL_shp.NAME == "De Soto",'NAME'] = "DeSoto"

FL_shp[['NAME','countyfips','geometry']].to_file("/arcgis/home
↪ /Flood_exposure_analysis/Results/FL_counties.shp",driver = "ESRI Shapefile")
```

Tip: Run the following commands in separate cells to prevent potential crashes from memory constraints.

9. In the next cell, type or copy the following code and click **Run**:

```
%%time
for i in ["2001","2019"]:
    FL_shp['fz_urban_'+i] = pd.DataFrame(zonal_stats(vectors =
    ↪ FL_shp['geometry'], raster = "/arcgis/home/Flood_exposure_analysis
    ↪ /Results/fz_urban_"+i+".tif", stats='sum', nodata = 0))['sum']
print(i)
```

You will now use **Zonal Statistics** to summarize urban lands in each county.

10. In the next cell, type or copy the following code and click **Run**:

```
%%time
for i in ["2001","2019"]:
    FL_shp['urban_'+i] = pd.DataFrame(zonal_stats(vectors =
    ↪ FL_shp['geometry'], raster = "/arcgis/home/Flood_exposure_analysis
    ↪ /Results/urban_"+i+".tif", stats='sum', nodata = 0))['sum']
print(i)
```

11. To visualize selected columns in the data frame, type or copy the following code and click **Run**:

```
FL_shp[['NAME','countyfips','fz_urban_2001','fz_urban_2019','urban_2001',
↪ 'urban_2019', 'geometry']]
```

The outcome should resemble the following table:

	NAME	countyfips	fz_urban_2001	fz_urban_2019	urban_2001	urban_2019	geometry
0	Okeechobee	93	32353.0	33984.0	117016.0	122517.0	POLYGON ((1.49e+06 6.28e+05, 1.49e+06 6.28e+05...
1	Orange	95	70282.0	89058.0	952935.0	1139468.0	POLYGON ((1.42e+06 7.44e+05, 1.42e+06 7.44e+05...
2	Osceola	97	51621.0	62952.0	320373.0	401657.0	POLYGON ((1.48e+06 7.05e+05, 1.48e+06 7.05e+05...
3	Palm Beach	99	177590.0	190611.0	1250210.0	1337684.0	MULTIPOLYGON (((1.59e+06 5.21e+05, 1.59e+06 5....
4	Putnam	107	29872.0	32044.0	297762.0	306581.0	POLYGON ((1.4e+06 8.34e+05, 1.4e+06 8.34e+05, ...
...	...	...	...	...	...	...	...
62	Bay	5	62265.0	74103.0	327734.0	360977.0	MULTIPOLYGON (((1.01e+06 8.23e+05, 1.01e+06 8....
63	Baker	3	17405.0	20736.0	115209.0	129294.0	POLYGON ((1.31e+06 9.28e+05, 1.31e+06 9.28e+05...
64	Walton	131	21842.0	26339.0	223563.0	248459.0	MULTIPOLYGON (((9.47e+05 8.65e+05, 9.47e+05 8....
65	Calhoun	13	17715.0	17770.0	101448.0	103007.0	POLYGON ((1.04e+06 8.96e+05, 1.04e+06 8.96e+05...
66	Nassau	89	21358.0	22959.0	164072.0	191667.0	POLYGON ((1.34e+06 9.6e+05, 1.34e+06 9.6e+05, ...

67 rows × 7 columns

You will now apply the flood exposure equation to calculate urban flood exposure. Review the top records and visualize them on the map for 2001 and 2019. You will create a graduated color map to represent urban flood exposure for both years.

12. In the next cell, type or copy the following code and click **Run**:

```
FL_shp['fz_urban_2001_ratio'] = FL_shp['fz_urban_2001']/FL_shp['urban_2001']
FL_shp['fz_urban_2019_ratio'] = FL_shp['fz_urban_2019']/FL_shp['urban_2019']
FL_shp['fz_urban_ratio_diff'] = FL_shp['fz_urban_2019_ratio'] -
↪ FL_shp['fz_urban_2001_ratio']
```

13. In the next cell, type or copy the following code and click **Run**:

```
FL_shp[['NAME','fz_urban_2001_ratio']].sort_values('fz_urban_2001_ratio')
↪ .head(5)
```

The outcome should resemble this table:

	NAME	fz_urban_2001_ratio
24	St. Lucie	0.036193
61	Gadsden	0.043474
47	Jackson	0.052321
49	Okaloosa	0.054745
45	Marion	0.062285

You will now sort the urban flood exposure ratio in descending order.

14. In the next cell, type or copy the following code and click **Run**:

```
FL_shp[['NAME','fz_urban_2001_ratio']].sort_values('fz_urban_2001_ratio',
↪ ascending = False).head(5)
```

The outcome should resemble this table:

	NAME	fz_urban_2001_ratio
42	Monroe	0.816585
33	Franklin	0.645481
6	Lafayette	0.573184
21	Dixie	0.510798
11	Gulf	0.497880

You will now create a graduated color map of Florida counties showing the ratio of flood exposure for urban land in 2001.

15. In the next cell, type or copy the following code and click **Run**:

```
fig, ax = plt.subplots()
FL_shp.plot(ax = ax,
            column = 'fz_urban_2001_ratio',
            cmap = "OrRd",
            legend = True,
            legend_kwds={"loc": "lower left"},
            scheme='user_defined',
            classification_kwds={'bins':[ 0.05, 0.1, 0.15, 0.2, 0.25, 1]})
plt.axis("off")
plt.show()
```

The outcome should resemble the following map:

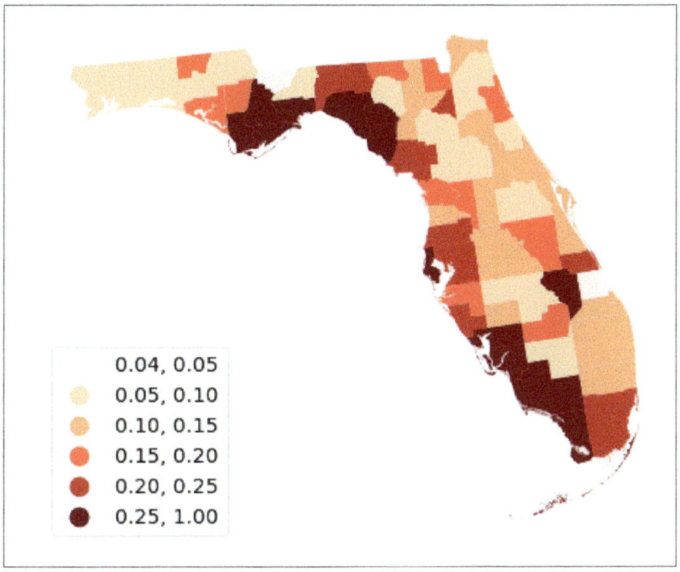

Next, you will calculate and visualize urban flood exposure in Florida counties in 2019 and create a graduated color map.

16. In the next cell, type or copy the following code and click **Run**:

```
FL_shp[['NAME','fz_urban_2019_ratio']].sort_values('fz_urban_2019_ratio')
↪ .head(5)
plt.show()
```

17. In the next cell, type or copy the following code and click **Run**:

```
FL_shp[['NAME','fz_urban_2019_ratio']].sort_values('fz_urban_2019_ratio',
↪ ascending = False).head(5)
```

You will now create a graduated color map of Florida counties showing the ratio of flood exposure for urban land in 2019.

18. In the next cell, type or copy the following code and click **Run**:

```
fig, ax = plt.subplots()
FL_shp.plot(ax = ax,
            column = 'fz_urban_2019_ratio',
            cmap = "OrRd",
            legend = True,
            legend_kwds={"loc": "lower left"},
            scheme='user_defined',
            classification_kwds={'bins':[ 0.05, 0.1, 0.15, 0.2, 0.25, 1]})
plt.axis("off")
plt.show()
```

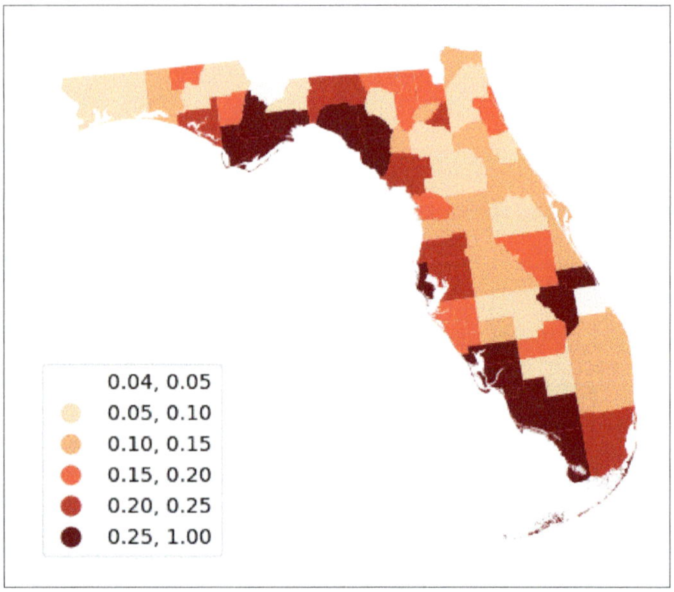

You will now examine changes in flood exposure between 2001 and 2019 and visualize the change on maps. You will review the difference in flood exposure between 2001 and 2019 and create a graduated color map with a divergent color scheme to view the change.

19. In the next cell, type or copy the following code and click **Run**:

```
FL_shp[['NAME','fz_urban_ratio_diff']].sort_values('fz_urban_ratio_diff')
↪ .head(5)
```

20. In the next cell, type or copy the following code and click **Run**:

```
FL_shp[['NAME','fz_urban_ratio_diff']].sort_values('fz_urban_ratio_diff',
↪ ascending = False).head(5)
```

21. In the next cell, type or copy the following code and click **Run**:

```
fig, ax = plt.subplots()
FL_shp.plot(ax = ax,
            column = 'fz_urban_ratio_diff',
            cmap='coolwarm',
            legend = True,
            legend_kwds={"loc": "lower left", 'fmt': "{:.3f}"},
            scheme='user_defined',
            classification_kwds={'bins':[ -0.01, -0.005, 0, 0.005, 0.01, 1]})
plt.axis("off")
plt.show()
```

The result will visualize changes in the ratio of urban land in flood zones by county in Florida between 2001 and 2019 using a range of blue (negative change) to red (positive change).

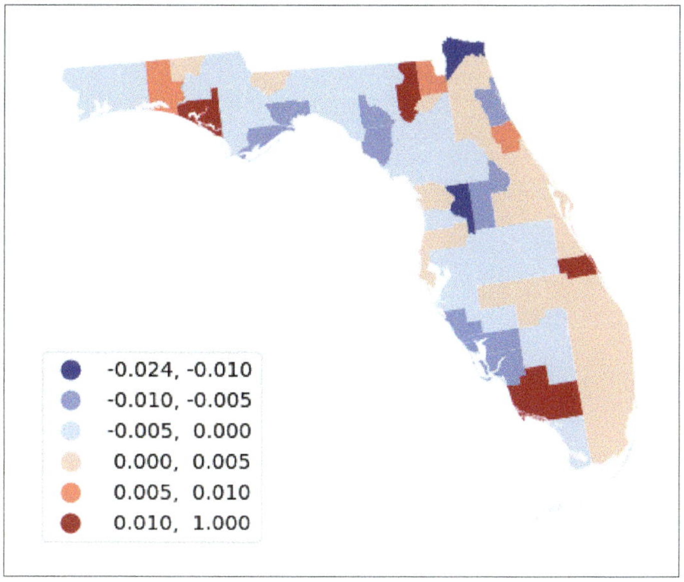

You have now completed the first half of the flood exposure assessment. In the next part, you will learn how to estimate the population exposed to flood hazards using dasymetric mapping. Dasymetric mapping is a type of thematic mapping that uses areal symbols to visualize a geographic field by refining a choropleth map with ancillary information about the distribution of the variable. Because the population is unevenly distributed within each county, merely multiplying the population by the ratio of urban land in flood zones to estimate populations living in flood zones may be arbitrary. So you will use dasymetric mapping with population data at the finest scale (census block groups). The population in each block group will then be aggregated to the county level to avoid disproportionate population distribution problems. This method will yield a more accurate estimation of people living in flood zones.

3 Use ArcGIS Notebooks to estimate exposed populations

You will first retrieve 2020 census demographic data at the block group level from ArcGIS Living Atlas and use this dataset in estimating populations with high flood exposure in Florida.

1. On the ribbon, click the **Add** button.
2. Click **My content** to open the menu and click **Living Atlas**. In the search bar, type Population by block group and press **Enter**.

   The top result should be the **USA 2020 Census Population Characteristics** feature layer by Esri.

   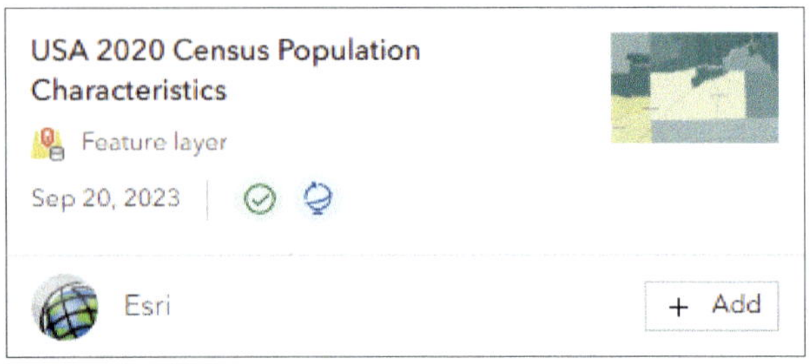

3. Click **Add** to add it to your notebook.

   After you add this ArcGIS Living Atlas layer to your notebook, a command cell will be added to your notebook.

```
# Item Added From Toolbar
# Title: USA 2020 Census Population Characteristics | Type: Feature Service |
↪ Owner: esri_demographics
item = gis.content.get("ebeb65deb5c14f4d8849fd68944b7ee6")
item
```

4. Click **Run**.

   You will now query the **USA 2020 Census Population Characteristics** dataset to show only Florida.

5. In the next cell, type or copy the following code and click **Run**:

```
item = gis.content.get("ebeb65deb5c14f4d8849fd68944b7ee6")
layer = item.layers[4]
query_result = layer.query(where="State = 'Florida'")
census_bg_2020_florida = query_result.sdf
```

Next, you will use **Zonal Statistics** to calculate the ratio of urban land in flood zones in each spatial unit (census block group). You need to confirm that the shapefile has the same coordinate system as the raster layer, or you may generate errors in the table. The coordinate system conversion can take some time—there are more than 10,000 block groups in Florida.

6. To convert the coordinate system before calculating statistics, in the next cell type or copy the following code and click **Run**:

```
census_bg_2020_florida.loc[:, ['GEOID', 'SHAPE']].spatial.to_featureclass
↪ ("/arcgis/home/Flood_exposure_analysis/Results/census_bg_2020_florida.shp")
FL_bg = gpd.GeoDataFrame.from_file('/arcgis/home/Flood_exposure_analysis
↪ /Results/census_bg_2020_florida.shp')
raster_urban01 = rasterio.open("/arcgis/home/Flood_exposure_analysis/Results
↪ /urban_2001.tif")
source_crs = Proj('epsg:3857')
target_crs = Proj(raster_urban01.crs.to_string())
transformer = Transformer.from_proj(source_crs, target_crs, always_xy=True)

def apply_transformation(geom):
    return transform(lambda x, y: transformer.transform(x, y), geom)

try:
    FL_bg['geometry'] = FL_bg['geometry'].apply(apply_transformation)
    FL_bg.crs = raster_urban01.crs
    print("Manual CRS transformation successful.")
except Exception as e:
    print("Manual CRS transformation failed:", str(e))
```

Now you will use **Zonal Statistics** to calculate urban population flood exposure. **Zonal Statistics** for larger datasets can take a long time.

Note: The following code may take more than 20 minutes to run.

7. In the next cell, type or copy the following code and click **Run**:

```
%%time
FL_bg['fz_ur_2019'] = pd.DataFrame(zonal_stats(vectors=FL_bg['geometry'],
↪ raster ="/arcgis/home/Flood_exposure_analysis/Results/fz_urban_2019.tif",
↪ stats = 'sum', nodata = 0))['sum']
FL_bg['ur_2019'] = pd.DataFrame(zonal_stats(vectors=FL_bg['geometry'],
↪ raster ="/arcgis/home/Flood_exposure_analysis/Results/urban_2019.tif",
↪ stats='sum', nodata = 0))['sum']
FL_bg[['fz_ur_2019','ur_2019']] = FL_bg[['fz_ur_2019','ur_2019']].fillna(0)
FL_bg['ur_fz_ratio'] = FL_bg['fz_ur_2019']/FL_bg['ur_2019']
FL_bg['ur_non_fz_ratio'] = 1 - FL_bg['ur_fz_ratio']
FL_bg[['geoid','fz_ur_2019','ur_2019','ur_fz_ratio','ur_non_fz_ratio']].head()
```

The outcome should resemble the following table:

	geoid	fz_ur_2019	ur_2019	ur_fz_ratio	ur_non_fz_ratio
0	120090621134	27.0	1735.0	0.015562	0.984438
1	120090621135	121.0	1147.0	0.105493	0.894507
2	120090621151	547.0	2857.0	0.191460	0.808540
3	120090621152	737.0	2570.0	0.286770	0.713230
4	120090623011	0.0	937.0	0.000000	1.000000

8. In the next cell, type or copy the following code and click **Run**:

```
census_bg_2020_florida_fz = census_bg_2020_florida.merge(FL_bg[['geoid',
↳ 'ur_fz_ratio','ur_non_fz_ratio']], left_on = "GEOID", right_on = "geoid")
```

Next, you will calculate the population inside or outside flood zones at the block group level.

9. In the next cell, type or copy the following code and click **Run**:

```
census_bg_2020_florida_fz[['GEOID','P0010001','ur_fz_ratio','ur_non_fz_ratio']]
census_bg_2020_florida_fz['pop_in_fz'] = round(census_bg_2020_florida_fz
↳ ['P0010001'].astype(int) * census_bg_2020_florida_fz['ur_fz_ratio'])
census_bg_2020_florida_fz['pop_in_non_fz'] = round(census_bg_2020_florida_fz
↳ ['P0010001'].astype(int) * census_bg_2020_florida_fz['ur_non_fz_ratio'])
census_bg_2020_florida_fz[['GEOID','County','pop_in_fz','pop_in_non_fz']]
```

In the code, `P0010001` is the total population, which will be used in the calculation. You can replace `P0010001` with other population estimates. To see other available variables, go to links.esri.com/2020CensusPopVariables.

Here, you may assume that the population is evenly distributed in urban developed areas for simplicity, but the reality is often different.

Next, you will summarize population data from census block groups to the county level by using the `groupby` function, which groups any row of a column with the same value.

10. To run the `groupby` function on the **countyfips** column (to summarize the population within each county), in the next cell type or copy the following code and click **Run**:

```
census_bg_2020_florida_fz_agg = pd.DataFrame(census_bg_2020_florida_fz
↳ [['County','pop_in_fz','pop_in_non_fz']].groupby(['County']).sum())
↳ .reset_index()
census_bg_2020_florida_fz_agg['NAME'] = census_bg_2020_florida_fz_agg.County
↳ .apply(lambda x: x[0:(len(x)-7)])
census_bg_2020_florida_fz_agg.head(5)
```

The outcome will have the columns specifying population inside and outside flood zones.

Next, you will merge the dataset with the block group dataset to visualize counties with vulnerable populations.

11. In the next cell, type or copy the following code and click **Run**:

```
FL_shp_popinfz = FL_shp[['countyfips','NAME','geometry']].merge
↪ (census_bg_2020_florida_fz_agg, on = 'NAME')
FL_shp_popinfz.head(5)
```

Finally, you will calculate the ratio of the population living in flood zones and visualize it on the map. You will use the `sort_values` function to check which county has the highest or lowest population living in flood zones.

12. In the next cell, type or copy the following code and click **Run**:

```
FL_shp_popinfz['pop_in_fz_ratio'] = FL_shp_popinfz['pop_in_fz']/
↪ (FL_shp_popinfz['pop_in_fz']+FL_shp_popinfz['pop_in_non_fz'])
FL_shp_popinfz.sort_values('pop_in_fz_ratio').head(5)
```

13. To create a map from this final dataset, in the next cell type or copy the following code and click **Run**:

```
fig, ax = plt.subplots()
FL_shp_popinfz.plot(ax = ax,
                    column = 'pop_in_fz_ratio',
                    cmap = "OrRd",
                    legend = True,
                    legend_kwds={"loc": "lower left"},
                    scheme='user_defined',
                    classification_kwds={'bins':[0.05, 0.1, 0.15, 0.2, 0.5,
                    ↪ 1]})
plt.axis("off")
plt.show()
```

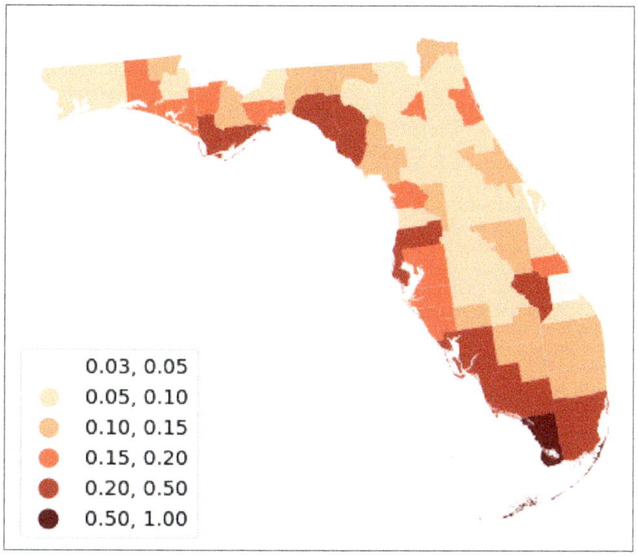

This task completes the workflow.

Interpretation of analysis

After completing this exercise, answer the following questions in detail. The answers to these questions, as well as the analysis results from the exercise workflow, will inform the intelligence brief.

1. What actionable information did the workflow yield that the public, local and state governments, and other stakeholders use to control flood exposure? See the sample response.

   Sample response: The results of the preceding workflow reveal that _____ (area) in Florida is faced with high flood exposure and ____ (area) has low flood risks. By comparing coastal and inland counties, we found that ____ (coastal/inland) counties faced higher flood exposure. Although we estimated the population living in flood zones, the findings ____ (support/disagree with) the outcomes derived from land-use data that people living in ____ (coastal/inland) counties faced higher flood exposure. Some examples of high flood exposure counties in both approaches include ______ (three counties with the highest flood exposure), whereas _____ (three counties with the lowest flood exposure) have a comparatively low flood risk.
2. Your analysis identifies certain communities as highly vulnerable to flooding. How do demographic factors such as income, age, and housing type contribute to flood exposure risk, and what policy implications arise from those findings?
3. Based on the results of this workflow, what kinds of mitigation strategies (zoning changes, infrastructure improvements, community education) would be most effective in reducing flood vulnerability in Florida?
4. How can your workflow be adapted to assess the impact of future climate change scenarios, such as sea level rise or increased hurricane intensity, on flood vulnerability? What additional datasets are necessary for this exploratory analysis?

Intelligence brief

Based on the analysis from the exercise workflow, produce a three- to five-page intelligence brief by responding to the following question prompts. Incorporate your geospatial outputs from the exercise to enrich and illustrate your conclusions.

- **BLUF:** List three to five primary observations from your analysis in bulleted-list format that summarize findings of this work.
- **Top-level analysis:** Report the primary data points supporting your BLUF.
- **Secondary analysis:** Provide the data sources and deeper analytic products from this analysis.
- **Impact:** Describe the impacts of your findings on the department or organization.
- **Confidence:** Report the level of confidence of your assessment based on available data, accuracy, and so on.
- **Bias:** Explain any background or information you have that could have affected your analysis.
- **Conclusions:** Determine what type of analytic conclusions there may be and how the data requirements, analysis, and brief writing may change based on what analytic framework you are following (descriptive, predictive, prescriptive).

Applying and extending your skills

This workflow includes the estimation of population living in flood-prone areas, and the outcome shows that coastal counties generally have a higher flood exposure than inland counties in Florida. Another finding is that Florida counties on the Atlantic have lower flood exposure when compared with those on the Gulf. This study is a preliminary assessment of flood exposure at the county level, but it has some limitations. For instance, although this method aims to accurately measure urban flood exposure, the assumption of population distributed evenly in each block group is not accurate. Some potential improvements may include using building footprints or human mobility data to redistribute population in each small spatial unit. Second, the FEMA 100-year flood map layer is often criticized for its outdated information and inaccurate measurement owing to the disproportionate distribution of gauges. In the future, flood maps may be validated with other flood damage datasets.

Overall, this workflow provides an example of visualizing flood exposure using land-use datasets and flood maps. All datasets used in this study are publicly available, but owing to the computing capacity, this tutorial focused only on Florida. However, this framework of assessing flood exposure can be extended to any region depending on the availability of land-use datasets and flood maps. The following data sources are used in this chapter:

- Source of land-use and landcover data (Multi-Resolution Land Characteristics Consortium): www.mrlc.gov/data
- Source of 100-year flood zone layer: https://msc.fema.gov/portal/advanceSearch
- Census Data API: www.census.gov/data/developers/guidance/api-user-guide.html

Additional resources

The following resources provide information about the flood hazard and human security challenges at hand.

- FEMA Flood Map Services: https://msc.fema.gov/portal/home
- FEMA National Flood Hazard Layer Viewer: www.fema.gov/flood-maps/national-flood-hazard-layer
- FEMA Geospatial Resource Center Flooding Portal: https://gis-fema.hub.arcgis.com/pages/flooding

For more information about dasymetric mapping, see the chapter references.

References

- Aerts, J. C. J. H., W. J. Botzen, K. C. Clarke, et al. 2018. "Integrating Human Behaviour Dynamics into Flood Disaster Risk Assessment." *Nature Climate Change* 8 (3), article 3. doi.org/10.1038/s41558-018-0085-1.

- Burningham, K., J. Fielding, and D. Thrush. 2008. "It'll Never Happen to Me: Understanding Public Awareness of Local Flood Risk." *Disasters* 32 (2): 216–38.
- Hauer, M. E., R. K. Saunders, and D. Shtob. 2022. "Research Note: Demographic Change on the United States Coast, 2020–2100." *Demography* 59 (4): 1221–32.
- Interagency Floodplain Management Review Committee. 1994. "Sharing the Challenge: Floodplain Management into the 21st Century." Report of the Interagency Floodplain Management Review Committee to the Administration Floodplain Management Task Force. https://irp.fas.org/agency/dhs/fema/sharing.pdf.
- Koks, E. E., B. Jongman, T. G. Husby, and W. J. W. Botzen. 2015. "Combining Hazard, Exposure, and Social Vulnerability to Provide Lessons for Flood Risk Management." *Environmental Science & Policy* 47:42–52. doi.org/10.1016/j.envsci.2014.10.013.
- Maantay, Juliana, and Andrew Maroko. 2009. "Mapping Urban Risk: Flood Hazards, Race, & Environmental Justice in New York." *Applied Geography* 29 (1): 111–24. doi.org/10.1016/j.apgeog.2008.08.002.
- Mennis, Jeremy. 2009. "Dasymetric Mapping for Estimating Population in Small Areas." *Geography Compass* 3 (2): 727–45. doi.org/10.1111/j.1749-8198.2009.00220.x.
- Mitchell, J. K. 2003. "European River Floods in a Changing World." *Risk Analysis: An International Journal* 23 (3): 567–74.
- Qiang, Yi, Nina S. N. Lam, Heng Cai, and Lei Zou. 2017. "Changes in Exposure to Flood Hazards in the United States." *Annals of the American Association of Geographers* 107 (6): 1332–50. doi.org/10.1080/24694452.2017.1320214.
- Schultz, B. 2001. "Irrigation, Drainage, and Flood Protection in a Rapidly Changing World." *Irrigation and Drainage: The Journal of the International Commission on Irrigation and Drainage* 50 (4): 261–77.
- Tate, E., M. A. Rahman, C. T. Emrich, and C. C. Sampson. 2021. "Flood Exposure and Social Vulnerability in the United States." *Natural Hazards* 106 (1): 435–57.
- Wing, O. E. J., P. D. Bates, A. M. Smith, et al. 2018. "Estimates of Present and Future Flood Risk in the Conterminous United States." *Environmental Research Letters* 13 (3): 034023. doi.org/10.1088/1748-9326/aaac65.
- World Meteorological Organization. 2021. "Weather-Related Disasters Increase over Past 50 Years, Causing More Damage But Fewer Deaths." https://wmo.int/media/news/weather-related-disasters-increase-over-past-50-years-causing-more-damage-fewer-deaths.
- Xu, Jinwen, and Yi Qiang. 2024. "Environmental Justice Implications of Flood Risk in the Contiguous United States: A Spatiotemporal Assessment of Flood Exposure Change from 2001 to 2019." *Cartography and Geographic Information Science* 5 (4): 492–512. doi.org/10.1080/15230406.2024.2328159.

About the authors

Jinwen Xu is a postdoctoral scholar in the Institute of Transportation Studies, University of California, Berkeley. He was research associate faculty at the GIS Center, Florida International University. He obtained his PhD in geography and environmental science and policy at the University of South Florida. He also has a master's degree in urban and environmental planning and a transportation certificate from Arizona State University. His research interests are in disaster resilience, geospatial data science, and GeoAI.

Levente Juhász is a research associate professor and assistant director of GIScience at the GIS Center, Florida International University, with a mix of academic and industry experiences. In academia, he held various roles at the University of Florida; Joint Research Centre of the European Commission in Ispra, Italy; the Carinthia University of Applied Sciences in Villach, Austria; and the University of Szeged, Hungary. In industry, he gained experience as a data scientist for a geospatial start-up, Mapillary, and as a full-stack GIS developer for a consulting firm. He has a PhD in geomatics.

Chapter 9

Evaluating shoreline properties threatened with future sea level rise in coastal Florida

Jinwen Xu, *University of California, Berkeley, Institute of Transportation Studies*
Levente Juhász, *Florida International University, GIS Center*

Overview

In this chapter, you will learn how to use GIST tools and methods to assess ownership of shoreline properties threatened with future sea level rise (SLR) in coastal Florida. The workflows and the final deliverable will show how SLR will affect coastal communities and provide actionable information to support various planning-related tasks that will be affected by changes in the current shoreline.

Learning objectives

- Generate a 500-meter geospatial buffer based on a projected SLR scenario to categorize and assess at-risk properties in the coastal zones of St. Lucie County, Florida.
- Transfer parcel ownership (polygon) to shoreline (polyline) using spatial analysis tools.
- Produce visualizations using interactive maps in ArcGIS Notebooks to ascertain patterns of shoreline ownership within St. Lucie County.
- Analyze and summarize changes in shoreline ownership across various SLR scenarios.

Technical components

- Software:
  - ArcGIS Notebooks in ArcGIS Online (Advanced level access)
- Datasets:
  - Sea level rise scenarios derived from NOAA SLR data
  - Parcel data from St. Lucie County's property appraiser
  - Shoreline data derived from Florida Geographic Information Office elevation data

Prerequisites

- Basic knowledge of or experience using GIS software or computer scripting language (Python) is required.
- Technical skill level: Intermediate GIS.
- Time to complete chapter: One to two hours without intelligence brief, three to five hours with intelligence brief.

Background and situational context

SLR poses a significant threat to coastal communities, with the potential to inundate coastal structures and inflict extensive damage. It is important to estimate and predict the impacts of future SLR to promote resilient communities and mitigate potential damage. Knowing the ownership of future shorelines affected by SLR is also important in some cases. Two such use cases are related to either infrastructure development or environmental conservation efforts. For example, local governments can establish better zoning regulations and land-use policies that consider risks associated with SLR. Public ownership knowledge can guide decisions on infrastructure investments—for example, building or reinforcing seawalls or flood protection systems. Similarly, ecosystem preservation and conservation decisions can help prioritize areas of restoration and preservation.

However, existing practices rely on current shoreline data, which may lead to overoptimistic assessments of future impacts from SLR. In addition, there appears to be a gap in the scientific literature and planning practice, which tends not to consider that the location of today's shoreline may well be different from that of the future under different SLR scenarios. This may stem from the lack of available data. Moreover, the ambiguous definition of shoreline property ownership complicates the identification of areas suitable for future development. This uncertainty impedes comprehensive and effective estimates of how many public and private properties might be at risk under varying SLR scenarios. A more accurate classification of property ownership is needed to plan effective mitigation strategies.

One of the leading shoreline restoration methods in Florida is known as living shorelines. Living shorelines are widely endorsed by conservation groups for their ability to improve estuarine water quality and aquatic ecosystem functionality, providing a crucial buffer for coastal communities. In both estuarine and freshwater systems, living shorelines can provide habitat for fish, birds, and other wildlife.

St. Lucie County, as one of the Florida counties with the greatest acreage of wetlands in the Treasure Coast region, has faced threats from SLR. According to Climate Central's Risk Finder, there is a 73 percent chance of at least one flood over three feet in St. Lucie Village between now and 2050 (see the chapter resources). With such a flood, more than 28 miles of roadways might be completely inundated. Several living shoreline projects have been conducted already within the Indian River Lagoon, such as in nearby Brevard Zoo.

When future living shorelines are located, it is important to determine whether the protected property is public or private. The project planner can then seek permission from the landowners or the government and identify potential beneficiaries.

The following exercise workflow will use data mining methods in Python to investigate parcel and shoreline data in St. Lucie County, especially the coastal areas near the Indian River Lagoon. By referencing property appraisers' records, this chapter will determine shoreline ownership and the proportion of public shoreline for siting future living shorelines.

Exercise workflow 9

Using various datasets (parcel information from St. Lucie County's property appraiser, five-foot SLR polygon from NOAA, current shoreline data), learners will perform a step-by-step workflow to create a buffer based on a certain sea level scenario, extract vulnerable communities within proximity of SLR inundation, assess property ownership according to property appraiser records, evaluate shoreline ownership, and locate potential sites for living shorelines.

Note: ArcGIS Notebooks does not have an auto-save function, so remember to save your work after each completed step or task.

1 Load the data

1. In your browser, navigate to www.arcgis.com and sign in to your ArcGIS organization account.
2. On the navigation bar, click **Notebooks**.
3. Click **New notebook** and select the **Advanced** notebook.

   Note: The Advanced Python notebook uses three credits per hour.

4. In the notebook, click **Files** and click the **home** folder.
5. Click **New folder**. Create a folder called Sea_level_rise_analysis.
6. In the new folder, create another folder called Results.
7. At the bottom of the Sea_level_rise_analysis folder, click **Choose file to upload**.

8. Navigate to the location of the **SecurityFirst_ExerciseData/Chapter09** folder. Upload the **Chapter09_data.zip** folder.

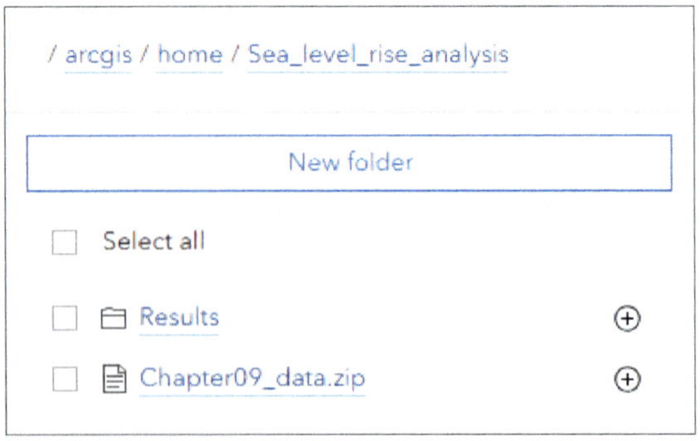

9. On the top right of the notebook, click **Save** > **Save As**.
10. In the **Save notebook as** window, apply the following settings:
    - **Title**: Ch09_SeaLevelRise
    - **Folder**: SecurityFirst

> **Tip:** If you do not already have a folder called SecurityFirst, click Create new folder and add the folder name.

11. Click **Save notebook**.

2 Run the code

1. In the notebook script, click the first cell and click **Run** to connect the notebook to your GIS.

   The following code will run the first few cells to load necessary packages (for example, `arcgis Python API` and `ArcPy`). ArcGIS Notebooks in ArcGIS Online uses a cloud-based environment, so you do not need to worry about installing Python packages on your own computer. After running the following cells, you will be able to install Python packages and load and test datasets.

2. In the next cell, type or copy the following code:

```
import numpy as np
import pandas as pd
import time
import arcpy
import arcgis
from arcgis import features
from arcgis.features import FeatureLayer
import requests
import json
import pyproj
import zipfile

arcpy.env.overwriteOutput = True
```

3. Click **Run** or use these shortcuts:

   - Run and Move to Next Cell: press Shift + Enter
   - Run and Stay in Current Cell: press Ctrl + Enter
   - Run and Insert a New Cell Below: press Alt + Enter

   The code will load the datasets. When the cell is running, the asterisk (*) appears in front of the cell. After finishing, it becomes a static number. You can monitor your progress this way.

4. In the next cell, type or copy the following code:

```
!unzip /arcgis/home/Sea_level_rise_analysis/Chapter09_data.zip -d
↪ /arcgis/home/Sea_level_rise_analysis/shapefiles
```

   Note: The arrow in the code block indicates a continuation of the code line. When you type the code, it will all be on one line.

5. Click **Run** to extract GIS files from the zip files.

   You will now create a 500-meter buffer to select vulnerable regions. Use the interactive map provided by ArcGIS API for Python to visualize the SLR extent and its buffer. By using the `create_buffers` function in ArcGIS API for Python, you will create a buffer based on the measurement units in the current coordinate system.

   The code will first upload the feature class as a feature layer and visualize the feature layer in an interactive map. This upload function will be used again later in the chapter. Next, it will create a 500-meter buffer for the 2100 intermediate–high SLR scenario and load it into your notebook. You want to create a 500-meter buffer for the layer. The process may take about five minutes.

6. In the next cells, type or copy the following code:

```
%%time
def featureclass_to_featurelayer(
    fc: str,
    gis: arcgis.gis.GIS,
    name: str="Feature Layer from Feature Class",
    folder: str=None
) -> arcgis.gis.Item:
    try:
        df = pd.DataFrame.spatial.from_featureclass(fc)
        fl_item = df.spatial.to_featurelayer(name, folder=folder)
        return fl_item
    except Exception as e:
        print(f"Could not convert feature class to feature layer because of the
        ↪ following error: {e}.")

item_SLR_extent = featureclass_to_featurelayer('/arcgis/home/Sea_level_rise_
↪ analysis/shapefiles/SLR_2100_int_high.shp',GIS, name="SLR_2100_int_high")
```

Note: Run the following commands in separate cells to prevent potential crashes from memory constraints.

```
%%time
polygon_lyr_buffer  = features.use_proximity.create_buffers(item_SLR_extent,
↪ distances=[500], units='Meters', ring_type='Rings',
↪ dissolve_type = 'Dissolve', end_type='Flat', output_name='SLR_500m_buffer')
```

Tip: If you encounter problems (for example, job failed) and the process stops during the first run, the notebook will prevent you from creating a layer with the same name. If this happens, use the following code to delete the layer. Otherwise, skip the following code and continue to the next code block.

```
def delete_layer_if_exists(layer_name):
    search_result = gis.content.search(query=f"title:{layer_name}",
    ↪ item_type="Feature Layer")
    if search_result:
        for item in search_result:
            if item.title == layer_name:
                item.delete()
                print(f"Deleted existing layer: {layer_name}")
delete_layer_if_exists("SLR_500m_buffer")
```

```
%%time
item_buffer = gis.content.get(polygon_lyr_buffer.id)
export_job = item_buffer.export(title="SLR_500m_buffer",
↪ export_format="Shapefile")
download_path = export_job.download(save_path="/arcgis/home/Sea_level_rise_
↪ analysis/Results")
```

Note: Run the following commands in separate cells to prevent potential crashes from memory constraints.

```
!unzip /arcgis/home/Sea_level_rise_analysis/Results/SLR_500m_buffer.zip -d
↪ /arcgis/home/Sea_level_rise_analysis/Results
```

Note: Run the following commands in separate cells to prevent potential crashes from memory constraints. This code will set up the function for revising the center and zoom level of interactive maps.

```
def compute_extent(center, zoom, width=800, height=600):
    resolution = 156543.03 / (2 ** zoom)
    half_width = (width / 2) * resolution
    half_height = (height / 2) * resolution
    transformer = pyproj.Transformer.from_crs("EPSG:4326", "EPSG:3857",
    ↪ always_xy=True)
    center_x, center_y = transformer.transform(center[1], center[0])
```

```
    return {
        "xmin": center_x - half_width,
        "ymin": center_y - half_height,
        "xmax": center_x + half_width,
        "ymax": center_y + half_height,
        "spatialReference": {"wkid": 102100}
    }
```

Note: Run the following commands in separate cells to prevent potential crashes from memory constraints.

```
interactive_map_1 = gis.map("Fort Pierce")
interactive_map_1.content.add(FeatureLayer(item_buffer.url+"/0"))
interactive_map_1.extent = compute_extent([27.45, -80.32], 12)
interactive_map_1
```

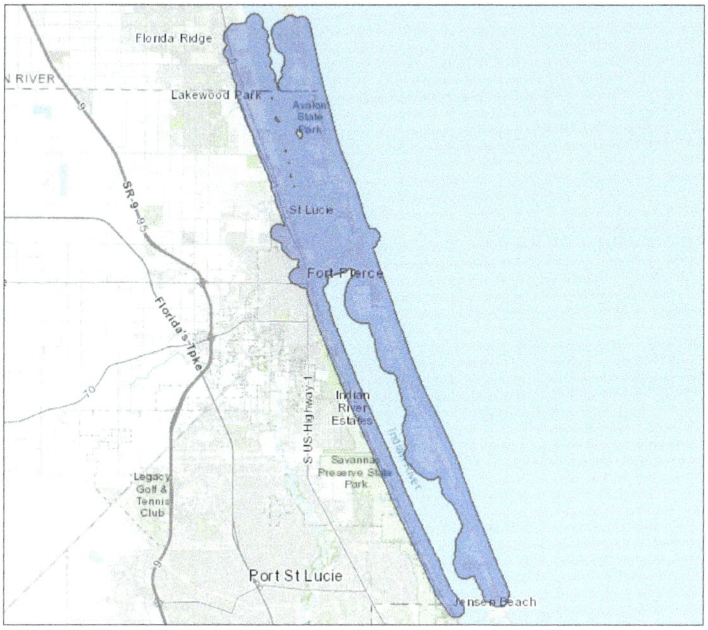

Note: The interactive map may sometimes experience delays. If the cell remains unresponsive for more than 30 seconds, interrupt the kernel to forcibly stop the cell. After waiting a few seconds, try to run the cell again.

When working in ArcGIS Pro, you can identify properties that intersect with the SLR extent by using the **Select by Location** tool. However, it is a bit different in ArcGIS Notebooks. To avoid selecting parcels that share only a part of their boundary with the SLR extent, you will define the `HAVE_THEIR_CENTER_IN` spatial relationship parameter within the `SelectLayerByLocation_management` function to find the parcels that face SLR risks.

First, you will extract parcels (polygon) with their centroids inside the buffer. Next, you will upload it to ArcGIS Online for visualization.

7. In the next cells, type or copy the following code and click **Run**:

```
%%time
Selection = arcpy.SelectLayerByLocation_management('/arcgis/home/Sea_level_
↪ rise_analysis/shapefiles/stl_ownership.shp', "HAVE_THEIR_CENTER_IN",
↪ '/arcgis/home/Sea_level_rise_analysis/Results/BufferedFeatures.shp')
arcpy.CopyFeatures_management(Selection,'/arcgis/home/Sea_level_rise_analysis
↪ /Results/stl_ownership_within_buffer.shp')
```

Note: Run the following command in separate cells to prevent potential crashes.

```
%%time
property_in_buffer = featureclass_to_featurelayer('/arcgis/home
↪ /Sea_level_rise_analysis/Results/stl_ownership_within_buffer.shp',GIS,
↪ name="property_in_buffer")
```

Note: Run the following command in separate cells to prevent potential crashes.

```
item_properties_in_buffer = gis.content.get(property_in_buffer.id)
property_in_buffer_featurelayer = FeatureLayer(item_properties_in_buffer
↪ .url+"/0")
```

Note: Run the following command in separate cells to prevent potential crashes.

```
interactive_map_2 = gis.map("Fort Pierce")
interactive_map_2.content.add(FeatureLayer(item_buffer.url+"/0"))
interactive_map_2.content.add(property_in_buffer_featurelayer)
interactive_map_2.extent = compute_extent([27.45, -80.32], 12)
interactive_map_2
```

Note: The interactive map may sometimes experience delays. If the cell remains unresponsive for more than 30 seconds, interrupt the kernel to forcibly stop the cell. After waiting a few seconds, run the cell again.

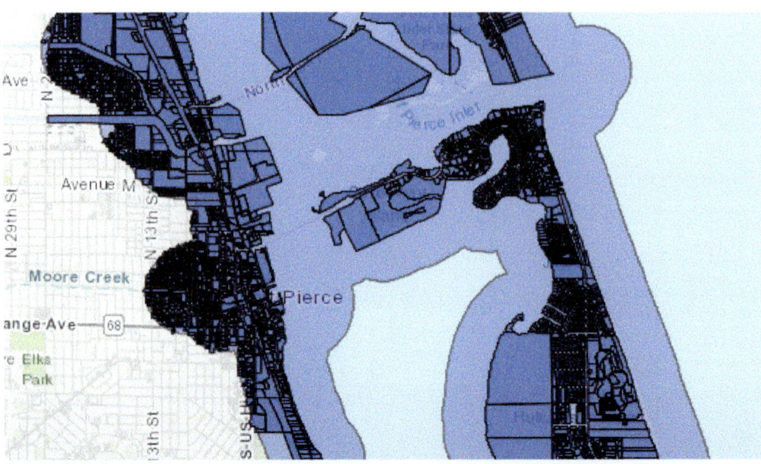

This map doesn't show ownership of parcels. However, this layer has a **land_type** field that can be used to distinguish property ownership. Use the following commands to set up different color schemes based on ownership for properties that are potentially vulnerable to SLR.

8. In the next cells, type or copy the following code and click **Run**:

```
request = requests.get(property_in_buffer_featurelayer.url +
↪ '?f=pjson&token={}'.format(gis._con.token))
property_in_buffer_featurelayer_json = request.json()
classification_polygon = {
    "type": "uniqueValueDef",
    "uniqueValueFields": ["land_type"],
    "fieldDelimiter": ",",
    "baseSymbol": {
        "type": "esriSFS",
        "style": "esriSFSSolid",
        "color": [0, 0, 0, 0],
        "outline": {
            "type": "esriSLS",
            "style": "esriSLSSolid",
            "color": [128, 128, 128, 255],
            "width": 1
        }
    },
    "colorRamp": {
        "type": "algorithmic",
        "fromColor": [115, 76, 0, 255],
        "toColor": [255, 25, 86, 255],
        "algorithm": "esriHSVAlgorithm"
    }
}
renderer = property_in_buffer_featurelayer.generate_renderer(classification_
↪ polygon)
renderer['uniqueValueInfos'][0]['symbol']['color'] = [241, 163, 64, 255]
renderer['uniqueValueInfos'][1]['symbol']['color'] = [153, 153, 153, 255]
renderer['uniqueValueInfos'][2]['symbol']['color'] = [128,0,128,255]
property_in_buffer_featurelayer_json['drawingInfo']['renderer'] = renderer
if 'editingInfo' in property_in_buffer_featurelayer_json:
    del property_in_buffer_featurelayer_json['editingInfo']
property_in_buffer_featurelayer.manager.update_definition(property_in_
↪ buffer_featurelayer_json)
```

Note: Run the following command in separate cells to prevent potential crashes.

```
property_in_buffer_featurelayer = FeatureLayer(item_properties_in_buffer
↪ .url+"/0")
interactive_map_3 = gis.map("Fort Pierce")
interactive_map_3.content.add(property_in_buffer_featurelayer)
interactive_map_3.extent = compute_extent([27.45, -80.32], 12)
interactive_map_3
```

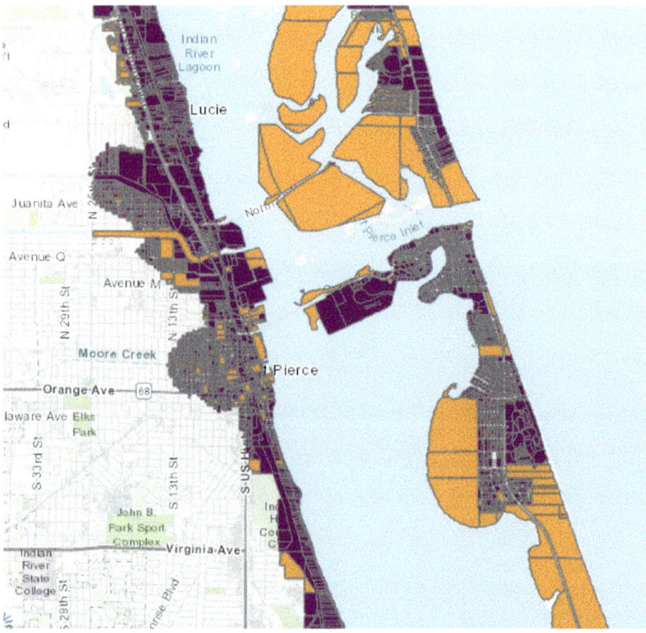

Now that your map distinguishes between public and private land, you can improve the color scheme. You will revise the color scheme to plot public land in green and private land in blue.

9. In the next cells, type or copy the following code and click **Run**:

```
renderer['uniqueValueInfos'][0]['symbol']['color'] = [161, 215, 106, 255]
renderer['uniqueValueInfos'][1]['symbol']['color'] = [153, 153, 153, 255]
renderer['uniqueValueInfos'][2]['symbol']['color'] = [145, 191, 219, 255]
property_in_buffer_featurelayer_json['drawingInfo']['renderer'] = renderer
if 'editingInfo' in property_in_buffer_featurelayer_json:
    del property_in_buffer_featurelayer_json['editingInfo']
property_in_buffer_featurelayer.manager.update_definition(property_in_
↪ buffer_featurelayer_json)
```

Note: Run the following command in separate cells to prevent potential crashes.

```
property_in_buffer_featurelayer = FeatureLayer(item_properties_in_buffer
↪ .url+"/0")
interactive_map_4 = gis.map("Fort Pierce")
interactive_map_4.content.add(property_in_buffer_featurelayer)
interactive_map_4.extent = compute_extent([27.45, -80.32], 12)
interactive_map_4
```

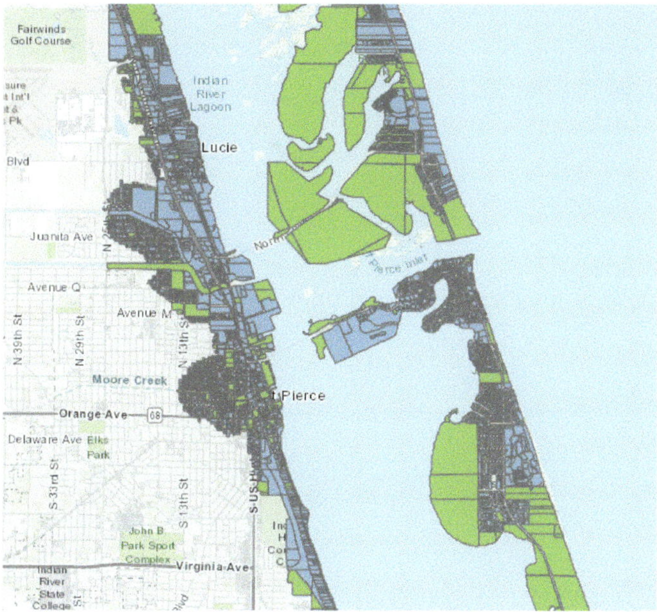

Your next goal is to transfer the ownership of classified properties to shoreline features. To do so, you will first break up the shoreline into five-meter segments to avoid arbitrarily assigning a single value to a long shoreline segment. Because polyline features cannot directly extract value from the nearest polygon, you will convert lines to points and obtain the ownership value from the nearest parcel. If the shoreline is too far from the parcel, you will set the search distance at 50 meters. After running cells in the entire section, you will upload the feature class to ArcGIS Online as a feature layer with a similar attribute table as shown in a following step.

10. In the next cells, type or copy the following code and click **Run**:

```
%%time
arcpy.management.GeneratePointsAlongLines(Input_Features='/arcgis/home
↪/Sea_level_rise_analysis/shapefiles/st_lucie_shoreline.shp',
Output_Feature_Class='/arcgis/home/Sea_level_rise_analysis/Results
↪/break_pts_5m.shp',
    Point_Placement="DISTANCE",
    Distance="5 Meters",
    Percentage=None,
    Include_End_Points="NO_END_POINTS",
    Add_Chainage_Fields="NO_CHAINAGE"
)
```

Note: Run the following command in separate cells to prevent potential crashes.

```
%%time
arcpy.management.SplitLineAtPoint(in_features="/arcgis/home/Sea_level_rise_
↪ analysis/shapefiles/st_lucie_shoreline.shp",
point_features='/arcgis/home/Sea_level_rise_analysis/Results/break_pts_5m.shp',
out_feature_class="/arcgis/home/Sea_level_rise_analysis/Results/st_lucie_
↪ shoreline_5m.shp",
search_radius="0.1 Meters")
```

Note: Run the following command in separate cells to prevent potential crashes.

```
%%time
arcpy.analysis.SpatialJoin(
target_features='/arcgis/home/Sea_level_rise_analysis/Results/st_lucie_
↪ shoreline_5m.shp',
join_features='/arcgis/home/Sea_level_rise_analysis/Results/stl_ownership_
↪ within_buffer.shp',
out_feature_class="/arcgis/home/Sea_level_rise_analysis/Results
↪ /shoreline_sj.shp",
    join_operation="JOIN_ONE_TO_ONE",
    join_type="KEEP_ALL",
    match_option="CLOSEST",
    search_radius="50 Meters",
    distance_field_name=""
)
```

Note: Run the following command in separate cells to prevent potential crashes.

```
%%time
item_shoreline = featureclass_to_featurelayer('arcgis/home/Sea_level_rise_
↪ analysis/Results/shoreline_sj.shp',GIS, name="shoreline_with_ownership")
sdf_shoreline_ownership = pd.DataFrame.spatial.from_layer(item_shoreline
↪ .layers[0])
```

11. In the next cell, type or copy the following code and click **Run**:

```
sdf_shoreline_ownership
```

Some values of the **land_type** column are None/empty, and remember that you also have an Unidentified category. These None/empty values are not related to parcel ownership; instead, it means that the shoreline is too far away from the nearest parcel (> 50 meters). You can keep them there.

	OBJECTID	FID	Join_Count	TARGET_FID	id	i	ORIG_FID	ORIG_SEQ	ParcelID	land_type	Shape__Length	SHAPE
0	1	0	1	0	1	36462	0	1	1034-800-0007-000-3	Private	5.64426	{"paths": [[[-8942130.07825543, 3193302.238095...
1	2	1	1	1	1	36462	0	7	1034-800-0007-000-3	Private	3.391691	{"paths": [[[-8942106.72577601, 3193281.766198...
2	3	2	1	2	1	36462	0	6	1034-800-0007-000-3	Private	5.652979	{"paths": [[[-8942110.69066883, 3193285.795565...
3	4	3	1	3	1	36462	0	5	1034-800-0007-000-3	Private	5.661703	{"paths": [[[-8942112.25823641, 3193290.807791...
4	5	4	1	4	1	36462	0	4	1034-800-0007-000-3	Private	5.652823	{"paths": [[[-8942116.22302125, 3193294.837046...
...	...	...	...	...	...	...	...	...	...	...	...	...
33445	33446	33445	0	33445	1	54	4781	6			2.896314	{"paths": [[[-8941031.91012999, 3179802.729861...
33446	33447	33446	0	33446	1	54	4781	5			5.66217	{"paths": [[[-8941031.94100577, 3179797.067775...
33447	33448	33447	0	33447	1	54	4781	4			5.662168	{"paths": [[[-8941031.97188147, 3179791.405691...
33448	33449	33448	0	33448	1	54	4781	3			5.662166	{"paths": [[[-8941032.0275711, 3179785.743610...
33449	33450	33449	0	33449	1	54	4781	2			5.662163	{"paths": [[[-8941032.03363267, 3179780.081531...

12. In the next cell, type or copy the following code and click **Run**:

```
%%time
shoreline_ownership_featurelayer = FeatureLayer(item_shoreline.url+'/0')
interactive_map_5 = gis.map("Cause Way Island")
interactive_map_5.content.add(shoreline_ownership_featurelayer)
interactive_map_5.extent = compute_extent([27.46, -80.308], 15)
interactive_map_5
```

You have now attached ownership value to the shoreline dataset. Next, you will visualize these results on the map. You can choose different colors to represent various types of shorelines, similar to what you did for parcels. Using the ArcGIS API for Python built-in function, you will be able to visualize these layers, pan around, and zoom within the interactive map.

13. In the next cells, type or copy the following code and click **Run**:

```
request = requests.get(item_shoreline.url +"/0"+ '?f=pjson&token={}'
↪ .format(gis._con.token))
shoreline_ownership_json =  request.json()
classification_line = {
    "type": "uniqueValueDef",
    "uniqueValueFields": ["land_type"],
    "fieldDelimiter": ",",
    "baseSymbol": {
        "type": "esriSLS",
        "style": "esriSLSSolid",
        "color": [0, 0, 0, 255],
        "width": 2
    },
    "colorRamp": {
        "type": "algorithmic",
        "fromColor": [115, 76, 0, 255],
        "toColor": [255, 25, 86, 255],
        "algorithm": "esriHSVAlgorithm"
    }
}
```

```
renderer = shoreline_ownership_featurelayer.generate_renderer(classification_
↪ line)
renderer['uniqueValueInfos'][0]['symbol']['color'] = [241, 163, 64, 255]
renderer['uniqueValueInfos'][1]['symbol']['color'] = [153, 153, 153, 255]
renderer['uniqueValueInfos'][2]['symbol']['color'] = [153, 153, 153, 255]
renderer['uniqueValueInfos'][3]['symbol']['color'] = [128,0,128,255]
shoreline_ownership_json['drawingInfo']['renderer'] = renderer
if 'editingInfo' in shoreline_ownership_json:
    del shoreline_ownership_json['editingInfo']
shoreline_ownership_featurelayer.manager.update_definition(shoreline_
↪ ownership_json)
```

Note: Run the following command in separate cells to prevent potential crashes.

```
interactive_map_6 = gis.map("Cause Way Island")
interactive_map_6.content.add(FeatureLayer(item_shoreline.url+'/0'))
interactive_map_6.extent = compute_extent([27.46, -80.308], 15)
interactive_map_6
```

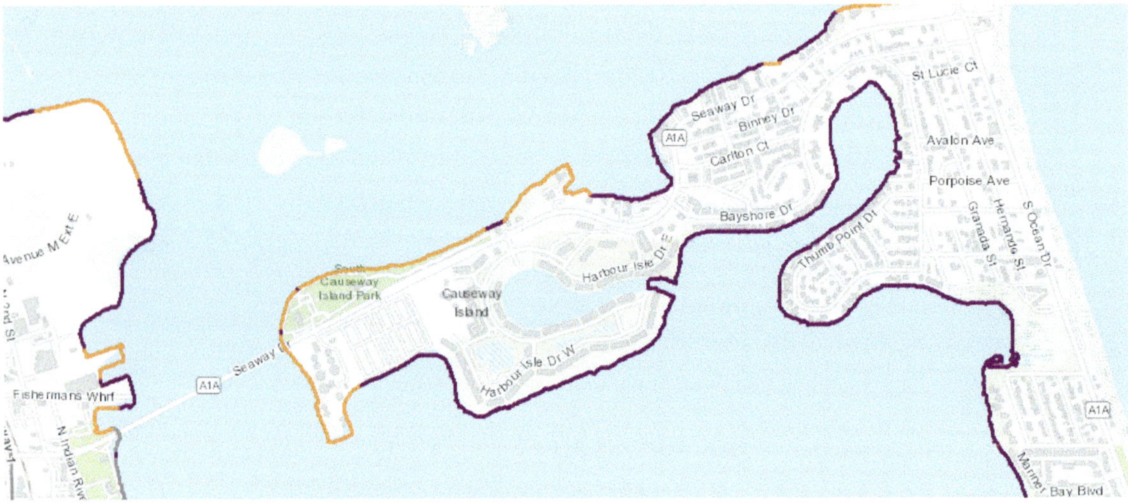

Note: Run the following command in separate cells to prevent potential crashes.

```
interactive_map_7 = gis.map("Cause Way Island")
interactive_map_7.content.add(FeatureLayer(item_properties_in_buffer.url+"/0"))
interactive_map_7.content.add(FeatureLayer(item_shoreline.url+'/0'))
interactive_map_7.extent = compute_extent([27.46, -80.308], 15)
interactive_map_7
```

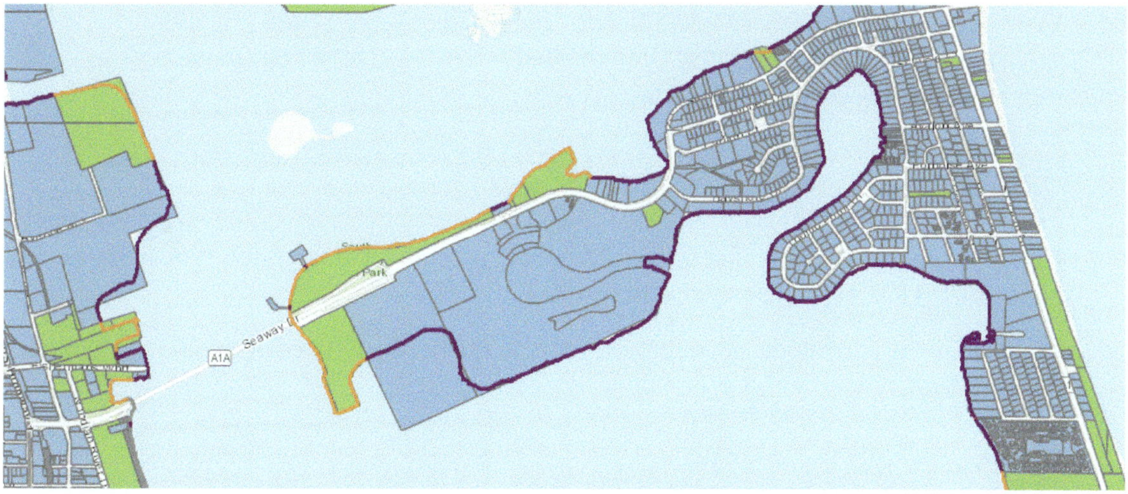

You have now visualized the segments of the shoreline according to public vs. private ownership. Next, you will perform some calculations in your notebook to find the proportion of publicly vs. privately owned shoreline. You can use length attributes to calculate the statistics. Remember that the length calculated is in meters, which corresponds to the measurement unit in the interactive map. You can convert meters to kilometers by dividing by 1,000. You can also calculate the proportion of public vs. private shoreline by using length as the numerator and total shoreline length as the denominator. Use the following code to complete these calculations and compare yours with the sample results code.

14. To calculate the length (in kilometers) of the part of the shoreline that is publicly owned, in the next cell type or copy the following code and click **Run**:

```
sum(sdf_shoreline_ownership[sdf_shoreline_ownership.land_type == "Public"]
↪ .Shape__Length) / 1000
```

The length of the shoreline that is publicly owned is 79.7638154980295 kilometers.

15. To calculate the length (in kilometers) of the part of the shoreline that is privately owned, in the next cell type or copy the following code and click **Run**:

```
sum(sdf_shoreline_ownership[sdf_shoreline_ownership.land_type == "Private"]
↪ .Shape__Length) / 1000
```

The length of the shoreline that is privately owned is `83.81084041791283` kilometers.

16. To calculate the length (in kilometers) of the part of the shoreline with unidentified ownership, in the next cell type or copy the following code and click **Run**:

```
sum(sdf_shoreline_ownership[(sdf_shoreline_ownership.land_type ==
↪ "Unidentified")|(sdf_shoreline_ownership.land_type == " ")].Shape__Length) /
↪ 1000
```

The length of the shoreline with unidentified ownership is 14.567163930047567 kilometers.

17. To calculate the length (in kilometers) of the part of the shoreline that is too far away from parcels, which results in a value of **Unidentified** in the **land_type** field, in the next cell type or copy the following code and click **Run**:

```
sum(sdf_shoreline_ownership[(sdf_shoreline_ownership.land_type ==
↪ "Unidentified")].Shape__Length)/1000
```

The length of the shoreline that is too far from the parcels and has a **land_type** field value of **Unidentified** is 0.5193017475902479 kilometers.

18. To calculate the percentage of public shoreline, in the next cell type or copy the following code and click **Run**:

```
print(str(round(sum(sdf_shoreline_ownership[sdf_shoreline_ownership.land_type
↪ == "Public"].Shape__Length)/sum(sdf_shoreline_ownership.Shape__Length) *
↪ 100,2)) + "%")
```

The percentage of public shoreline is 44.78 percent.

19. To calculate the percentage of private shoreline, in the next cell type or copy the following code and click **Run**:

```
print(str(round(sum(sdf_shoreline_ownership[sdf_shoreline_ownership.land_type
↪ == "Private"].Shape__Length)/sum(sdf_shoreline_ownership.Shape__Length) *
↪ 100,2)) + "%")
```

The percentage of private shoreline is 47.05 percent.

Challenge step

Now you can explore the future shoreline (2050 intermediate–high, approximately 1.35 feet, Shoreline_SLR_2050_int_high.zip data). Using the provided blocks of code, apply the necessary revisions for the code to run for the 2050 shoreline prediction.

Interpretation of analysis

After completing this exercise, answer the following questions in detail. The answers to these questions, as well as the analysis results from the exercise workflow, will inform the intelligence brief.

1. What is the most important takeaway from the analysis workflow? What actionable information can urban planners and coastal specialists use to site living shorelines from these steps?

   Sample response: The results of the preceding workflow reveal that ___ km current shoreline in St. Lucie County is public and ___ km is private. Public shoreline occupies ___ percent of the total shoreline length, and private shoreline occupies ___ percent. In a 2050 intermediate–high SLR scenario, the proportion of public shoreline changes to ___ percent and that of private shoreline changes to ___ percent, indicating that more ____ (public/private) lands become vulnerable to SLR in the future.

2. What are the limitations of using historical shoreline data to predict future SLR impacts, and how might these limitations be addressed?
3. How can socioeconomic factors, such as population density and infrastructure investment, influence the prioritization of areas for mitigation or adaptation strategies?

Intelligence brief

Based on the analysis from the exercise workflow, produce a three- to five-page intelligence brief by responding to the following question prompts. Incorporate your geospatial outputs from the exercise to enrich and illustrate your conclusions.

- **BLUF:** List three to five primary observations from your analysis in bulleted-list format that summarize findings of this work.
- **Top-level analysis:** Report the primary data points supporting your BLUF.
- **Secondary analysis:** Provide the data sources and deeper analytic products from this analysis.
- **Impact:** Describe the impacts of your findings on the department or organization.
- **Confidence:** Report the level of confidence of your assessment based on available data, accuracy, and so on.
- **Bias:** Explain any background or information you have that could have affected your analysis.
- **Conclusions:** Determine what type of analytic conclusions there may be and how the data requirements, analysis, and brief writing may change based on what analytic framework you are following (descriptive, predictive, prescriptive).

Applying and extending your skills

Different approaches can be used to model SLR scenarios, including GIS and hydrological methods. The most common method is the "bathtub" model, which simulates how the ocean fills with water. The bathtub model assumes that sea level changes are uniform and areas with lower elevations will be inundated first. The data used in this study is from NOAA and was computed from the bathtub model. However, the bathtub model has some drawbacks, including ignoring hydrological connections, sediment supply, offshore geology, and tidal effects.

Most SLR datasets are free of charge, but the volume of data is large, and processing such datasets may require high-performing computers and additional processing resources. Instead of providing raw data, agencies usually host data in their portals and build a web interface for viewing and exploring these datasets. The following lists a few SLR datasets and viewers:

- NOAA Sea Level Rise Viewer: https://coast.noaa.gov/slr
- University of Florida Sea Level Rise Viewer: https://sls.geoplan.ufl.edu/viewer
- Sea Level Impact Projection (SLIP) Study Tool: https://floridadep-slip.org/map.aspx
- NOAA Sea Level Rise Viewer Data Download: https://coast.noaa.gov/slrdata

Additional resources

The following additional resources provide information about the Indian River Lagoon and SLR inundation modeling:

- Indian River Lagoon National Estuary Program (NEP): https://onelagoon.org/irlnep
- Datums for Trident Pier, Port Canaveral, FL: https://tidesandcurrents.noaa.gov/datums.html?datum=NAVD88&units=0&epoch=0&id=8721604&name=Trident+Pier%2C+Port+Canaveral&state=FL
- Marine and Estuarine Habitats: https://myfwc.com/wildlifehabitats/habitat/ahcr/marine-estuarine
- NOAA 2022 Sea Level Rise Technical Report: https://oceanservice.noaa.gov/hazards/sealevelrise/sealevelrise-tech-report.html
- "Detailed Method for Mapping Sea Level Rise Inundation": https://coast.noaa.gov/data/digitalcoast/pdf/slr-inundation-methods.pdf
- Climate Central Risk Finder: www.climatecentral.org/tool/tool-risk-finder
- Surging Seas Risk Finder: https://riskfinder.climatecentral.org
- Florida Peninsular Lidar Dashboard by the Florida Geographic Information Office: www.floridagio.gov/pages/lidar-resources

References

- Juhász, Levente, Jinwen Xu, and R. W. Parkinson. 2023. "Beyond the Tide: A Comprehensive Guide to Sea Level Rise Inundation Mapping Using FOSS4G." *Geomatics* 3 (4): 522–40.
- Saint Lucie County Property Appraiser: www.paslc.gov/data-requests (accessed September 2023)

About the authors

Jinwen Xu is a postdoctoral scholar in the Institute of Transportation Studies, University of California, Berkeley. He was research associate faculty at the GIS Center, Florida International University. He obtained his PhD in geography and environmental science and policy at the University of South Florida. He also has a master's degree in urban and environmental planning and a transportation certificate from Arizona State University. His research interests are in disaster resilience, geospatial data science, and GeoAI.

Levente Juhász is a research associate professor and assistant director of GIScience at the GIS Center, Florida International University, with a mix of academic and industry experiences. In academia, he held various roles at the University of Florida; Joint Research Centre of the European Commission in Ispra, Italy; the Carinthia University of Applied Sciences in Villach, Austria; and the University of Szeged, Hungary. In industry, he gained experience as a data scientist for a geospatial start-up, Mapillary, and as a full-stack GIS developer for a consulting firm. He has a PhD in geomatics.

Part 3

Social justice and human rights

Chapter 10

Introduction to environmental justice monitoring: The case of air pollution in Los Angeles

Bita Minaravesh, *University of Southern California, Spatial Sciences Institute*

Overview

This chapter introduces applications of GIST to monitor environmental justice concerns, such as exposure to poor air quality at the census tract level. Comparing these aggregations against specific points of interest, such as the locations of major emitters, allows users to assess the impact of industrial sources of air pollution on local communities and the potential extent of their reach to identify key areas for interventions.

Learning objectives

- Perform exploratory data visualization of multiple environmental justice indicators.
- Layer two forms of vector data.
- Draw a set of buffers around point data sources and aggregate values within the buffers.
- Create a bivariate map.
- Critically analyze the implications of trends found in the data to spatially assess relationships and provide recommendations for environmental interventions.

Technical components

- Software: ArcGIS Online (at least Publisher level access)
- Datasets:
  - CalEnviroScreen 4.0 (2021)
  - EPA's Greenhouse Gas Reporting Program (2021)

Prerequisites

- Prior knowledge of or experience using GIS software or spatial data is helpful but not required.
- Technical skill level: Introductory GIS.
- Time to complete chapter: One to two hours without intelligence brief, three to five hours with intelligence brief.

Background and situational context

Environmental justice requires transparent monitoring and reporting of the exposures and inequities experienced by the community's most vulnerable residents. Spatial analysis through GIST is essential in this pursuit because it exposes the spatial patterns of environmental affluence and risk often materialized in inequitable burdens felt by low-income communities of color. To ensure that aggregated data does not mask the experiences of communities within the direct reach of industrial sites contributing to water and air pollution, GIS analysts can assess multiple sources of information to produce a thorough representation of those living by major emitters.

Los Angeles County and the City of Los Angeles both plan to keep the region on track to meet the Paris Agreement's goal of reducing greenhouse gas (GHG) emissions across multiple sectors and sources. LA County is drafting a 2045 Climate Action Plan (CAP), which includes strategies to convert all vehicles in unincorporated LA County to zero carbon emissions, reduce diesel fuel sales, and reduce the use of diesel-powered vehicles across the county. The City of LA's 2019 Green New Deal, also known as the pLAn, similarly seeks to aggressively cut GHG emissions to reach carbon neutrality by 2050, recognizing that 20 to 23 percent of GHG emissions consistently stems from transportation, and zero emission fleets must be prioritized. Both plans go beyond environmental considerations to recognize the sociodemographic patterns of inequity. They refer to environmental justice and equity as driving factors to improve community conditions and share a reliance on public health considerations, either as tools to designate disadvantaged communities or as a target to be met, such as reducing asthma-related emergency room visits. Through these considerations, both the city and the county recognize that a multidimensional approach to environmental justice is necessary. Together, they provide coverage for most of LA County, meaning that virtually all neighborhoods will be positively impacted by these initiatives.

For urban planners, an initial step in introducing environmental justice through plans such as CAP or pLAn is to assess the current distribution of exposures and experiences across the region to identify key areas for environmental interventions. This chapter provides a guide to assessing environmental equity across LA County by using two complementary environmental justice data sources. The first, CalEnviroScreen 4.0 (2021), provides a range of composite and specific ecological measures across California, aggregated to the census tracts, with a collection of demographic characteristics collected from the American Community Survey (ACS). The second is a national dataset developed by the Environmental Protection Agency (EPA), which collects and monitors annual GHG data from large emitting facilities, suppliers of fossil fuels, and industrial gases (GHGRP). For this chapter, this dataset has been converted from a table of coordinates into a shapefile limited to California. Both data sources were published in 2021.

Taking the role of an urban planner tasked with identifying key communities for environmental justice interventions within LA County, your objective is to assess the relationship between the following:

- The distribution of major GHG emitters and measures of diesel particulate matter (PM)
- The distribution of major GHG emitters and rates of emergency room visits for asthma symptoms
- The distribution of major GHG emitters and the percentage of the population with income less than two times the federal poverty level

The workflow is presented in two sections. First, you will visualize major GHG emitters in the county against poor air quality through measures of diesel PM, the solid particles that come from the exhaust of transportation vehicles using a diesel engine. Not only is contact with diesel PM harmful to respiratory health, but it can also be used to assess the presence of diesel-fuel-operated vehicles that produce additional air toxins and GHGs. You will establish a spatial range of influence and summarize values within that buffer. Buffers will be drawn at 1.0 mile around the facilities classified as direct emitters of greenhouse gases to represent a conservative reach of the emissions. This step will prepare you to assess the influence that facilities hold over local communities, the relationship between these sites and socioeconomic indicators, and the implications for environmental justice in the region.

The second section of the workflow relates the measured exposure to community indicators, such as measures of asthma (a sensitive population indicator) and poverty (socioeconomics). You will explore the relationship between the points and CES datasets individually before creating a bivariate map for asthma and poverty. Bivariate maps combine two choropleth maps to show the sequence of combinations, such as high poverty and high asthma simultaneously experienced by a tract. The workflow provides an opportunity to replicate this analysis with an alternative focus on one of CalEnviroScreen's other environmental exposures, such as drinking water contaminants, or alternative health indicators, such as the rate of cardiovascular disease. Mapping two indicators simultaneously allows you to consider the experiences of community members through an intersectional angle, where compounding characteristics such as race, income, and exposures are considered to reflect disadvantage.

By completing this chapter, you will be able to answer the following:

1. Is there a clear relationship between the source, the GHG emitters, and a single indicator? Should a combination of indicators be considered?
2. How does the consideration of multiple sources of environmental justice influence the understanding of the distribution of social and environmental inequities in a given space?
3. How can urban planners from CAP or pLAn develop an intersectional perspective to environmental justice when selecting key zones for interventions?

Exercise workflow 10

In the first part of this exercise, you will work with major GHG emitters.

1 Prepare the exposure assessment

1. In your browser, navigate to www.arcgis.com and sign in to your ArcGIS organization account.
2. On the navigation bar, click the **Map** tab to open a new map in **Map Viewer**.
3. On the **Contents toolbar** (dark), click **Save and open** > **Save as**.
4. In the **Save map** window, apply the following settings:
   - **Title**: Environmental Justice Ch. 10 <your initials>
   - **Folder**: SecurityFirst
   - **Tags**: Security First, Chapter 10
   - **Summary**: Environmental justice case study of air pollution in Los Angeles.

   Tip: If you need to create the folder, click the down arrow, choose **Create new folder**, and type the folder name.

5. Click **Save**.
6. On the **Contents toolbar** (dark), navigate to **Layers** and click **Add** > **Browse Layers.**

   Note: On the bottom left of the page, click the double arrows to expand the Contents toolbar.

7. Click **My content** and, from the menu, click **ArcGIS Online**.
8. In the search bar, type owner:SecurityFirst EPA_GHGRP.
9. Navigate to the resulting layer and click **Add**.
10. Repeat this process, typing owner:SecurityFirst and the necessary layer name, and add these layers:
    - **LA_County_Boundary**
    - **CES 4.0**
11. Click the **Back** button to close the **Add layer** panel.

12. Click the **Options** button next to the **LA_County_Boundary** layer and click **Zoom to.**
13. Hover over the **EPA_GHGRP** layer and click the **Visibility** button to hide the layer for now.

2 Update the symbology

Next, you will change the symbology of LA's boundary.

1. On the **Settings** toolbar, click **Styles** to open the **Styles** pane.
2. Confirm that **LA_County_Boundary** is selected in the **Layers** pane.
3. In the **Styles** pane, under **Pick a style**, click **Style options.**
4. Under **Symbol style**, click the **Edit** button.
5. For **Current symbol**, click **Vector polygon** to open the **Change symbol** menu. For **Category**, click **Basic shapes** and select **Basic polygon.**
6. For **Fill color**, select **No color.**
7. Increase the **Outline width** to 2 px.

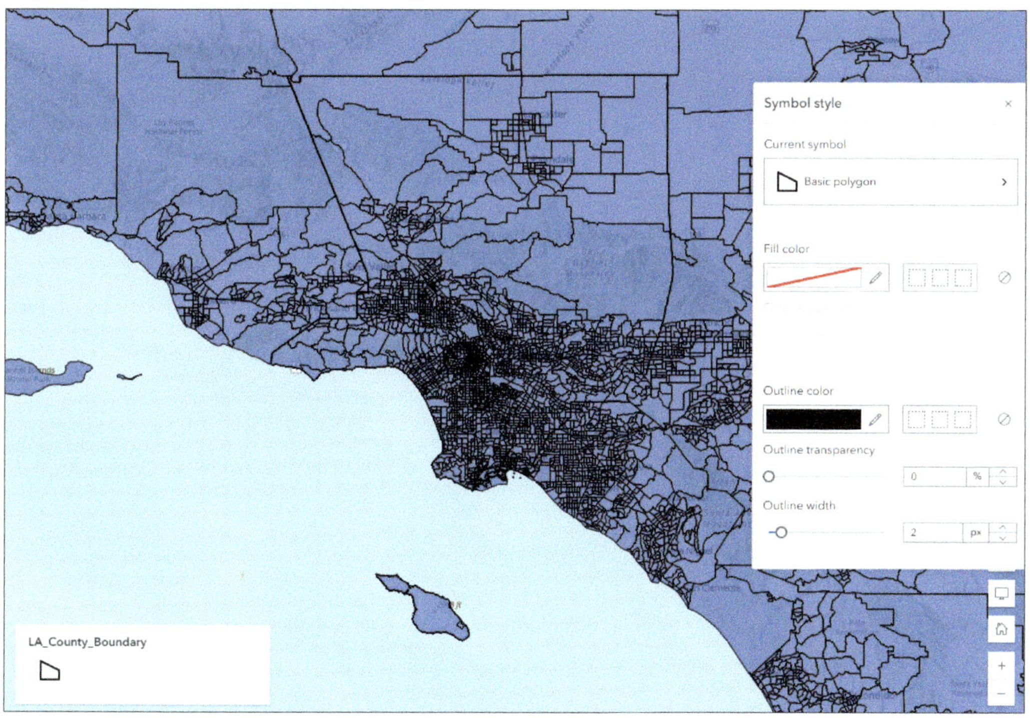

8. Click **Done** and **Done** again. Close the **Style** pane.
9. On the **Contents** toolbar, click **Save and open** > **Save** to save these changes.

3 Explore and modify the CES 4.0 dataset to examine diesel PM

Next, you will examine the CES data layer, note the variables, and update the symbology to reflect diesel PM measures across the state.

1. In the **Layers** panel, click the **Options** button next to **CES 4.0** and click **Show table.**

   Spatial scale: The data is assessed to census tracts but provides information on the respective zip codes, county, and approximate locations.

   Data types: For each of the environmental justice indicators, the values are available as a raw measurement and a standardized percentile rating for all tracts in California. This chapter will use the percentile ratings because they provide a consistent range for all the indicators.

   Demographic data: CES incorporates data from the ACS. These values are at the far right of the table and can be used to understand socioeconomic patterns.

2. Review the attributes and then close the table.
3. Confirm that **CES 4.0** is selected in the **Layers** pane and open its **Styles** pane.
4. Under **Choose attributes**, click **Field.**
5. In the **Select fields** menu, scroll down and check the box for **DieselPM_P**. Click **Add.**

   Note: Diesel PM_P provides the standardized percentile ratings for diesel particulate matter across all census tracts in California.

   Now there are more options available in the **Styles** pane.

6. Under **Pick a style**, click **Counts and Amounts (color).**
7. Confirm that the **Theme** is set to **High to low**. Click **Style options.**

   Note: Your colors may default to a different color ramp. This will not alter your progress.

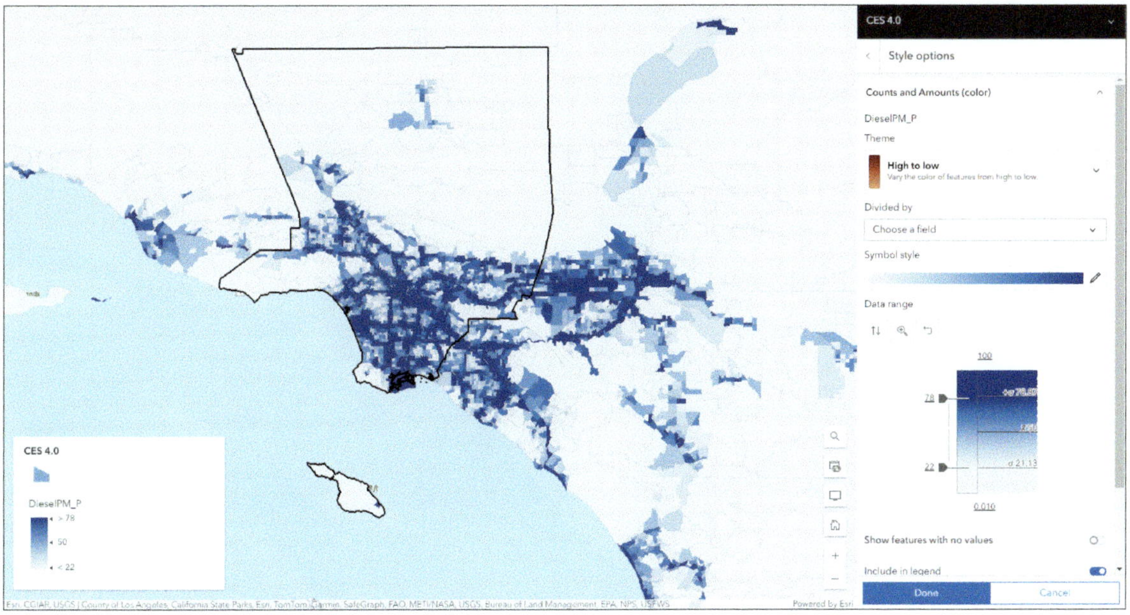

8. Under **Symbol style**, click the **Edit** button.

   Note: Higher diesel PM percentile implies higher exposure, which you are conveying with darker colors.

9. Confirm your changes and close the **Styles** pane. Save your changes.

   What do you observe about the distribution and concentration of diesel PM throughout LA County? You can zoom to different parts of the county or zoom out to see the distribution across California.

   Do any regions seem to have notably high or low values?

4 Explore and modify the EPA GHGRP dataset

Next, you will explore the EPA GHGRP dataset.

1. In the **Layers** pane, next to the **EPA_GHGRP** layer, click the **Visibility** button.
2. Explore the data's attribute table as you did with the **CES 4.0** dataset.

   Note: Observe the information available for each point. Explore the range of values under Industry_1. Also, note the emissions value—available for methane, nitrous oxide, and others.

3. Close the table after reviewing the attributes.

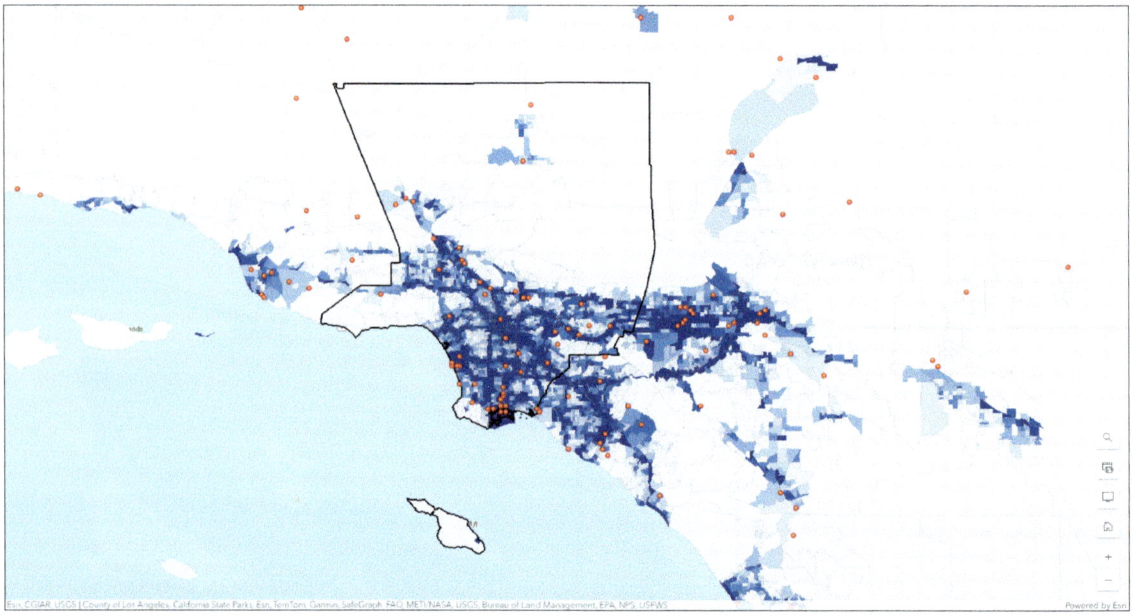

Your map now shows the major greenhouse gas emitters across Los Angeles County against standardized measures of annual diesel PM exposure. What do you observe about the distribution of the points against the color ramp of the **CES 4.0** layer?

- Does the distribution of the points seem to match the diesel PM trends?
- Do the points accurately reflect the range of emissions from the sites?

5 Analyze the GHG emission buffer

To better represent the range of emissions, you will draw buffers around the points.

1. Confirm that **EPA_GHRGP** is selected in the **Layers** pane.
2. On the **Settings** toolbar, click **Analysis**.
3. Click **Tools.** In the search bar, type Create Buffers and select the tool.
4. In the **Create Buffers** tool, enter the following parameters:
    - **Input layer**: EPA_GHGRP
    - **Distance values:** 1.0
    - **Output name:** GHG_1_Mile_Buffer
    - **Save in folder**: SecurityFirst

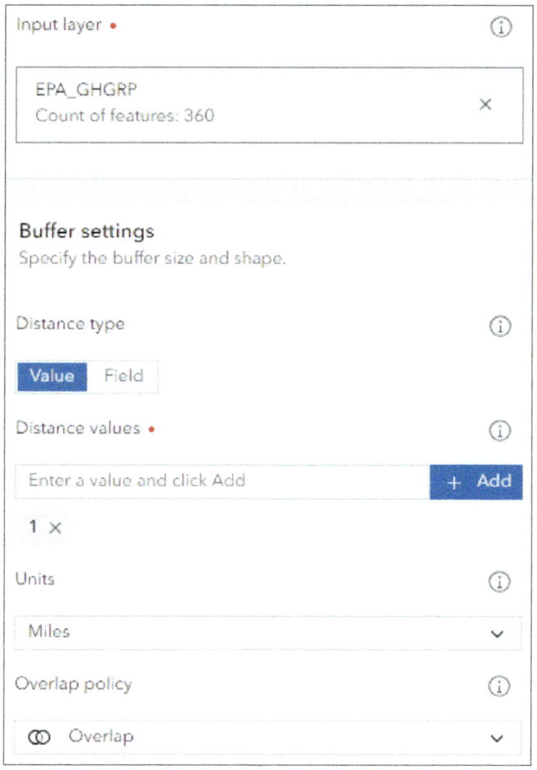

5. Click **Run**.
6. When the tool has finished running successfully, close the **Analysis** pane.

   Note: The buffer should be automatically added to your layers. If it does not appear, repeat these steps and find it under Your content.

7. In the **Layers** pane, drag **GHG_1_Mile_Buffer** under **EPA_GHGRP** to reorder the layers.

   Note: Although these buffers present an equal likelihood of the diffusion of GHGs across the space, predominant wind patterns will lead to varying exposure levels of air pollution, with residents downwind from a factory bearing the greatest burden.

6 Update the buffer analysis symbology

You need to adjust the buffer symbology to help illustrate the distribution of GHG emissions.

1. Change the buffer symbology, as you did with the LA boundary, to give the buffers a single symbol style with no fill.
2. For **Outline color**, select **White**.
3. Increase the **Outline width** to 1.5 px.
4. Confirm and save your changes.

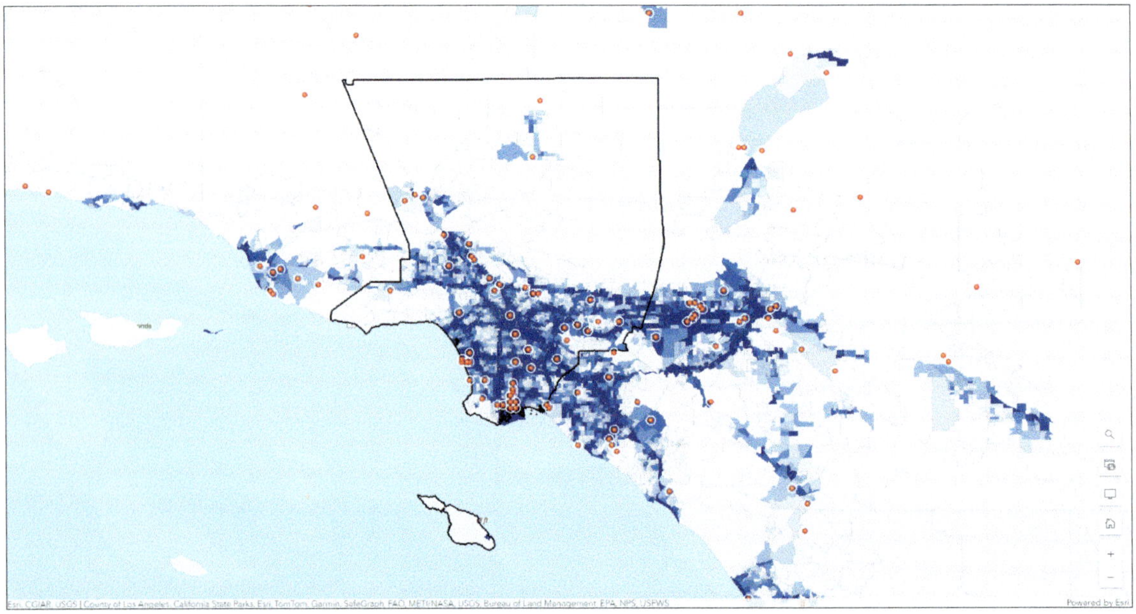

Your map now shows the major greenhouse gas emitters with a one-mile radius of influence.

5. Take a screenshot of this map.

   Do the buffers, which extend the spatial context of your observations, allow you to better evaluate the range of emissions near a GHG site? How do you expect these values to change as the buffers increase or decrease in distance from the emission site?

7 Calculate GHG emissions

Building on the visual assessment of GHG emissions using one-mile buffers that you just completed, you will now calculate the average values within them.

1. On the **Settings** toolbar, open the **Analysis** pane.
2. In **Tools**, search for Summarize Within and open the tool.
3. Add the following parameters:
   - **Input features:** CES 4.0
   - **Summary polygon layer:** GHG_1_Mile_Buffer
   - **Include summarized geometry of features:** toggle off
   - **Field Statistics:** DieselPM_P
   - **DieselPM_P:** Select all
   - **Output name:** Diesel_Aggregations
4. Click **Run**.

   Note: This step uses 8.392 credits. If credits are limited, clip the buffers to the county boundary first and then run with the clipped buffers as the summary polygon layer.

5. When the tool has finished, close the **Analysis** pane.

8 Update GHG emission calculations symbology

You will now update the symbology of the **Diesel_Aggregations** layer.

1. Confirm that **Diesel_Aggregations** is selected in the **Layer** pane.
2. On the **Settings** toolbar, open the **Styles** pane.
3. Under **Choose attributes**, click **Field** and add **Mean DieselPM_P**.
4. Under **Pick a style**, click **Counts and Amounts (color)**.
5. For **Theme**, choose **Above and Below** and select **Style options**.
6. Under **Symbol style**, click the **Edit** button and then click the **Color Ramp**.
7. Click **Flip ramp colors** so that red represents the higher values and blue represents the lower values. Click **Done**.
8. In the **Style options** pane, click the toggle to turn on **Classify data**.
9. Set the **Method** to **Natural breaks**.
10. Set the **Number of classes** to 5.
11. Confirm and save your changes.

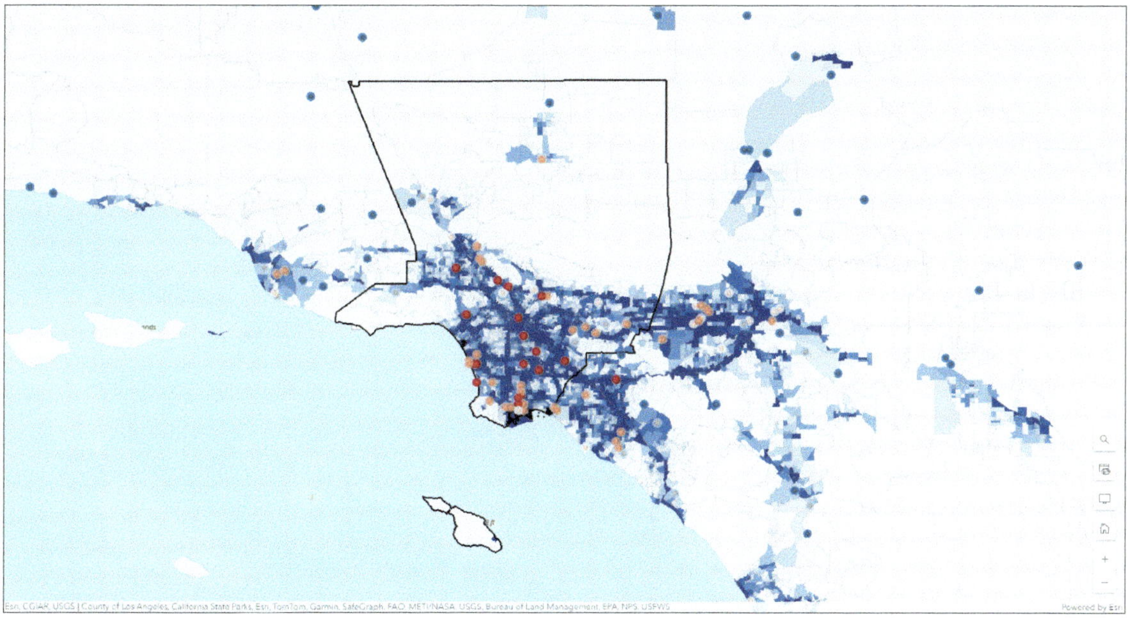

Your map now shows the average diesel PM values within a one-mile radius of major GHG emitters in Los Angeles County.

12. Take a screenshot of this map.

9 Review the results

You have successfully analyzed the data and updated the symbology. What do you observe from these results?

1. Adjust the visibility of your map layers so that only **Diesel_Aggregations** and **LA_County_ Boundary** are displayed on the map.

   What patterns do you observe?

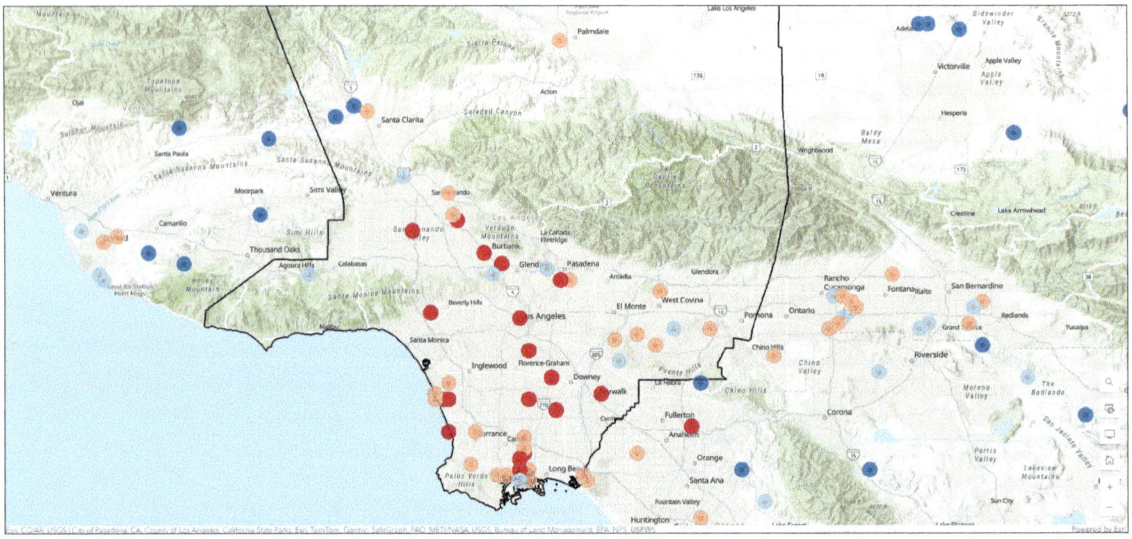

2. Repeat the preceding styling steps with an alternative classification method for **Mean DieselPM_P** (for example, use **Standard Deviation**).

   Note the differences in distribution.

3. Repeat the steps to run the **Summarize Within** tool with an alternative aggregation, such as the **Minimum** or **Maximum** observed value within the buffer.

   Note the differences in distribution.

4. If credits allow, repeat the steps to use the **Create Buffers** tool to create another buffer distance of your choosing, such as 0.5 miles or three miles. How do your results compare with one-mile buffers?

   Note the differences in aggregations and conclusions based on different distances.

   Now that you have seen the spatial distribution and aggregate values of emissions from major GHG sites, you will explore the connections to other elements of environmental justice.

   You will visualize the GHG emissions against rates of asthma by changing the symbology.

10 Visualize GHG emissions against asthma rates

1. In the **Layers** pane, turn off the visibility for **Diesel_Aggregations** and make **CES 4.0** visible.
2. Confirm that **CES 4.0** is selected in the **Layers pane** and open the **Styles** pane.
3. Under **Choose attributes**, remove **DieselPM_P** as a field and replace it with **AsthmaP**.
4. Under **Pick a style**, click **Counts and Amounts (color)**.
5. Change the **Symbol style** to a different color ramp that is appropriate for the scenario.

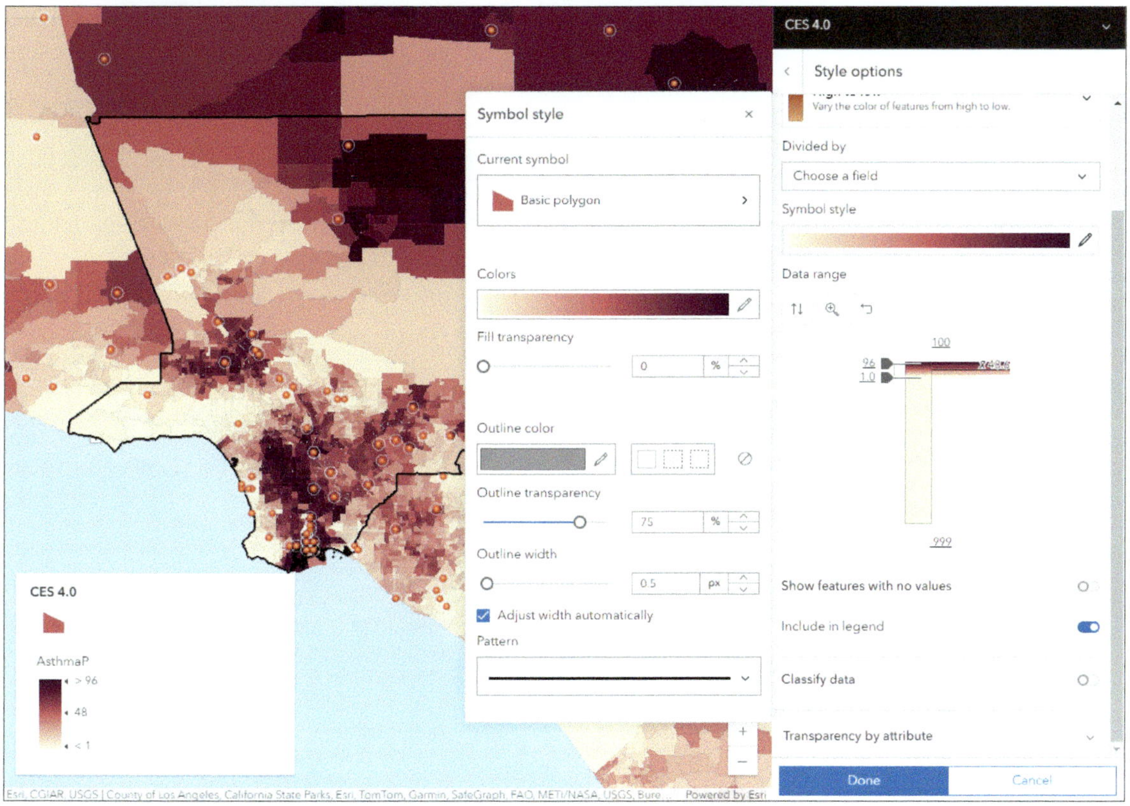

6. Confirm and save your changes.
7. Pan around the county to better assess the trends of asthma rates.
   - What patterns do you see across the county?
   - What patterns do you see against the GHG buffers?

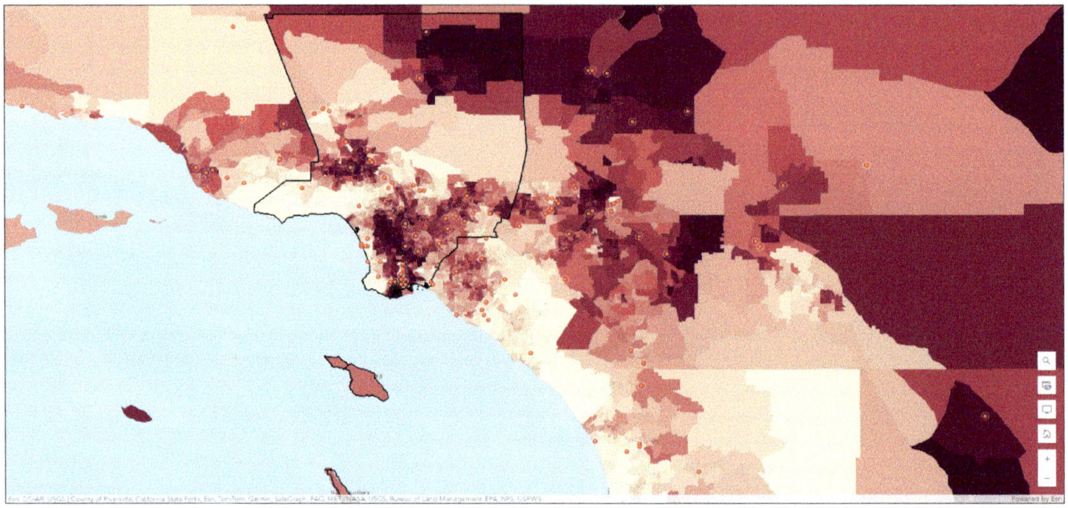

Your map now shows the major GHG emitters with a one-mile radius of influence against rates of emergency room visits for asthma attacks.

8. Take a screenshot of this map.

11 Visualize GHG emissions against poverty rates

You will repeat the preceding steps to explore the relationship against rates of poverty.

1. Replace the field with **PovertyP**.
2. Choose a new color ramp.
3. Confirm and save your changes.

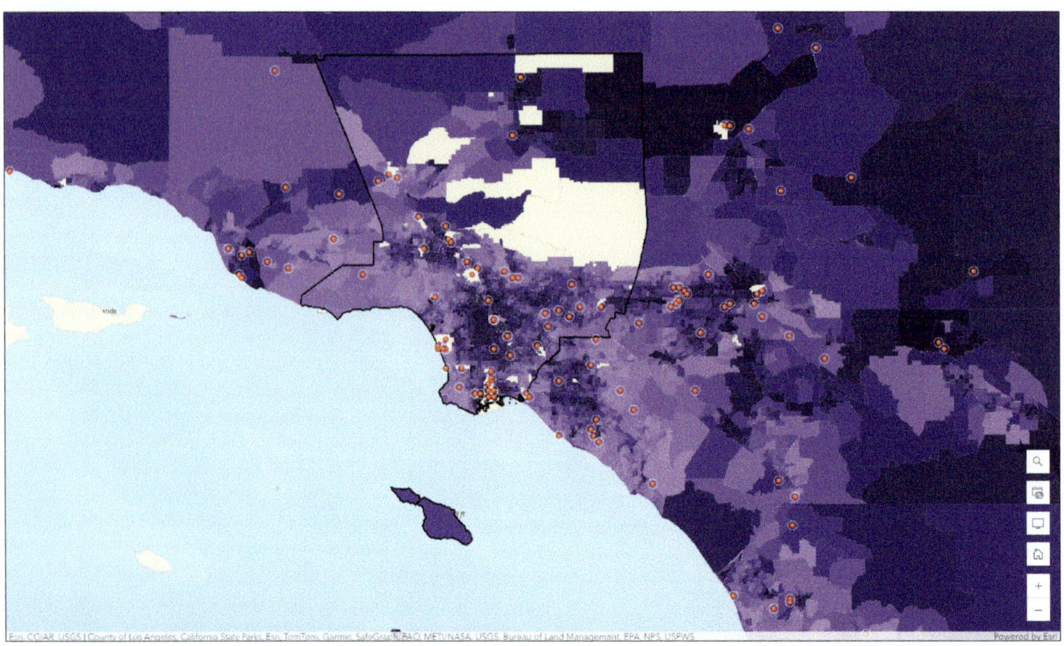

Your map now shows the major GHG emitters with a one-mile radius of influence against a standardized measure of residents living under the federal poverty level.

4. Take a screenshot of this map.

12 Create a bivariate map

Now that you have assessed the variables separately, you will visualize them together in a bivariate map.

1. With **CES 4.0** selected in the **Layers** pane, return to the **Styles** pane.
2. Under **Choose attributes**, for **Field**, add **AsthmaP**.

   Both **PovertyP** and **AsthmaP** are selected attributes.
3. Under **Pick a style**, scroll through the style options and select **Relationship**.
4. Confirm and save your changes.

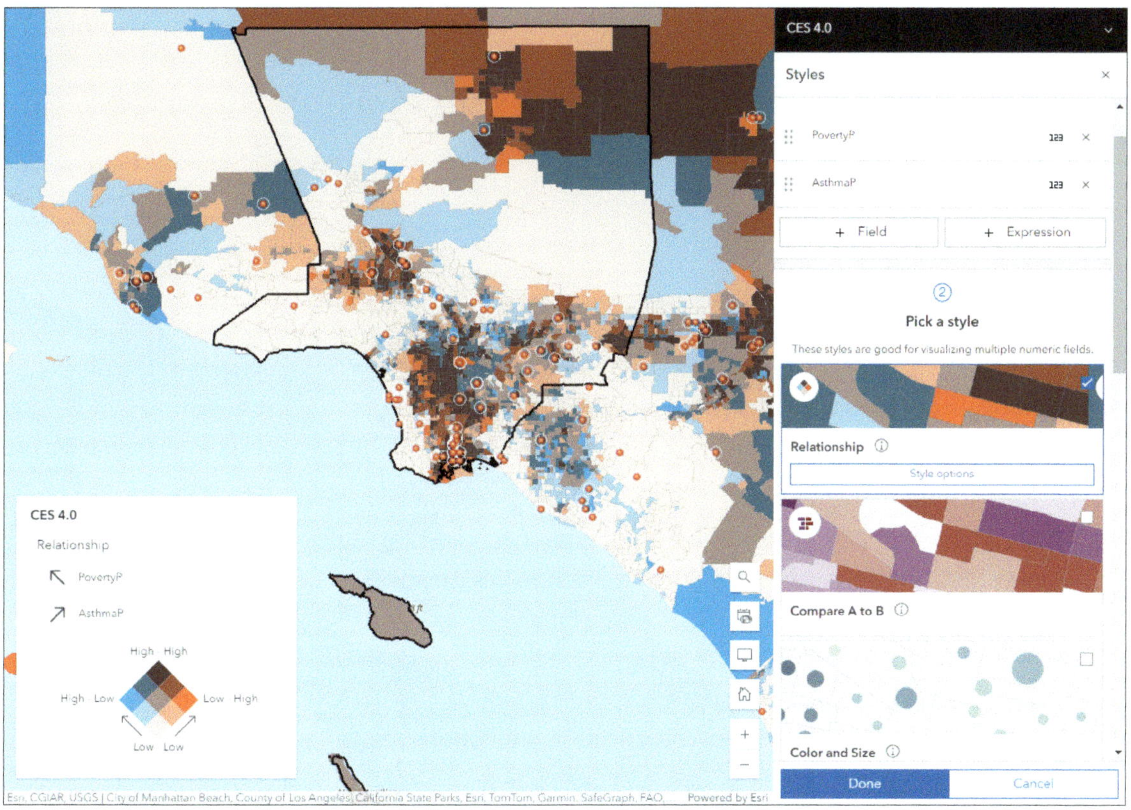

Your map now shows the major GHG emitters with a one-mile radius of influence against both measurements of asthma and poverty at the census tract level.

Note that the color shown as **High-High** (dark brown) represents census tracts with high burdens in both categories, whereas **Low-Low** represents low burdens in both categories.

5. Take a screenshot of this map.
6. Turn on the visibility for the **Diesel_Aggregations** layer.
7. Confirm and save your changes.

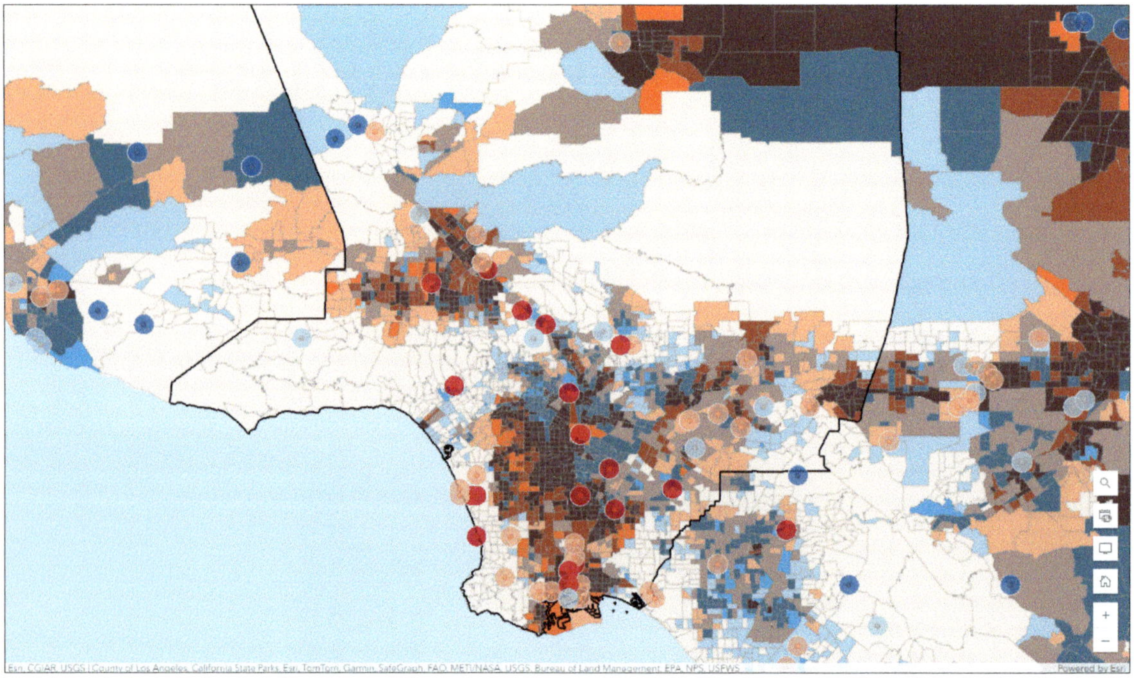

Interpretation of analysis

After completing this exercise, answer the following questions in detail. The answers to these questions, as well as the analysis results from the exercise workflow, will inform the intelligence brief.

1. Based on this analysis, what patterns do you observe when viewing the aggregated diesel emissions against the bivariate map of poverty and asthma rates?
2. Are diesel emissions evenly distributed across Los Angeles County? If not, which residents are exposed to the highest diesel concentrations?
3. What actionable information can be derived from this analysis? How might Los Angeles County use this information to develop a response plan?
4. How does your choice of data classification alter your visual analysis of the values? Which aggregation best reflects the disparity in exposures?
5. How does a bivariate map change your interpretations of the relationship between emitters and the sociodemographic of the communities that house them?
6. What other factors should be explored to understand the full scope of environmental equity across this space?

Intelligence brief

Based on the analysis from the exercise workflow, produce a three- to five-page intelligence brief by responding to the following question prompts. Incorporate your geospatial outputs from the exercise to enrich and illustrate your conclusions.

- **BLUF:** List three to five primary observations from your analysis in bulleted-list format that summarize findings of this work.
- **Top-level analysis:** Report the primary data points supporting your BLUF.
- **Secondary analysis:** Provide the data sources and deeper analytic products from this analysis.
- **Impact:** Describe the impacts of your findings on the department or organization.
- **Confidence:** Report the level of confidence of your assessment based on available data, accuracy, and so on.
- **Bias:** Explain any background or information you have that could have affected your analysis.
- **Conclusions:** Determine what type of analytic conclusions there may be and how the data requirements, analysis, and brief writing may change based on what analytic framework you are following (descriptive, predictive, prescriptive).

Applying and extending your skills

Repeat the workflow for another county in California and compare the distribution of GHG emitters and environmental indicators against those for Los Angeles County.

- What similarities do you see? What patterns are unique to the county?

To extend the workflow, additional CalEnviroScreen indicators can be observed independently or in relationship with one another. Suggestions for exploring:

- Ozone, PM 2.5, Toxic Releases from Facilities
- Drinking Water Contaminants, Groundwater Threats, Impaired Water Bodies
- Hazardous Waste Generators and Facilities
- Cardiovascular Disease, Housing Burden
- Race and Ethnicity

Additional points of interest that can contribute to the analysis:

- Brownfields
- Airports
- Landfills

Alternatively, a concentration on parks, open spaces, and green spaces serves as a contrast by highlighting ecological investments in low-income communities of color.

Additional resources

The following additional resources provide information about the GIST approach to monitoring environmental justice concerns:

- California Office of Environmental Health Hazard Assessment, CalEnviroScreen 4.0: https://oehha.ca.gov
- City of Los Angeles, Green New Deal: https://plan.mayor.lacity.gov/get-involved/faq
- Los Angeles County, Draft 2045 Climate Action Plan: https://planning.lacounty.gov/long-range-planning/climate-action-plan
- United States Environmental Protection Agency, Facility Registry Service (FRS): www.epa.gov/frs
- United States Environmental Protection Agency, Greenhouse Gas Reporting Program (GHGRP): www.epa.gov/ghgreporting

About the author

Bita Minaravesh, PhD, is a lecturer at the Spatial Sciences Institute of the University of Southern California, with a focus on environmental equity and sustainability research. She received her doctoral degree from the interdisciplinary Population, Health, and Place program and has a master's degree in public health.

Chapter 11

Using geospatial information dashboards to visualize multidimensional poverty

Sarbeswar Praharaj, *University of Miami, Department of Geography and Sustainable Development; University of Miami, School of Architecture*

Oumayma Moufid, *Hassania School of Public Works, Morocco*

Overview

The rise in geospatial data availability in cities has encouraged authorities to invest in ways to explore, operationalize, and visually communicate vast amounts of information. There has been a growing trend to use geospatial information dashboards as instruments for policymakers to integrate and visualize multiagency data to monitor and respond to urban issues (Praharaj et al. 2022; Mattern 2015). Dashboards are software tools enabling information visualization and data storytelling through maps, graphs, diagrams, indicators, and other interactive widgets, consolidated and arranged on a single screen or a web page (Few 2006). Dashboards are dynamic visualizations because they are programmed to update as new data is released (Batty 2015), helping users track and compare over time and space to support near-real-time decision-making.

This chapter outlines the technical workflow for developing a geospatial information dashboard, with an example of its application in urban poverty mapping and socioeconomic vulnerability assessment. By completing this exercise, you will learn how to use ArcGIS Dashboards to integrate spatial and quantitative datasets to design actionable dashboards. The tools used in this tutorial can be repurposed to set up dashboards for a variety of uses, ranging from urban transport monitoring to disaster management. Dashboard skills gained through this training can be beneficial for addressing public policy issues and communicating data to the public.

Learning objectives

- Join spatial data layers (for example, shapefiles) with statistical information (for example, Microsoft Excel and CSV) to combine, query, and visualize information.
- Create visual analytics products, including maps, charts, and diagrams, using ArcGIS Dashboards.
- Design interactive geospatial information dashboards that can be deployed on the web.
- Identify how to deploy actionable dashboards to manage specific urban problems with an example of green infrastructure monitoring.

Technical components

- Software:
  - ArcGIS Pro 3.4
  - ArcGIS Dashboards
  - ArcGIS Online (at least Publisher-level access)
- Datasets:
  - A sample of the poverty assessment data obtained from the HEIGHT Poverty Dashboard (https://resilience.asu.edu/HEIGHT-poverty-dashboard)
  - Zip Code shapefile from the US Census Bureau

Prerequisites

- Basic knowledge of GIS shapefiles and cartographic techniques.
- Technical skill level: Intermediate GIS.
- Time to complete chapter: Two to three hours without intelligence brief, three to five hours with intelligence brief.

Background and situational context

With the advancements in digital technology, organizations today are generating near-real-time data feeds at a high volume, velocity, veracity, and variety—technically referred to as big data (Praharaj 2020; 2025). Geospatial big data and digital technologies are increasingly playing a vital role in the design of government policies, including urban planning, management, and disaster response (Moufid et al. 2024; 2025). The ability of cities to collect, process, and use data to enforce a logic of control enacted through digital technologies shows immense potential to transform how we address and respond to emerging challenges and shocks. Building on the ongoing revolution in data and analytics to enable evidence-based decision-making, the practice of "dashboard governance" (Few 2006) has proliferated, where data visualization interfaces play a vital role in everyday decision-making, acting as a potential channel of communication between decision-makers and community stakeholders.

Stephen Few (2006) defines dashboards as a "visual display of the most important information needed to achieve one or more objectives." Dashboards embody the many ways of visualizing a variety of data that is representable, translatable, and intelligible to both experts and nonexpert target audiences and can be used as a decision support tool by governments, corporations, urban stakeholders, and communities (Batty 2015; Kitchin 2014; Mattern 2015). Although many dashboard project implementations are coming to fruition, there is still considerable research and development potential yet to be explored and exploited. Not all dashboards are created equal in layout, structure, visual flair, content, and navigability (Praharaj and Wentz 2022). The multiplicity of technical approaches and software applications also determines diverse dashboard outcomes.

In this context, the current chapter aims to provide a template for dashboard design using ArcGIS Dashboards to emphasize the importance of spatial thinking and the effectiveness of cartographic visualizations. The approaches laid out in this section focus on augmenting the technological capacity of researchers and experts to integrate and visualize complex datasets through digital interfaces to address urban issues in a timely and efficient manner. This work is grounded in the fact that designing responses to emerging challenges requires a combination of situational data and easy-to-use decision support tools (Praharaj et al. 2022) that can inspire and organize communities and policymakers to initiate action. This chapter particularly demonstrates the application of dashboards in measuring and visualizing multidimensional poverty.

The US Census uses a federal poverty threshold to determine an individual's ability to live within a certain standard of living. However, poverty thresholds do not account for living costs beyond a basic food budget. In the last few decades, the costs of housing, transportation, childcare, and health care have risen far more rapidly than food costs. The HEIGHT poverty measure addresses these gaps to build a novel econometric approach accounting for the costs of food, housing, and utilities, health care, childcare, transportation, taxes, and other essential needs to capture the real nature of poverty. Analyzing those local costs of living with the household's income data, the tool determines how many households are struggling to meet basic living costs across neighborhoods. The data provides insights about families with different characteristics—whether the household includes children, whether the head of household is male or female, the number of adults in the household, and the ethnicity of householders—to enable communities to design interventions targeted to their specific needs. Using poverty and demographic data, users can explore how various datasets (spatial, statistical) can be collected, combined, analyzed, and disseminated through public dashboards to enable our communities to gain critical knowledge, decipher complex evolving issues, and prepare responses to critical social issues.

Exercise workflow 11

1 Download and review the data

You will first check to see whether your ArcGIS organization account includes the ArcGIS Dashboards app.

1. In your browser, navigate to www.arcgis.com and sign in to your ArcGIS Online account.
2. On the top right, click the **App Launcher** (nine dots) and click **Dashboards**.

If the **Dashboards** app is not visible, make sure that it is part of your license. You can check this detail under your account name.

3. Click **My settings** > **Licenses** > **Essential Apps**.

You should find **ArcGIS Dashboards**. If not, contact your organization to activate your license.

Next, you will download the datasets provided with this chapter and familiarize yourself with key fields and attributes. This exercise involves three datasets: (1) two spatial data layers containing public use microdata (PUMA) boundaries and the county boundaries in Arizona and (2) one Microsoft Excel field containing multiple sheets.

Sheets named "C_" relate to county-level data, and sheets named "P_" relate to PUMA-level data. It is important to study and recognize the different attributes in the spreadsheets (for example, county, PUMA, household type, and so on) before starting the analysis.

4. In File Explorer, navigate to **SecurityFirst_ExerciseData/Chapter11**. Open each dataset and analyze the data.

2 Prepare a map in ArcGIS Pro

To add maps to your dashboard in ArcGIS Dashboards, you will first create those maps in ArcGIS Pro, share them online, and finally import them into the dashboard.

You will create three maps and share them to your online account:

- Map 1 represents the percentage of households struggling to meet the household cost per counties per household type (White, African-American, and so on).
- Map 2 represents the percentage of households struggling to meet the household cost per PUMA per household type.
- Map 3 represents the percentage of households below the Resilience Capacity Index (RCI) per PUMA (total families, regardless of household type).

Create and share map 1

1. Start **ArcGIS Pro**. Under **New Project**, click **Map.**
2. In the **New Project** window, apply the following settings:
    - **Name**: Chapter11Map1
    - **Location**: C:\SecurityFirst

    Note: This tutorial will use C:\SecurityFirst.

3. Click **OK.**
4. On the **Map** tab, in the **Layer** group, click **Add Data.**
5. From the **SecurityFirst_ExerciseData/Chapter11** folder, add **Counties.shp**.
6. In the **Contents** pane, right-click the **Counties** layer and click **Copy.** Paste the layer eight times so that you have nine layers total of the **Counties** shapefile.
7. Click **Add Data** and double-click **Families_RCI_Race.xlsx.** Import the nine sheets in the Excel field that represent Counties data per household type. Each name starts with **C_**.

Name	Type
C_AfAm$	Excel Table
C_AmInAl$	Excel Table
C_Asian$	Excel Table
C_HisLat$	Excel Table
C_NatHaw$	Excel Table
C_NotHisOrLat$	Excel Table
C_OtherRace$	Excel Table
C_TwoOrMoreRaces$	Excel Table
C_White$	Excel Table

Next, you will join each **Counties** shapefile layer with one table.

8. Right-click the first **Counties** layer and click **Joins and Relates** > **Add Join.** In the **Add Join** pane, apply the join using the common attribute reflecting the name of the county with the following parameters:
    - **Input Table**: Counties
    - **Input Field**: NAMELSAD
    - **Join Table**: C_AfAm$
    - **Join Field**: Name

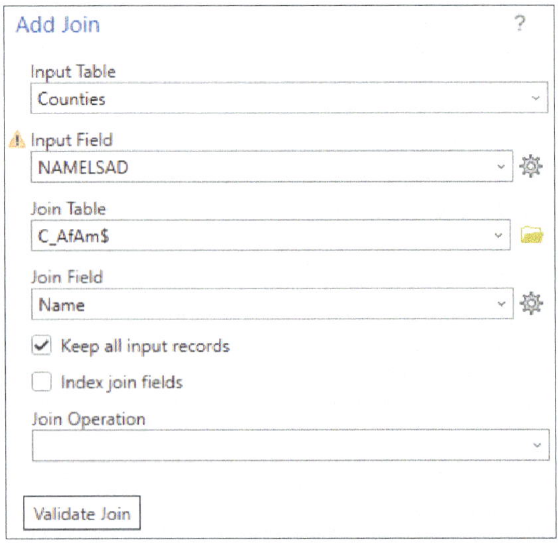

9. Click **OK**.
10. Repeat the preceding steps for each **Counties** layer.

    Each type of **C_** Excel table will be joined with its own **Counties** layer.

    Now that you have all the data joined to the **Counties** spatial layer, you need to export it as a feature class so that you can create maps based on the joined attributes.

11. Right-click the first **Counties** layer and click **Data** > **Export Features**. Name the output feature class with the name of the household type:
    - **Input Features**: Counties
    - **Output Feature Class**: AfricanAmerican_Counties
12. Click **OK**.
13. Repeat the preceding steps for each **Counties** layer.

    Each household type will have its own feature layer. When exporting each feature layer, remember to check the attribute table to identify the appropriate household type.

    Tip: If you are unsure of the names of each household type, right-click the Excel table and click Open to view its attribute table. The name can be found in the Household Type column.

    Next, you will create a choropleth map to show the percentage of households struggling to meet household cost per household type at the county scale.

14. Drag the **AfricanAmerican_Counties** layer to the top of the layer list.
15. Right-click **AfricanAmerican_Counties** and click **Symbology**.
16. In the **Symbology** pane, change the **Primary symbology** to **Graduated Colors**.
17. Change the **Field** to **% Households Struggling to Meet Household Cost**. Change the **Color scheme** to **Oranges (Continuous)**.

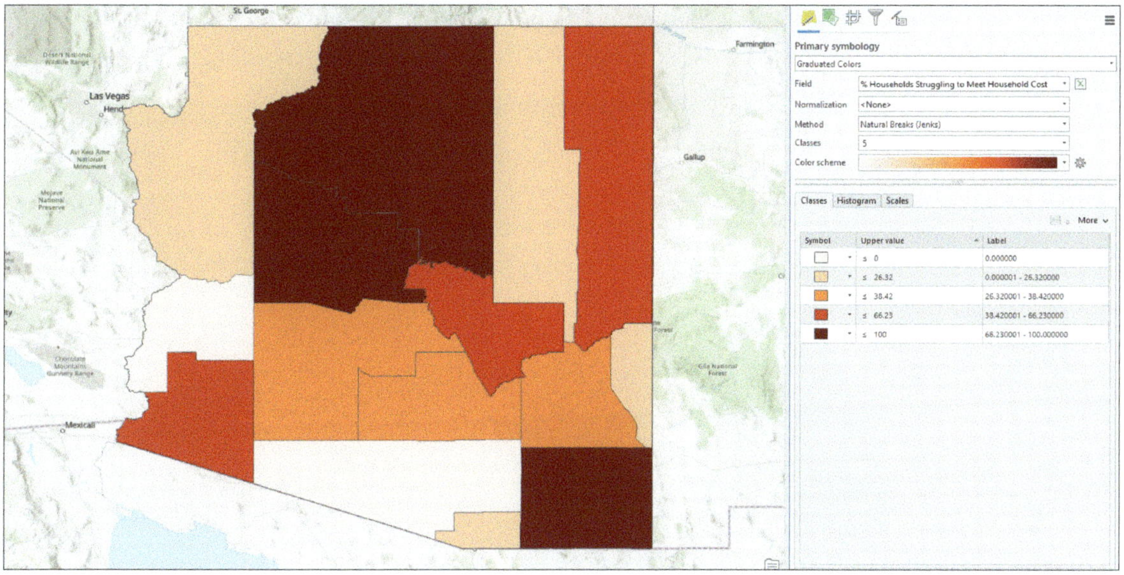

The color of each county on the map reflects an interval of household percentage. However, it would be better to display this percentage on the map. To do this, you will add a label to the map and connect it with the percentage attribute.

18. In the **Contents** pane, right-click **AfricanAmerican_Counties** and click **Label.**

    At this point, you will not see any change in the map because you have not connected the attribute yet.

19. Right-click **AfricanAmerican_Counties** and click **Labeling Properties.**
20. In the **Label Class** pane, uncheck the box for **Expression** if it is not empty.
21. Under **Fields**, double-click **% Households Struggling to Meet Household Cost**. The expression should read:

    ```
    $feature.F__Households_Struggling_to_Mee
    ```

22. Click **Apply.**

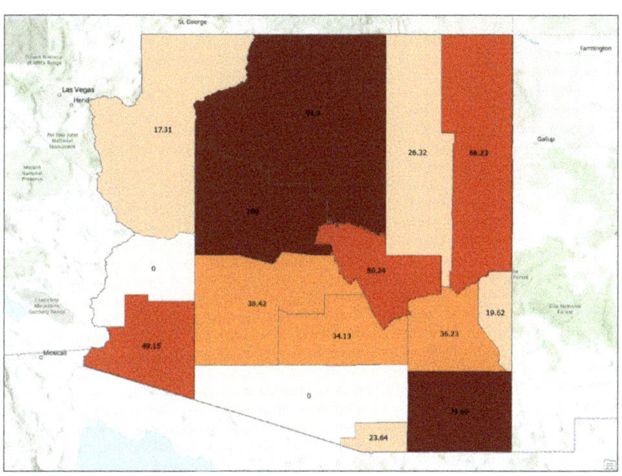

The map has the percentage attribute, but it does not have the percentage symbols. You will change the expression to see the percentage on the map with the percentage symbol.

23. In the **Label Class** pane, change the **Expression** as follows:

```
Concatenate($feature.F__Households_Struggling_to_Mee, "%")
```

24. Click **Apply**.

You should now see the percentage attribute with the percentage symbol for each county.

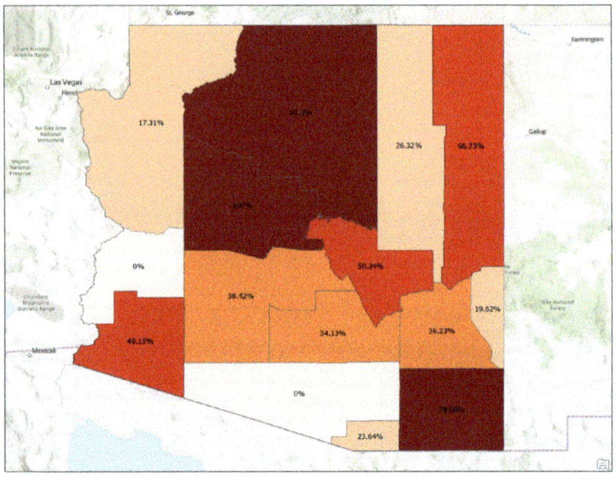

25. Repeat the preceding steps for the other eight layers.

You will have nine feature classes total, with each layer representing the nine household types.

26. In the **Contents** pane, remove all the **Counties** layers.

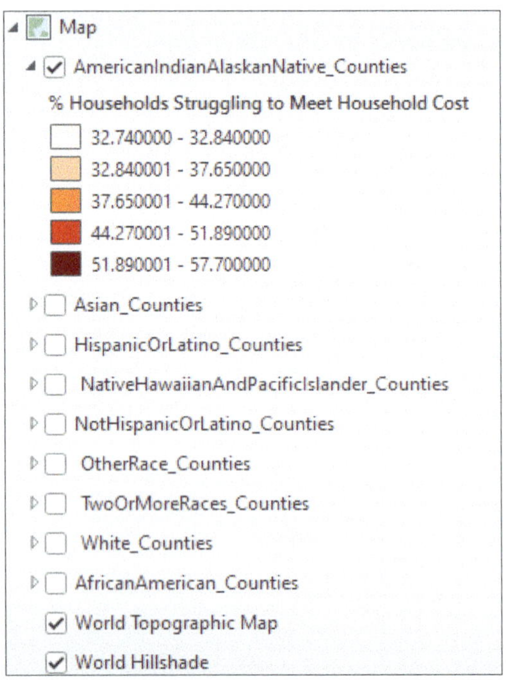

The next step is to share the map.

27. On the **Share** tab, in the **Share As** group, click **Web Map**.
28. In the **Share As Web Map** pane, change the **Name** to Map_Counties, and click **Analyze**.

    A few errors will pop up on the **Messages** tab.

29. For **Error 00230**, right-click the **error** and click **Update Map to Use Basemap's Coordinate System.**
30. For Error **00374**, right-click the **error** and click **Auto-Assign IDs Sequentially**.
31. Click **Analyze** again.
32. When all the errors have been resolved, click **Share.**
33. When the **Share** process is complete, sign in to your ArcGIS Online account in a browser. Click the **Content** tab to see your web map.

    The web map should contain the nine layers that you just created and mapped at the county scale.

Create and share map 2

To create map 2, you will create a new ArcGIS Pro project and follow steps that are similar to those used for map 1. The key difference is that instead of using the **Counties** shapefile, you will use the **PUMA** shapefile, and in place of the **Counties** sheets (**C_**), you will use the **PUMA** sheets (**P_**). At the end of the process, you will have an online map containing nine layers, each layer representing data per household type at the PUMA scale.

1. Repeat all the steps that you used for map 1 but use the **PUMA** shapefile instead of the **Counties** shapefile and use the **PUMA** sheets instead of the **Counties** sheets.
2. When the **Share** process is complete, open ArcGIS Online in a browser. Click the **Content** tab to see your web map. It should contain the nine layers you just created and mapped at the PUMA scale.

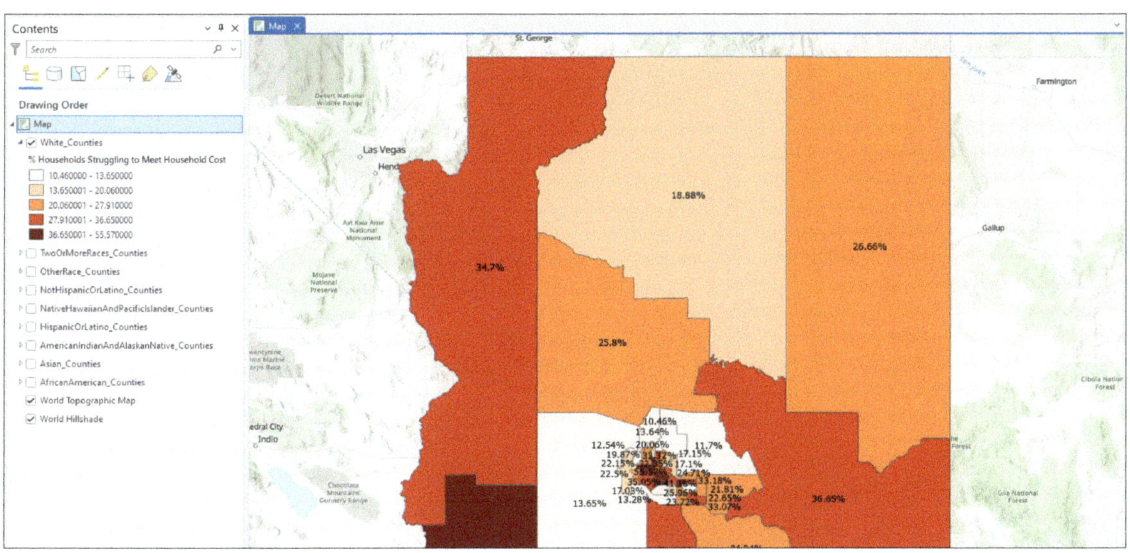

Create and share map 3

Map 3 will be created at the PUMA scale and will contain only one layer.

To build this map, you will create a new project in ArcGIS Pro, and you will use the **PUMA** shapefile and the sheet named **'P_Total Families'**. You will apply the join based on the **GEOID** attribute and make a choropleth map using the **Percentage Families below RCI_total** attribute.

1. Create a new **Map Project** called Chapter11Map3 and save it to the same folder as **Map1** and **Map2**.
2. Click **Add Data**. Add **tl_2018_04_puma10.shp**. From **Families_RCI_Race.slsx**, add **'P_Total Families$'**.
3. Right-click the PUMA layer (**tl_2018_04_puma10.shp**) and click **Joins and Relates** > **Add Join**. Apply the join using the following parameters:
    - **Input Table**: PUMA
    - **Input Field**: GEOID10
    - **Join Table**: P_Total Families$
    - **Join Field**: GEO_ID
4. Export the features and name the output PUMA_RCI.
5. Using the steps for the previous maps, create a choropleth map using the Percentage Families below RCI_total attribute.
6. Add percentage labels using the **Percentage Families below RCI_total** field. Adjust to two decimal points.
7. Remove all layers from the **Contents** pane except **PUMA_RCI**.
8. Share the map to your ArcGIS Online account and name the map Map_PUMA_RCI.
9. When the **Share** process is complete, open ArcGIS Online in a browser. Click the **Content** tab to see your web map.

    It should contain the layer you just created and mapped for the RCI at the PUMA scale.

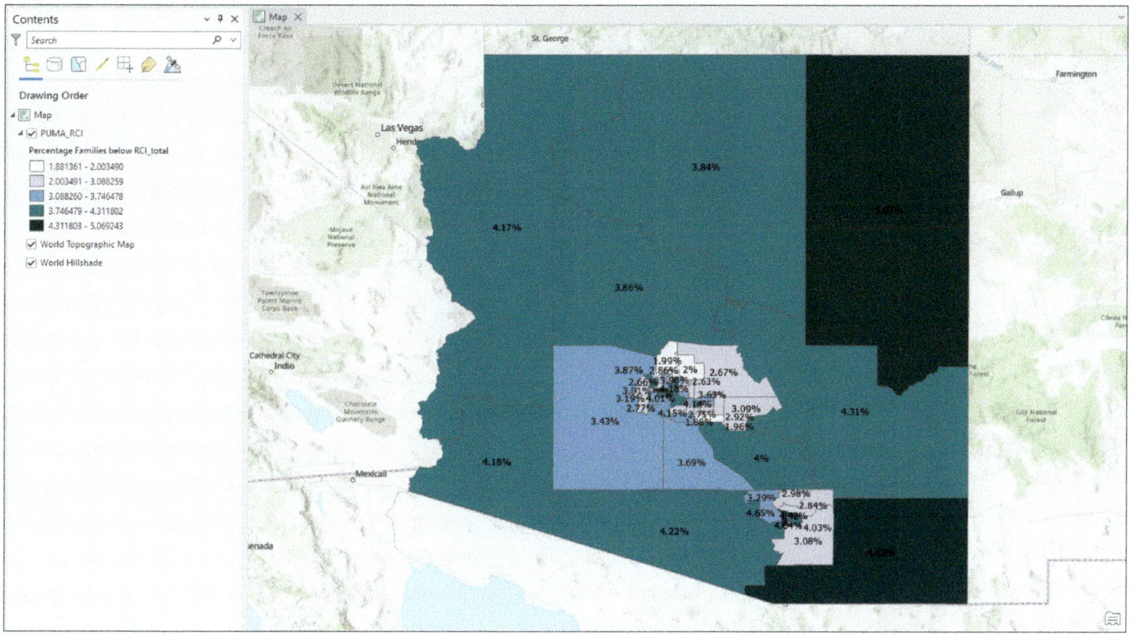

3 Build dashboards and share data

Earlier, you created and shared three maps online. Now you will build dashboards in ArcGIS Online to share and visualize data.

Build dashboard 1

You will begin by sharing the third map you created earlier that shows the percentage of households below the RCI per PUMA (total families, regardless of household type).

1. In your browser, navigate to www.arcgis.com. Sign in to your ArcGIS organization account, click the **App Launcher**, and select **Dashboards**.
2. In ArcGIS Dashboards, click **Create dashboard**.
3. In the **Create new dashboard** window, apply the following settings:
    - **Title**: Dashboard 1 Arizona Households
    - **Tags**: Arizona, Counties, PUMAs
    - **Summary**: A dashboard showing the percentage of Arizona households struggling to meet the household cost per household types in Counties and PUMAs.
    - **Folder**: SecurityFirst

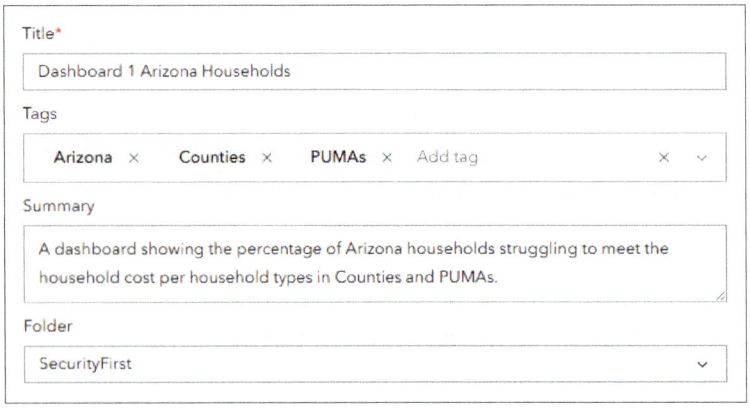

4. Click **Create dashboard**.

   The **Add** button is used to add different elements to your dashboard. First, you will add map 1 to the dashboard.

5. Click the **Add** button and select **Map**.
6. From **My content**, select **Map_Counties**.
7. On the **Settings** tab, turn on the following features:
   - **Search**
   - **Initial view and bookmarks**
   - **Layer visibility**
   - **Basemap switcher**
   - **Zoom in/out**

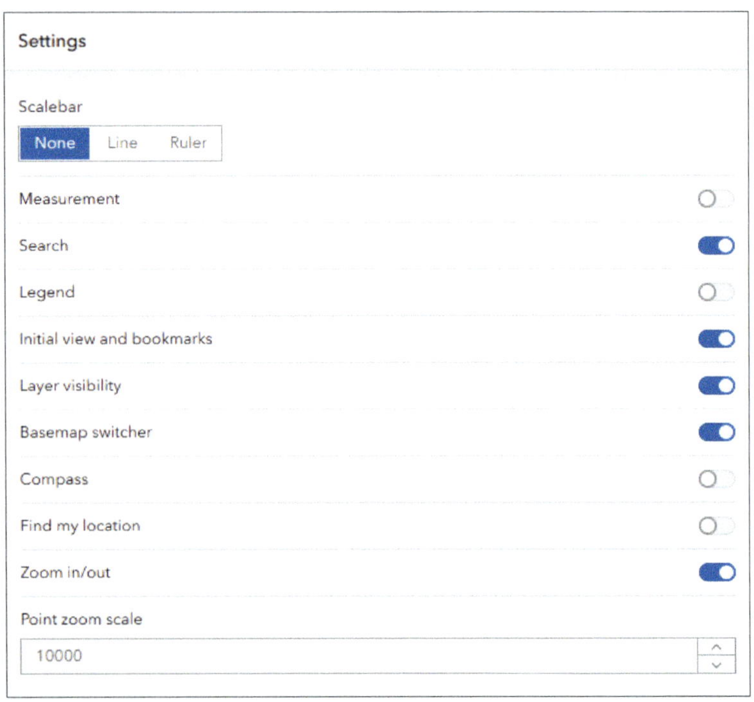

8. Click **Done**.

At this point, your dashboard should have **Map_Counties** loaded on a basemap. When you click the **Layers** icon, you can see the nine layers you created earlier in ArcGIS Pro. Each layer in the map represents the percentage of households struggling to meet the household cost, for each household type. You can turn layers on and off using the **Visibility** icon.

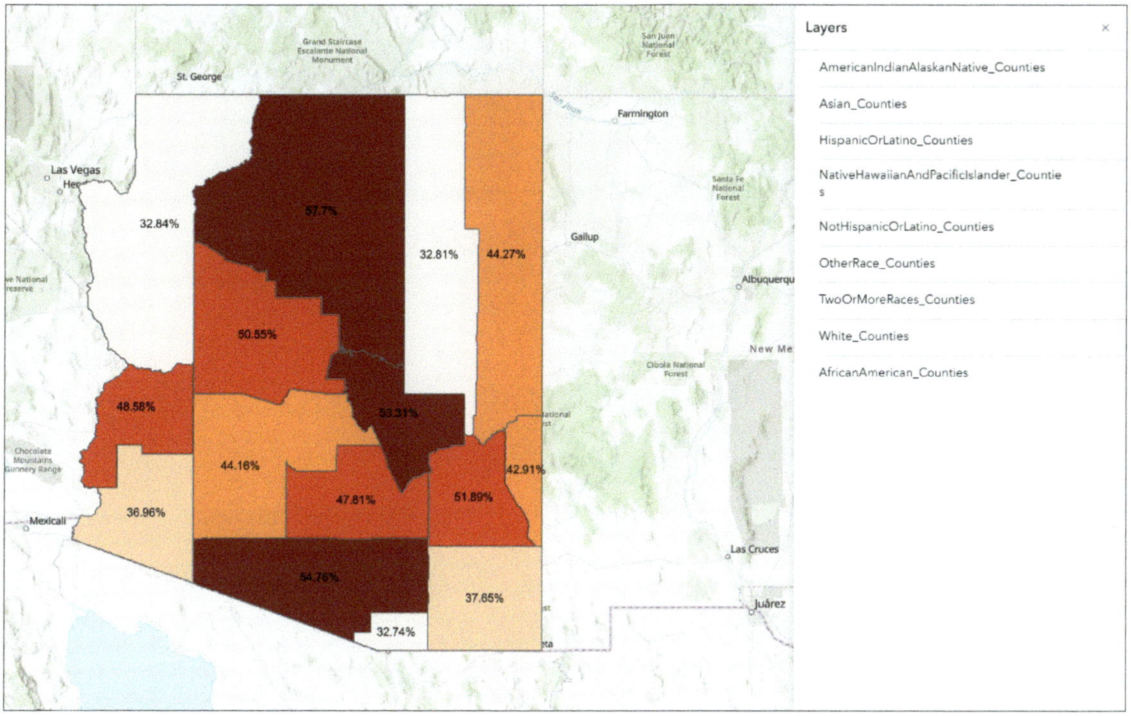

Next, you will add a list.

9. On the left toolbar, click the **Add element** button. Click the plus button above the map and select the **List** element.

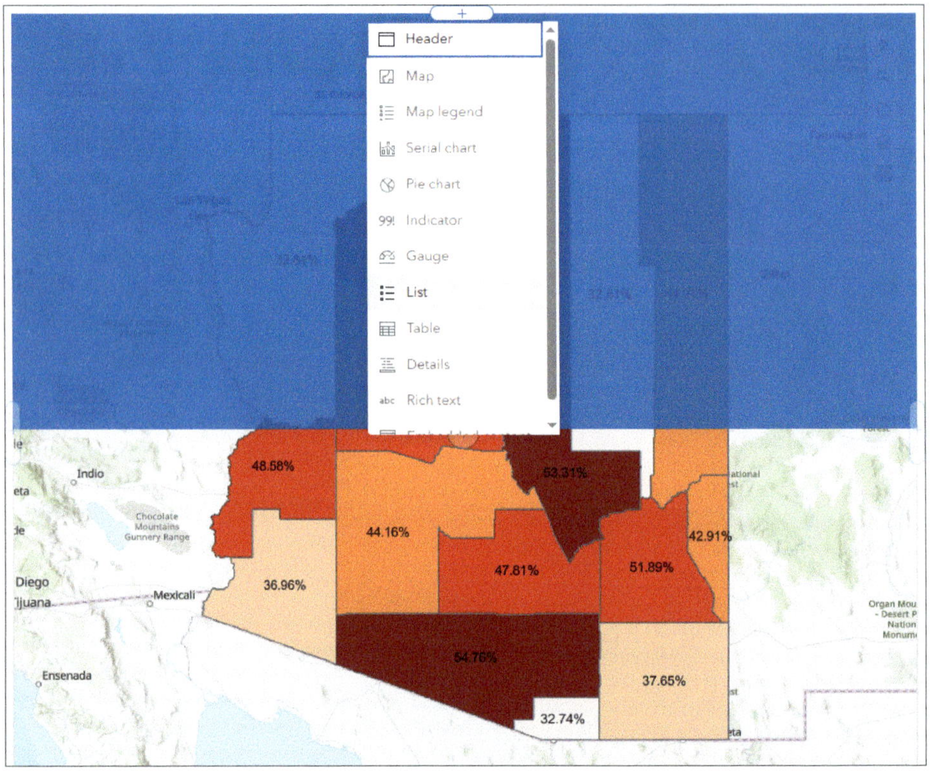

You are adding this **List** element to use as a filter. The **List** element can perform actions on elements in the dashboard. At this point, you will manage these actions on map 1. It will allow you to select and zoom to a county (or multiple counties) of interest.

10. In the **Select a layer** window, select any layer (for example, **White_Counties**).

    Now you need to customize the **List** settings.

11. On the **Data** tab, for **Sort by**, select **NAME**.
12. Click the **List** tab. For **Template**, ensure that the expression reads {field/NAME}. Turn on Show search.
13. Click the **General** tab. Change the **Title** to County name.
14. Click the **Actions** tab. Apply the following settings:
    - **Selection mode**: Multiple
    - **Filter**: Turn on all layers; for each layer, change the **Source field** and the **Target Field** to **NAME**
    - **Flash**: Turn on **Map_Counties**
    - **Show pop-up**: Turn on **Map_Counties**
    - **Zoom**: Turn on **Map_Counties**
15. Click **Done**.

16. On the dashboard, drag the bottom edge of the **List** element so that it takes up only the top part of the view.

    Next, you will search a county.

17. In the search bar, type Coconino. Click the search result and see the result on the map.

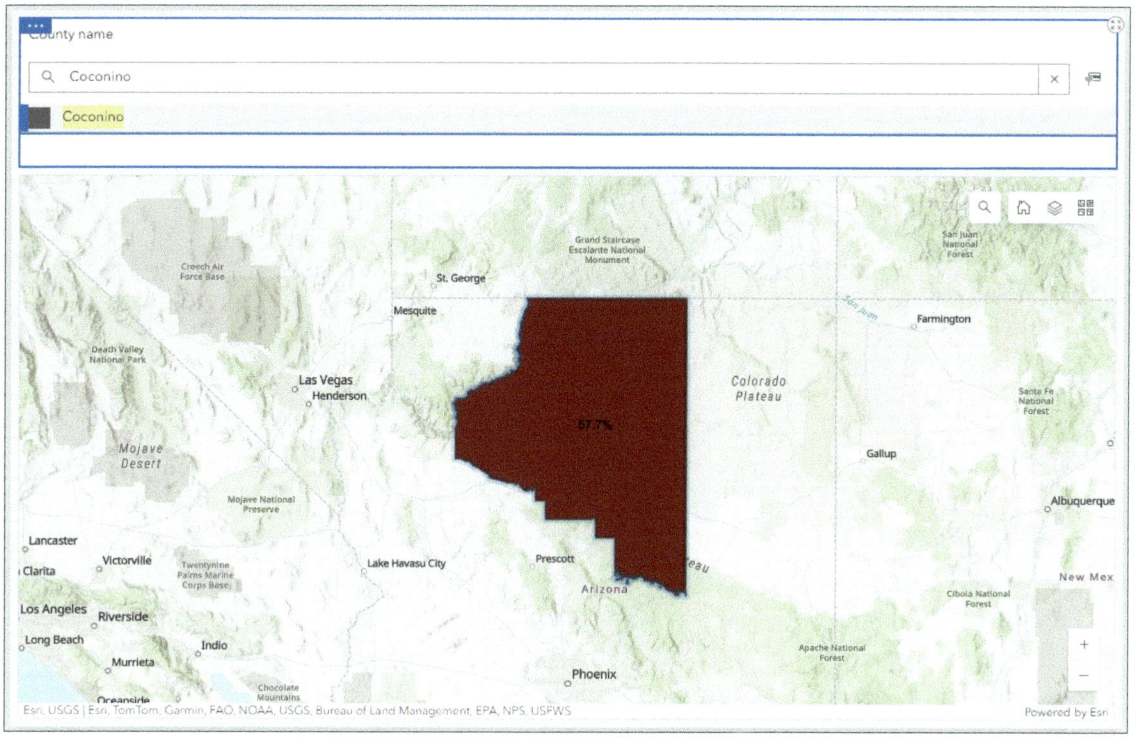

Tip: If the highlighted section is difficult to see, you can change the color in the map's settings on the General tab. Change the Selection color.

Next, you will add map 2.

18. Follow the steps used for map 1, but this time, add **Counties_Map2** related to PUMA.
19. Add the new element to the right of map 1.

    Now you will add list 2.

20. Following the same steps used for list 1, create list 2. Add the new element to the right of list 1. Connect the list to map 2 using the same household type used in map 1 (for example, **White_Counties**). However, make the following modifications:
    - On the **Data** tab, increase the **Maximum features displayed** to 54 (because the total number of PUMAs in the map is 54).
    - On the **General** tab, change the **Title** to PUMA name.
    - On the **Actions** tab, apply the same settings from list 1, but apply **Flash** and **Zoom** actions to both map 2 and map 1.

When searching a PUMA in list 2, both maps highlight the outline of the resulting PUMA and are zoomed to the location.

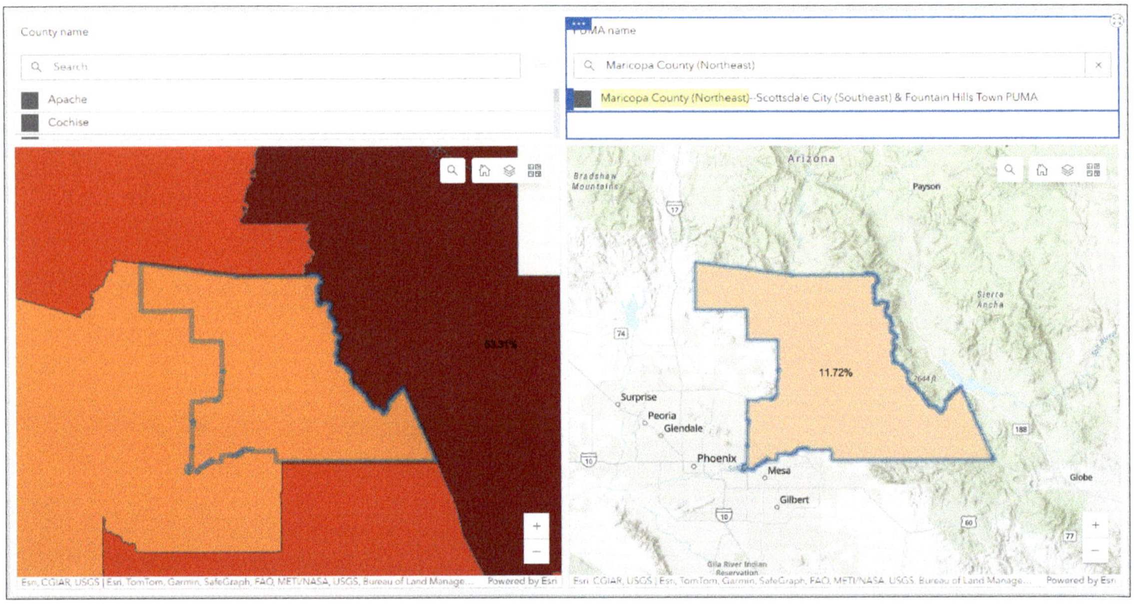

A gauge is a type of chart that you will use to express the total of households struggling to meet the household cost per each type of household. So, you will create nine gauges—one for each of the t household types. To create these gauges, you can connect them with data in the **Counties** map or the **PUMA** map. In either case, you will have the same total of households.

21. Click the **Add element** button. Below map 1 and map 2, add a **Gauge** element. Select a layer (for example, **White_Counties**).
22. On the **Data** tab, apply the following settings:
    - **Statistic**: Sum
    - **Field**: Estimated Households Struggling to Meet Household Cost
    - **Maximum value** > **Value type**: Statistic
    - **Maximum value** > **Statistic**: Sum
    - **Maximum value** > **Field**: Total Households
23. On the **Gauge** tab, change the appearance by changing the **Band** color to a color of your choice.
24. On the **General** tab, set the **Title** as the layer's household type (for example, White). Adjust the text so that it is clear to see on the dashboard.
25. Click **Done**.
26. Repeat the creation of the **Gauge** element with the same steps for each household type. Adjust the size of the gauges so that they fit on the dashboard.

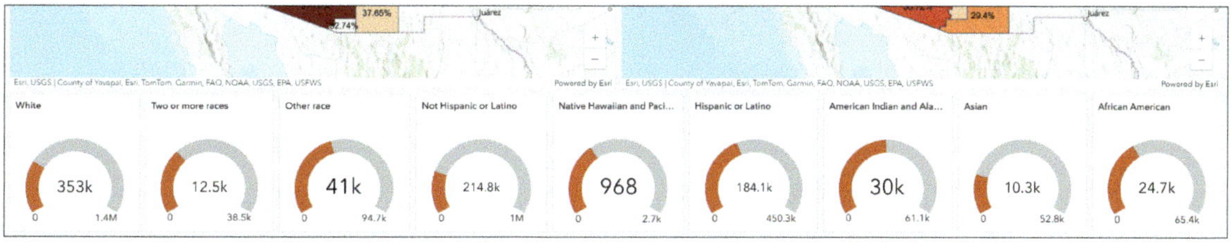

You will add a header next.

27. On the left toolbar, click the **View** button. On the **Header** tab, click **Add Header**.
28. Choose either a **Title** or a **Subtitle**. (Title text size is bigger.) Type Households Struggling to Meet Household Costs in Arizona. Add a logo or a background image.
29. Click **Done**.
30. On the left toolbar, click the **Save** button to save your dashboard.
31. Search for the dashboard in your ArcGIS Online content and open it.

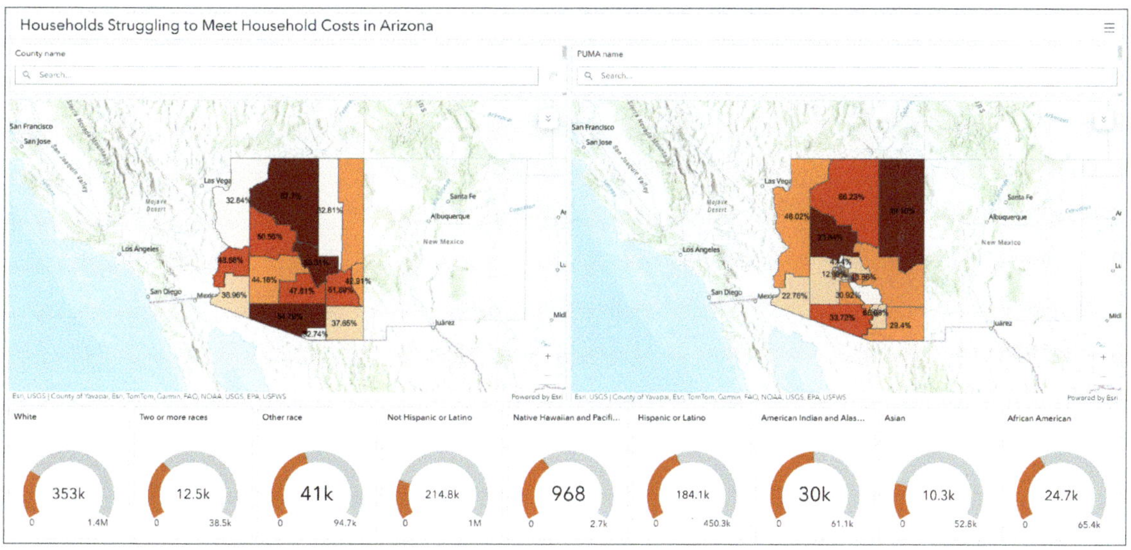

Build dashboard 2

To build the second dashboard, you will use **Map_PUMA_RCI** that you created earlier.

1. Create a new dashboard using the following settings:
   - **Title:** Dashboard 2 PUMA RCI
   - **Tags:** Arizona, RCI
   - **Summary:** A dashboard showing the percentage of households below the Resilience Capacity Index (RCI) in Arizona by PUMAs.
   - **Folder:** SecurityFirst

   Now you will start building the dashboard using a map and other elements.

2. Add **Map_PUMA_RCI** to the dashboard.
3. For **Settings**, turn on the following:
    - **Search**
    - **Initial view and bookmarks**
    - **Layer visibility**
    - **Basemap switcher**
    - **Zoom in/out**

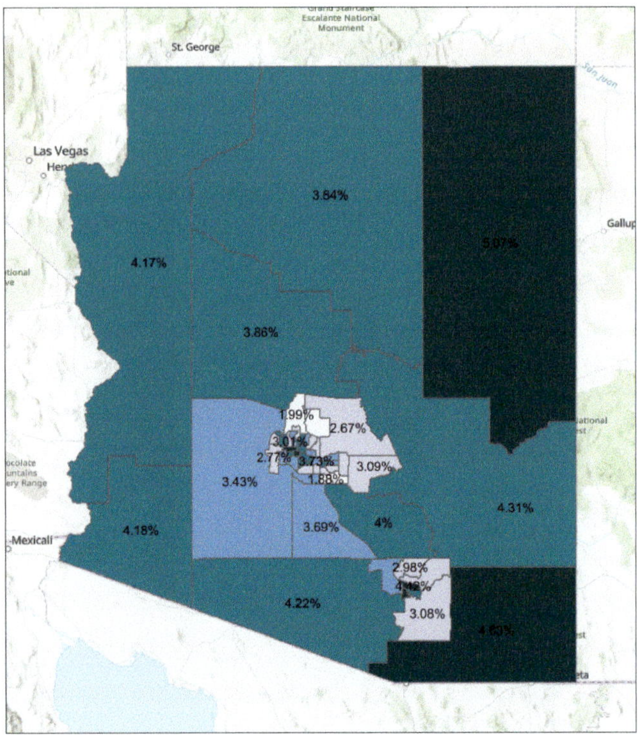

The goal is to create a chart that represents the total amount of families below RCI for each county on the map.

4. Click **Add element**. Add a **Serial chart** element to the left of the map.
5. Select the **PUMA_RCI** layer.
6. On the **Data** tab, adjust the settings as follows:
    - **Categories from**: Grouped values
    - **Category field**: County
    - **Statistic**: Sum
    - **Field**: Families below RCI_total
7. On the **Chart** tab, change the orientation to **Horizontal.**
8. On the **Value axis** tab, change the **Title** to Total of families below RCI.

9. On the **Series** tab, change the chart's bar color to #0f7b85.
10. On the **Actions** tab, apply **Flash** and **Zoom** on the map.
11. Click **Done.**

    When selecting a county on the chart, the map highlights and zooms to the PUMAs in this county.

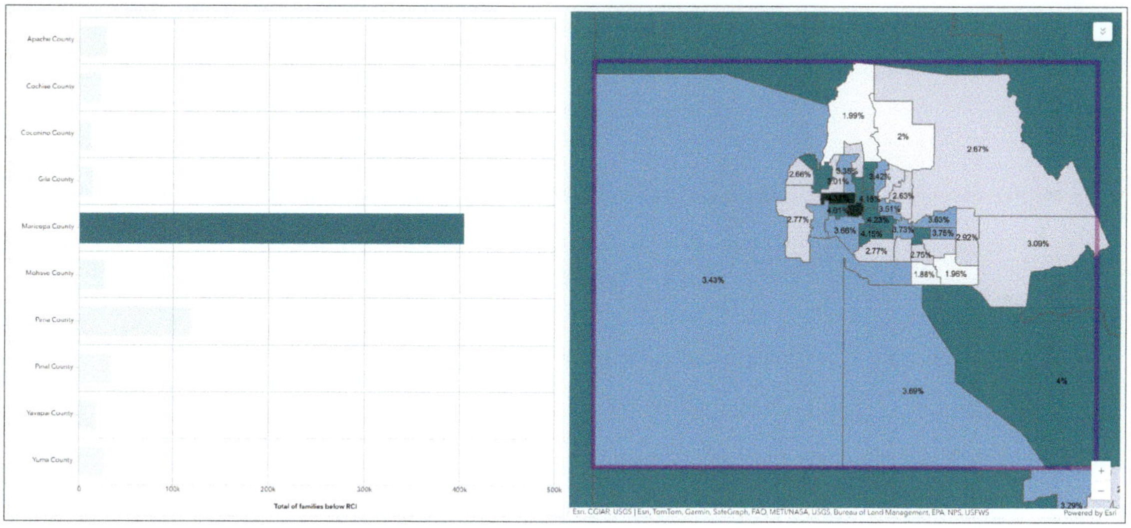

Next, you will use a pie chart to present the percentage of families below RCI in counties.

12. Click **Add element.** Add a **Pie chart** below the serial chart.
13. Select the **PUMA_RCI** layer.
14. On the **Data** tab, apply the following settings:
    - **Categories from**: Grouped values
    - **Category field**: County
    - **Statistic**: Sum
    - **Field**: Percentage Families below RCI_total
15. On the **General** tab, change the **Title** to Percentage of Families below RCI Per County.
16. On the **Actions** tab, apply **Filter** and **Zoom** on the map.
17. Click **Done.**

    When selecting a county in the pie chart, the map makes a filter and zooms to the PUMAs in this county.

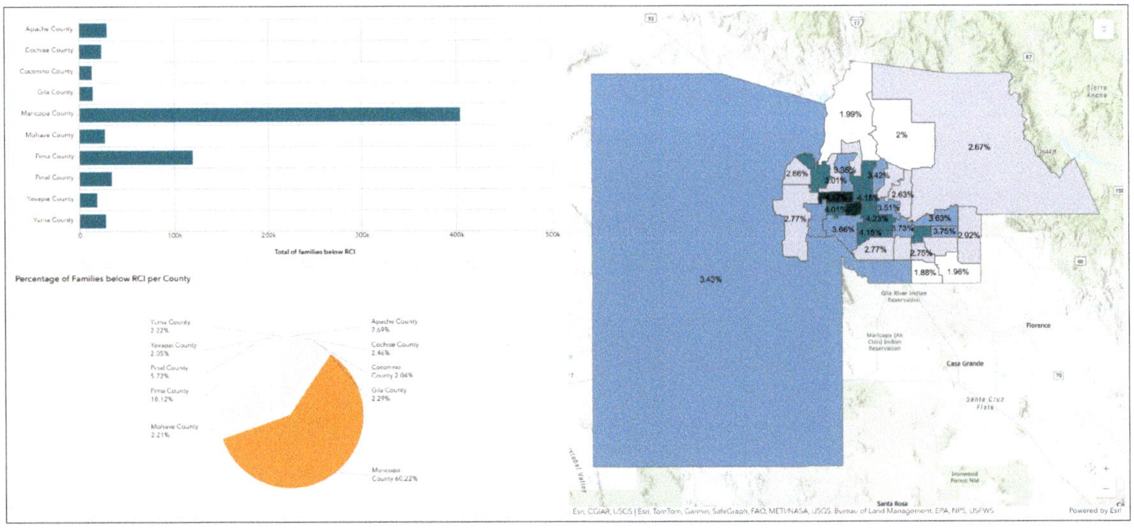

You will now add a header and modify the title.

18. Add a **Header** element to the dashboard. Change the **Title** to Families below RCI in Arizona State.

19. Click the **Save** button to save your work, and then close your dashboard.

20. In ArcGIS Online, under **Content**, find **Dashboard 2** and open it.

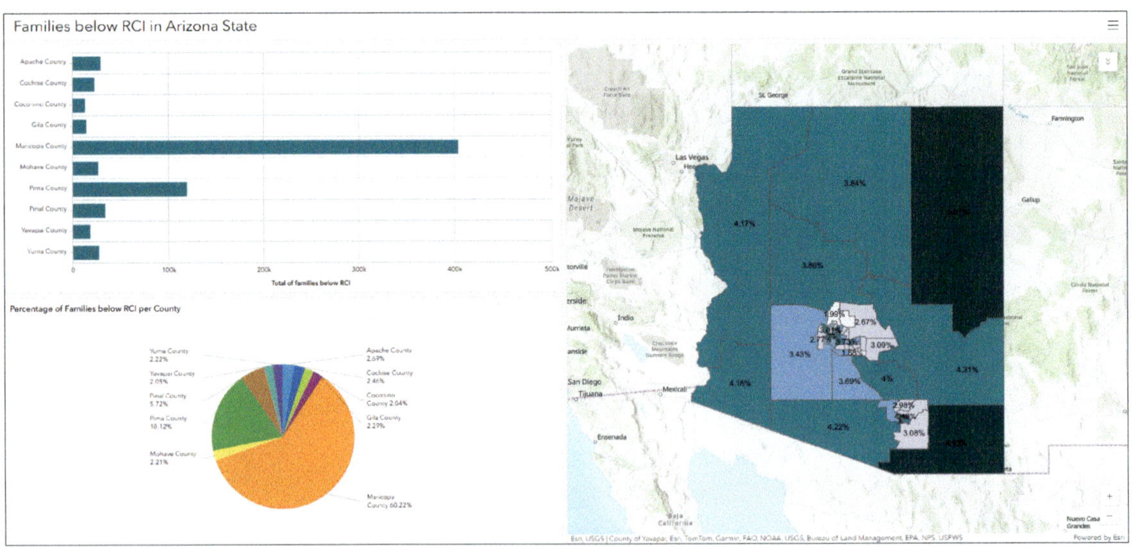

Interpretation of analysis

After completing this exercise, answer the following questions in detail. The answers to these questions, as well as the analysis results from the exercise workflow, will inform the intelligence brief. This may require additional research to understand the geospatial distribution of multidimensional poverty in the state of Arizona.

1. From the overall dashboard analysis, what can you infer about the general distribution of poverty rates in Arizona?
2. Do you recognize any clusters of poverty indicators in Arizona? By zooming in to counties and PUMAs, do you recognize a correlation between these two scales?
3. What are the county and the PUMA that have the highest percentage of families below RCI? What is the spatial relation between them?
4. How could these indicators at the two scales support data-driven policymaking for enhancing socioeconomic resilience?
5. What are the other attributes in the given database that you think are valuable to logically expand on this analysis?

Intelligence brief

Based on the analysis from the exercise workflow, produce a three- to five-page intelligence brief by responding to the following question prompts. Incorporate your geospatial outputs from the exercise to enrich and illustrate your conclusions.

- **BLUF:** List three to five primary observations from your analysis in bulleted-list format that summarize findings of this work.
- **Top-level analysis:** Report the primary data points supporting your BLUF.
- **Secondary analysis:** Provide the data sources and deeper analytic products from this analysis.
- **Impact:** Describe the impacts of your findings on the department or organization.
- **Confidence:** Report the level of confidence of your assessment based on available data, accuracy, and so on.
- **Bias:** Explain any background or information you have that could have affected your analysis.
- **Conclusions:** Determine what type of analytic conclusions there may be and how the data requirements, analysis, and brief writing may change based on what analytic framework you are following (descriptive, predictive, prescriptive).

Applying and extending your skills

The following suggestion will help enhance your data visualization and dashboard skills to generate insights on pressing community issues:

- Discuss how your findings may influence multidimensional poverty risk-related decisions in the study area and how streamlining data analysis through a dashboard platform might support the community's progress toward its sustainability goals and climate or heat action plans.

Additional resources

Fit-for-purpose dashboards

You can build many types of dashboards to tackle various kinds of problems. Dashboards are situated in a wide range of contexts, including business, academia, government, and public information. For example, strategic dashboards help executives track key performance indicators, and tactical dashboards help business managers analyze historical data and visualize trends. Operational dashboards help with event monitoring in real time, and informational dashboards help organizations in community outreach. Knowing the design principles and purpose of these instruments (Sarikaya et al. 2019) will help you design more fit-for-purpose dashboards.

Application in green infrastructure decision-making and policy

To learn more about how dashboards are used in policymaking settings to advance poverty reduction and urban equity initiatives, review the "Integrative Spatial Framework and Co-Design Toolkit to Measure and Visualize Multidimensional Poverty in the United States," published in *Geographical Analysis* (Praharaj 2025).

For more information about ArcGIS Dashboards, including a gallery and online Help resources, visit www.esri.com/en-us/arcgis/products/arcgis-dashboards/overview.

References

- ArcGIS Dashboards, https://www.esri.com/en-us/arcgis/products/arcgis-dashboards/overview.
- Batty, Michael. 2015. "A Perspective on City Dashboards." *Regional Studies, Regional Science* 2 (1): 29–32. doi.org/10.1080/21681376.2014.987540.
- Few, Stephen. 2006. *Information Dashboard Design: The Effective Visual Communication of Data*. O'Reilly Media Inc.
- Kitchin, Rob. 2014. "The Real-Time City? Big Data and Smart Urbanism." *GeoJournal* 79:1–14. doi.org/10.1007/s10708-013-9516-8.
- Mattern, Shannon. 2015. "Mission Control: A History of the Urban Dashboard." *Places Journal*. Accessed April 17, 2025. doi.org/10.22269/150309.

- Moufid, O., S. Praharaj, and H. J. Oulidi. 2024. "Digital Technologies in Urban Regeneration: A Systematic Review of Literature." *Journal of Urban Management*. https://doi.org/10.1016/j.jum.2024.11.002.
- Moufid, O., S. Praharaj, H. J. Oulidi, and K. Momayiz. 2025. "A Digital Twin Platform for the Cocreation of Urban Regeneration Projects: A Case Study in Morocco." *Habitat International* 161, 103427. https://doi.org/10.1016/j.habitatint.2025.103427.
- Praharaj, S. 2020. "Development Challenges for Big Data Command and Control Centres for Smart Cities in India." *Data-driven Multivalence in the Built Environment* 75–90. https://doi.org/10.1007/978-3-030-12180-8_4.
- Praharaj, S. 2025. "An Integrative Spatial Framework and Co-Design Toolkit to Measure and Visualize Multidimensional Poverty in the United States." *Geographical Analysis*. https://doi.org/10.1111/gean.12420.
- Praharaj, Sarbeswar, and Elizabeth A. Wentz. 2022. "Building Community Resilience Through Geospatial Information Dashboards." *International Archives of the Photogrammetry, Remote Sensing and Spatial Information Sciences* XLVIII-4/W5-2022: 151–57. doi.org/10.5194/isprs-archives-XLVIII-4-W5-2022-151-2022, 2022.
- Praharaj, Sarbeswar, Patricia Solis, and Elizabeth A. Wentz. 2022. "Deploying Geospatial Visualization Dashboards to Combat the Socioeconomic Impacts of COVID-19." *Environment and Planning B: Urban Analytics and City Science 50* (5). doi.org/10.1177/23998083221142863.
- Sarikaya, Alper, Michael Correll, Lyn Bartram, Melanie Tory, and Danyel Fisher. 2019. "What Do We Talk About When We Talk About Dashboards?" *IEEE Transactions on Visualization and Computer Graphics* 25 (1): 682–92. doi.org/10.1109/TVCG.2018.2864903.

About the authors

Sarbeswar Praharaj, PhD, is an assistant professor (tenure track) at the University of Miami Department of Geography and Sustainable Development and the School of Architecture. He recently served as the associate director of the Knowledge Exchange for Resilience at Arizona State University. He has a PhD and postdoctoral research experience in urban planning and analytics from the University of New South Wales, Australia. His research and teaching interests combine geospatial technologies for advancing community resilience.

Oumayma Moufid, PhD, earned her doctorate at Hassania School of Public Works, Morocco, and is a Fulbright Research Scholar supervised by Professor Praharaj at the University of Miami. She conducts research in geography and urban planning in the Department of Geography and Sustainable Development and the School of Architecture. Oumayma has an academic background in geospatial data science and surveying engineering and professional experience in urban planning for the National Agency of Urban Regeneration and Rehabilitation in Morocco.

Chapter 12

Cultural impacts of the war in Ukraine

| ***Madeline Rouse,*** *University of Southern California, Spatial Sciences Institute*

Overview

In this chapter, you will learn how to use ArcGIS Pro to investigate and visualize how the war in Ukraine, since 2022, has affected various cultural groups in Ukraine. Understanding the cultural impacts of conflict can provide a variety of benefits, including informing humanitarian efforts and interventions and analyzing the intentions of combative groups.

Learning objectives

- Select appropriate geospatial data to represent cultural variables.
- Transform and analyze imperfect vector data in ArcGIS Pro.
- Create a variety of cartographic products to analyze patterns and inform decision-makers.
- Critically analyze how data limitations affect the confidence level of your findings.

Technical components

- Software: ArcGIS Pro 3.4
- Datasets: Ukrainian language, Ukrainian ethnicity, other Ukrainian cultural variables, attacks on civilians in Ukraine

Prerequisites

- Prior knowledge or experience with GIS software or spatial data is helpful but not required.
- Technical skill level: Introductory GIS.
- Time to complete chapter: One to three hours without intelligence brief, two to four hours with intelligence brief.

Background and situational context

On February 22, 2022, Russian president Vladimir Putin announced a "special military operation" to take place in Ukraine, citing concerns about NATO expansion and skepticism about the legitimacy of Ukrainian statehood following the collapse of the USSR. The subsequent war in Ukraine is still ongoing as of spring 2025 and involves many political, military, and cultural factors.

The cultural impacts of conflict are not commonly considered in the GEOINT workflow. However, understanding these impacts can provide a variety of benefits:

- Informing humanitarian efforts and interventions (for example, What language do those affected speak? Are certain ethnic groups in more or less need of assistance?)
- Analyzing the intentions of combative groups (for example, Are certain cultural groups being targeted or avoided by a given state? Are a group's actions matching its leader's rhetoric?)

Although Putin has stated that one of Russia's goals in Ukraine is to protect and reassimilate linguistically and ethnically Russian communities that reside within Ukraine, data about Russian attacks on civilians and cultural groups in Ukraine can provide insight into whether Russian actions are matching Putin's words.

In this chapter, you will use cultural variables to determine which groups in Ukraine are being most affected by Russian attacks on civilians and whether certain groups are being targeted or protected from attacks.

Exercise workflow 12

1 Add data to a map

You will first add data to a new map in ArcGIS Pro.

1. In File Explorer, navigate to this chapter's data folder, **SecurityFirst_ExerciseData/Chapter12**.
2. Double-click **Ch12_Ukraine.ppkx**, an ArcGIS Pro project package file, to open the preloaded ArcGIS Pro project.

   The map contains the following layers:

   - **Russian Attacks on Ukrainian Civilians**
   - **Support for Euromaidan**
   - **Most Common Language Spoken at Home**

- **Most Common Native Language**
- **Ukrainian Ethnicity**
- **Ukrainian Oblasts**

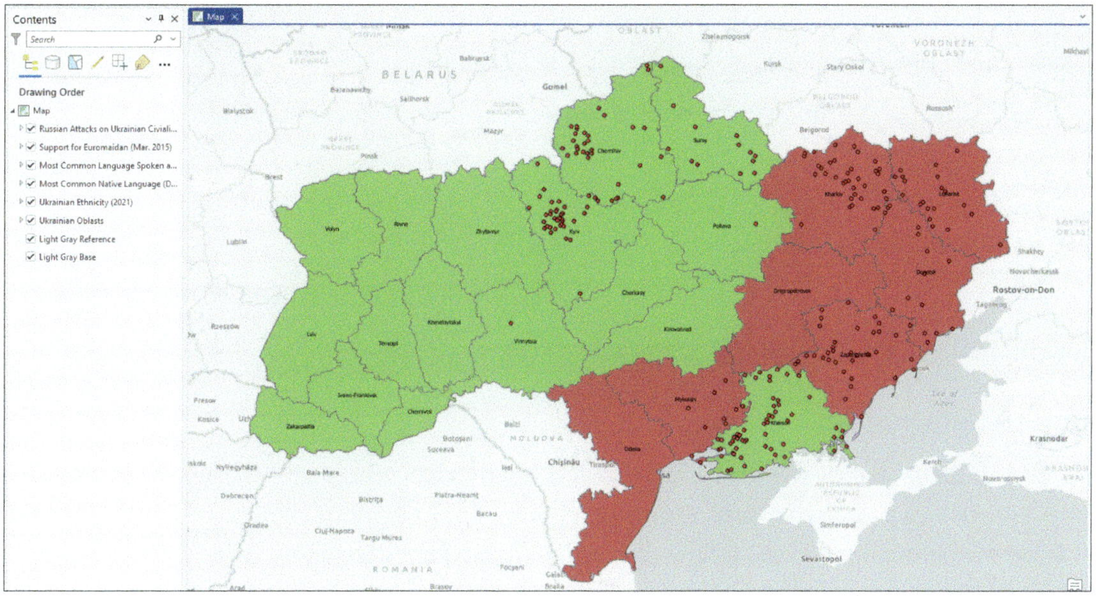

In Ukraine, oblasts are first-level administrative divisions, similar to US states and Canadian provinces.

3. On the ribbon, click the **Project** tab. Click **Save Project As**. Save your project as Chapter_12 to the **C:/SecurityFirst** folder.

2 Explore and select cultural variables for analysis

A variety of feature classes have been prepared for you in the geodatabase for this project. The first step is to determine which variables you want to analyze in relation to civilian attacks. Culture is complex, with a unique combination of linguistic, ethnic, political, religious, and other factors contributing to an individual's sense of cultural identity. What cultural factors play the largest role in your identity? What cultural factors do you think may be most important in the context of the invasion of Ukraine? Use these questions to guide your selection of cultural variables.

1. One at a time, view each of the following layers in ArcGIS Pro. Explore the data, metadata, and symbology for each layer to familiarize yourself with them.
    - **Most Common Native Language** (Dec. 2015)
    - **Most Common Language Spoken at Home** (Dec. 2015)
    - **Support of Euromaidan** (Mar. 2015)
    - **Ukrainian Ethnicity** (2021)

Tip: To view a layer's information, right-click the layer in the Contents pane and select the feature you are interested in (for example, symbology, properties, and so on).

What patterns do you already see in these data layers? Ukraine is often considered a country divided between east and west. What characteristics are more prevalent in the east vs. the west? Why?

For this activity, you will analyze ethnicity and one other cultural variable as they relate to each other and ongoing violence against civilians in Ukraine.

2. Decide which of the following variables you will use in addition to ethnicity:
   - Most common native language
   - Most common language spoken at home
   - Support of Euromaidan

   This tutorial will analyze ethnicity with the most common native language.

3. Remove the two layers you will not be using from your map.

Explore attack data

If you also chose to analyze the most common native language, you can follow along directly with these two variables. Otherwise, substitute your choice of most common language spoken at home or support for Euromaidan in the following steps. Before proceeding, familiarize yourself with the variables you have selected and their general role in Ukrainian culture.

1. In the **Contents** pane, turn off all layers except **Russian Attacks on Ukrainian Civilians (Feb 22 2022–Jun 30 2023)** and **Ukrainian Oblasts**.

   Each red dot on the map represents an attack by a Russian military or police entity on civilians in Ukraine from the start of the war (February 22, 2022) through June 30, 2023.

2. Right-click the **Russian Attacks on Ukrainian Civilians** layer and click **View Metadata** to understand its source and temporal resolution.

3. On the map, click a few dots and explore the layer's attribute table to understand the data provided by Armed Conflict Location and Event Data (ACLED).

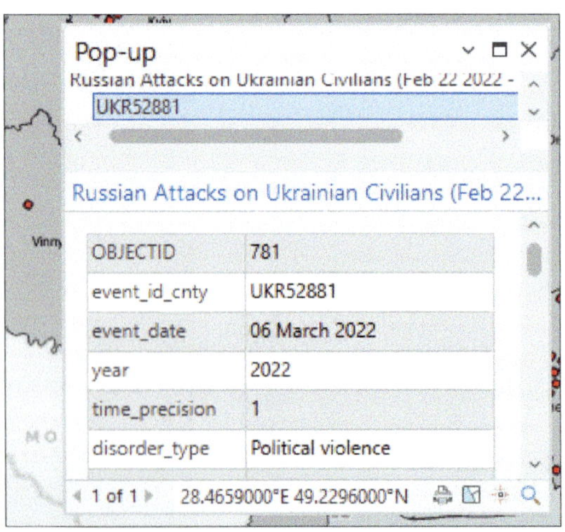

3 Transform and visualize your data

Real-world data is often messy and complex. To transform the data into actionable information, you must simplify and analyze it to discover patterns. The linguistic and political layers provided have already been simplified and summarized by oblast; the next step for you will be to transform the ethnicity and attack data to match.

Transform ethnicity data

1. Turn on the **Ukrainian Ethnicity (2021)** layer. Right-click the layer and click **Symbology**.
2. In the **Symbology** pane, at the top of the pane, click the **Symbol layer drawing** button.
3. Turn on **Enable symbol layer drawing**.

   This allows you to manually change the order that the five ethnicity sublayers are drawn in.

   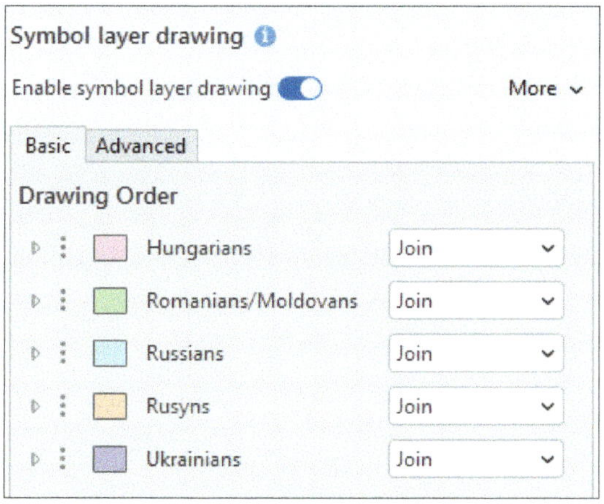

4. Experiment with changing the **Drawing Order** by dragging the **Hungarians** ethnicity sublayer below the **Romanians/Moldovans** sublayer.

   On the map, the layers will be drawn in order from bottom to top.

   What do you notice displayed on the map? Which areas are covered by multiple polygons, and what does this mean?
5. Return the sublayers to their original order and turn off **Enable symbol layer drawing**.

Use the Tabulate Intersection tool

Next, you will use the **Tabulate Intersection** tool to calculate the percentage by area of each ethnicity within each oblast. For more information about the tool, see links.esri.com/SF-TabIntersectTool.

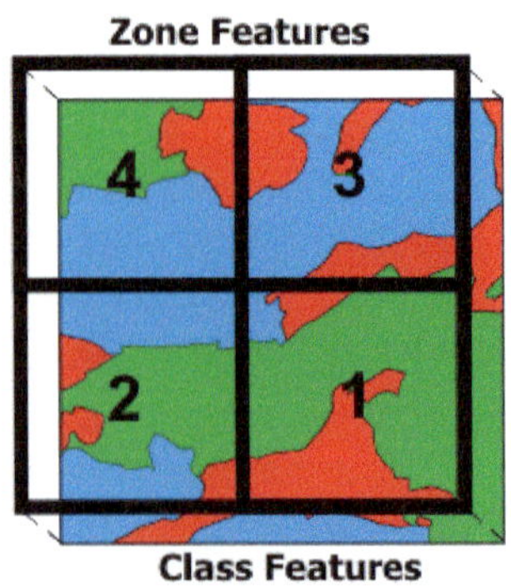

Output

Zone	Color	AREA	PERCENTAGE
1	Blue	0.7	7.9
1	Green	6.3	67.5
1	Red	2.3	24.6
2	Blue	3.4	37.1
2	Green	4.1	44.3
2	Red	1.7	18.7
3	Blue	5.2	56.1
3	Green	1.3	13.9
3	Red	2.8	30.0
4	Blue	5.5	59.3
4	Green	2.0	21.5
4	Red	1.8	19.1

1. On the ribbon, click the **Analysis** tab. In the **Geoprocessing** group, click **Tools.**
2. In the **Geoprocessing** pane, search for the Tabulate Intersection tool and open it.
3. In the **Tabulate Intersection** tool pane, apply the following settings:
   - **Input Zone Features:** Ukrainian Oblasts
   - **Zone Fields:** Name
   - **Input Class Features:** Ukrainian Ethnicity (2021)
   - **Output Table:** EthnicityPctByOblast
   - **Class Fields:** group

   Note: The filter for Input Class Features is set up to show only the five ethnicities present in Ukraine for this dataset.

   The **Input Zone Features** and **Zone Fields** values specify that you want to divide the area into oblasts, which can be defined by their unique names. The **Input Class Features** and **Class Fields** values specify that you want to analyze the ethnicity layer within each oblast and that polygons with the same group value should be considered part of the same ethnicity. A table will be generated.

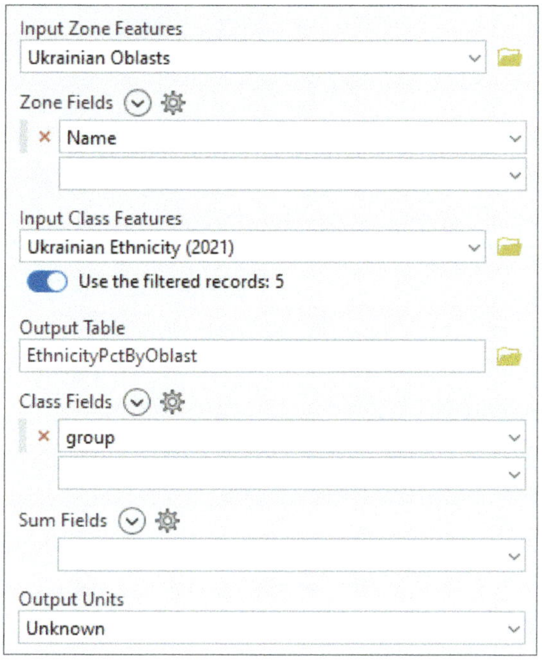

4. Click **Run**.
5. In the **Contents** pane, right-click the new table and click **Open**.

   How does it compare with the visualization of ethnicity on the map? Do the ethnicities for each oblast add up to 100 percent? Why or why not?

Apply a definition query

Because of the Russian rhetoric about protecting and reassimilating ethnic Russians within Ukraine, you will focus on Russian ethnicity for this analysis.

1. Right-click the **EthnicityPctByOblast** table and click **Properties**.
2. Click the **Definition Query** tab and click **New definition query**.

   > **Tip:** A definition query allows you to filter a layer or table without deleting the rest of the underlying data.

3. Build your new query: **Where group_ is equal to Russians.**

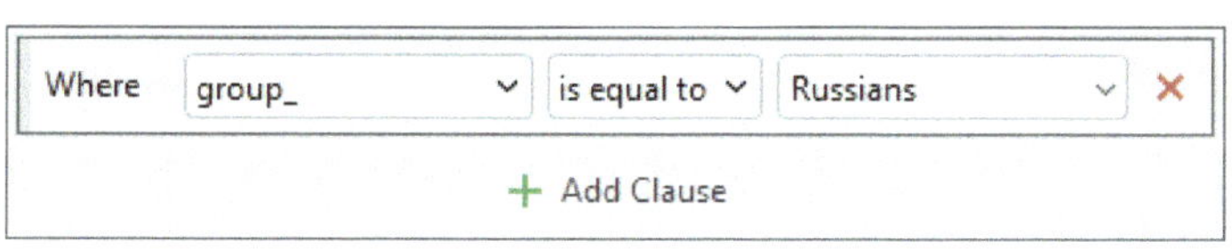

4. Click **Apply** and then click **OK** to close the definition query window.

5. Confirm that the **EthnicityPctByOblast** table shows only rows for Russian ethnicity.

   Note: Not every oblast will be represented in this table because some oblasts have 0 percent Russian ethnicity.

Use the Add Join tool

Now that you have calculated the Russian ethnicity area percentage by oblast, you will add it to the map. You will match the oblast names in the table to oblast names in the **Ukrainian Oblasts** layer with the **Add Join** tool. For details about the tool, see links.esri.com/SF-AddJoinTool.

INPUT

OBJECT ID#	Landuse Code
1	2
2	0
3	1

← Join Fields →

\+

Landuse Code	Landuse Type
0	Unclassified
1	shrub
2	water

OUTPUT

OBJECT ID#	Landuse Code	Join Table Landuse Code	Join Table Landuse Type
1	2	2	water
2	0	0	Unclassified
3	1	1	shrub

1. In the **Contents** pane, turn off the **Ukrainian Ethnicity (2021)** layer.
2. Right-click the **Ukrainian Oblasts** layer and click **Copy** (or press **Ctrl+C**). Paste (or press **Ctrl+V**) the layer into the pane.
3. Double-click the new layer and change its **Name** to Ethnicity by Oblast (2021).
4. Right-click the **Ethnicity by Oblast (2021)** layer and click **Label** to turn off labeling.
5. Open the **Geoprocessing** pane. Search for the Add Join tool and open it.
6. In the **Add Join** tool pane, apply the following settings:
   - **Input Table**: Ethnicity by Oblast (2021)
   - **Input Join Field**: Name
   - **Join Table**: EthnicityPctByOblast
   - **Join Table Field**: Name
7. Click **Run.**

   The **Input Table** value specifies that you want to modify the **Ethnicity By Oblast (2021)** attribute table. The **Join Table** value specifies that you want to bring in attributes from the **EthnicityPctByOblast** table. **Input Join Field** and **Join Table Field** specify which attributes both match the two tables.
8. Open the attribute table for the **Ethnicity by Oblast (2021)** layer.

   This table now includes only 15 of the 24 oblasts—the same 15 that were in the **EthnicityPctByOblast** table when you filtered to data for only Russian ethnicity.

9. Scroll to the far right of the attribute table.

   The percentages you calculated and joined to this layer are listed as a decimal number in an attribute titled **PERCENTAGE**. You will update this name to better describe the data.

10. On the ribbon, click the **Data** tab. In the **Data Design** group, click **Fields**.

11. Scroll to the bottom of the **Fields** table.

    You will change the alias for the **EthnicityPctByOblast.PERCENTAGE** field.

12. Double-click the **Alias** box that currently reads **PERCENTAGE** and type the new name: Percent Russian Ethnicity by Area. Press **Enter**.

13. On the **Data** tab, in the **Manage Edits** group, click **Save**. Save your project.

14. Confirm that the title of the final column has changed.

Percent Russian Ethnicity by Area
7.507435
10.5357
16.069049
1.815176
1.30286
1.474016
2.549961
13.407648

15. Close the **Fields** table and the attribute table.

Visualize ethnicity data

Now that the ethnicity data is cleaned, calculated, and represented on your map, you will choose appropriate symbology.

1. Open the **Symbology** pane for **Ethnicity by Oblast (2021)**.

2. Change the **Primary symbology** to **Graduated Colors**.

3. For **Field**, select **Percent Russian Ethnicity by Area**.

   Because the oblasts missing from this layer are those with 0 percent Russian ethnicity, your map currently shows oblasts with no Russian ethnicity in gray because the **Ukrainian Oblasts** layer is visible underneath.

   You will choose a color scheme that works with this existing symbology.

4. For **Method**, select **Equal Interval**. For **Classes**, type 4.

   This will break the data into four groups (for a total of five with the 0 percent category) where each group spans the same size range of the data distribution.

5. For **Color scheme**, select **Reds (Continuous)**.

> **Tip:** In the Color scheme list, check the box for Show names to see the names of each color scheme.

This color scheme is close to the desired look, but the gray does not fit with the gradient as nicely as it could—rather than the gray looking like the bottom end of the gradient, it stands out as its own color. You can adjust the chosen color scheme to make the gray look like part of the overall symbology rather than a stand-alone symbol.

6. At the bottom of the **Color scheme** list, click **Format color scheme**.
7. On the top right of the **Color Scheme Editor** window, use the **Remove color** button to configure the colors so that you have only three colors remaining:
    - The color at the far left
    - The color at the 50 percent position
    - The color at the far right
8. Click the color tab on the far left. Under **Settings**, click **Color**. At the bottom of the color picker, click **Color Properties**. In the **Hex #** box, type E2CCCC. Click **OK**.
9. Using the same procedure, configure the colors for the following color tabs:
    - 50 percent position: #E06565
    - Far right: #A21F21

Now the gray fits better into the general color scheme.

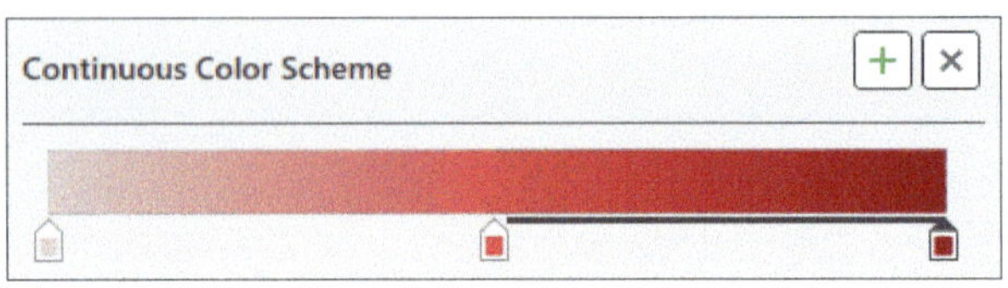

10. Click **OK**.

The final step is to adjust the classes and labels slightly to be more reader-friendly.

11. In the **Symbology** pane, in the **Classes** table, adjust **Upper value** and **Label** using the following values:

Upper value	Label
4	Under 4%
8	4–8%
12	8–12%
16.069049	Over 12%

12. Save your project.
13. Turn off the **Ethnicity by Oblast (2021)** layer.

Transform attack data

Currently, each attack is shown as an individual red dot on the map. You will show a summary of these attacks by oblast using the **Summarize Within** tool. You can find more information about the tool at links.esri.com/SF-SummarizeWithinTool.

1. In the **Contents** pane, turn on the **Russian Attacks on Ukrainian Civilians** layer.
2. Open the **Geoprocessing** pane. Search for the Summarize Within (Analysis Tools) tool and open it.
3. In the **Summarize Within** tool pane, apply the following settings:
   - **Input Polygons**: Ukrainian Oblasts
   - **Input Summary Features**: Russian Attacks on Ukrainian Civilians (Feb 22 2022–Jun 30 2023)
   - **Output Feature Class**: RussianAttacksSummarized

   Note: The filter on Input Summary Features is set up to show attacks only on Ukrainian soil during the set time.

   The **Input Polygons** value consists of the polygons that you want to summarize within—oblasts. The **Input Summary Features** value consists of the features that you want to summarize in the oblasts—attacks. The **Output Feature Class** will be the name of your new layer.
4. Click **Run**.
5. Turn off the **Russian Attacks on Ukrainian Civilians** layer.
6. Right-click the new **RussianAttacksSummarized** layer and click **Label** to turn off labeling.
7. Open the layer's **Properties** window and rename it Russian Attacks Summarized (Feb 22 2022–Jun 30 2023).
8. Open the attribute table for the new layer.

   Note the new field at the end of the table. This column represents the number of attacks counted in each oblast.
9. Open the **Fields** table to change the **Alias** of this new field to Number of Attacks.
10. Save your changes and close the **Fields** and the attribute table.

Visualize attack data

Now that you have a field storing the number of Russian attacks on Ukrainian civilians, you can visualize attack frequency by oblast.

1. Open the **Symbology** pane for **Russian Attacks Summarized**.
2. Change the **Primary symbology** to **Graduated Colors**.
3. For **Field**, select **Number of Attacks**.

   In this layer, all 24 oblasts are represented, even those with no recorded attacks. You want to choose a color scheme that accurately represents attack severity.

4. For **Method**, confirm that **Natural Breaks** is selected. For **Classes**, confirm that **5** is set.

   You will manually modify the category ranges later to be intuitive and minimize misrepresentation.

5. For **Color scheme**, select **Grays (Continuous)**.

   Although this is a good way of visualizing the attack data—white oblasts for no attacks and darker oblasts for more attacks—the use of pure black as the darkest color erases your oblast layers. Lightening the color scheme will improve the visualization.

6. Open the **Color Scheme Editor** window.
7. Use the **Remove color** button to configure the colors so that you have only three colors:
   - The color at the far left
   - The color at the 50 percent position
   - The color at the far right
8. Using **ArcGIS Colors**, set each color tab to the following color:
   - Far left: **Arctic White**
   - 50 percent position: **Gray 40 percent**
   - Far right: **Gray 70 percent**

   Now you have a gray gradient that can still show black labels on top.

   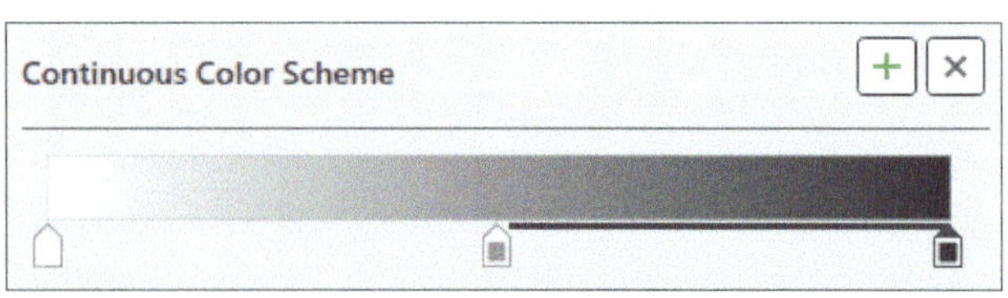

   Note: In this cartographic scenario, if the data analyst's goal is to show relative differences in the number of attacks across oblasts, counts should be normalized by area or population; however, if the analyst's goal is to show the magnitude only, then raw counts are recommended.

9. Click **OK**.

   Finally, you will adjust the classes and labels slightly to be more reader-friendly.

10. In the **Symbology** pane, in the **Classes** table, adjust **Upper value** and **Label** using the following values:

Upper value	Label
0	0
25	25 or fewer
50	50 or fewer
100	100 or fewer
250	Over 100

11. Save your project.

4 Create cartographic products

Create a map of Ukrainian ethnicity

1. On the ribbon, click the **Insert** tab. In the **Project** group, click **New Layout**.
2. From the list, select **ANSI–Landscape Letter**.
3. On the **Insert** tab, in the **Map Frames** group, click **Map Frame**. Select **Map**, which is the map you have been working on.
4. Click the top right of the layout and drag it to the bottom right so that the map spans the whole page.
5. In the **Contents** pane, turn off all layers except:
    - **Ethnicity by Oblast (2021)**
    - **Ukrainian Oblasts**
    - **Light Gray Reference**
    - **Light Gray Base**

   As you choose what to add to your map and how to format it, consider your intended audience. For example, should your scale bar be in miles for an American audience or in kilometers for a European or international audience?
6. On the **Insert** tab, in the **Graphics and Text** group, add a **Rectangle** text element to the top right of the layout. Type Percent Russian Ethnicity by Ukrainian Oblast (2021).
7. On the ribbon, click the **Text** tab. In the **Text Symbol** group, adjust the **Text Symbol Font Size** so that the text is big enough to be a title. For example, change the size to 24 pt.

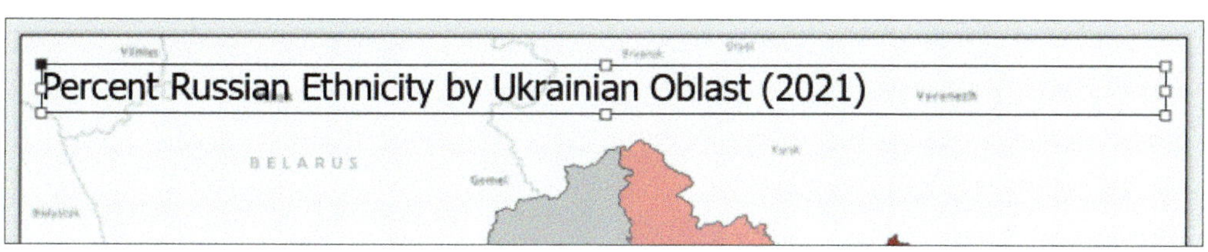

8. Add a smaller **Text** element below the title to call out data credits. Type Data from ETH Zurich GeoEPR Project.

9. On the **Insert** tab, in the **Map Surrounds** group, click **Legend**. Select any legend. For example, **Legend 1**. Place it at the bottom right of the layout.

> **Tip:** There are multiple ways to customize a legend. Right-click the legend and select Convert to Graphics to modify and delete individual elements of a legend in ArcGIS Pro.

10. On the map, add more elements that you see fit:
    - Scale bar (optional)
    - North arrow (optional)
    - Change the base layer (optional)
    - Other elements and modifications as you see fit

11. On the ribbon, click the **Share** tab. In the **Output** group, click **Export Layout** to export your layout as a high-quality JPEG file. Give it an appropriate name, such as Percent Russian Ethnicity by Ukrainian Oblast (2021).

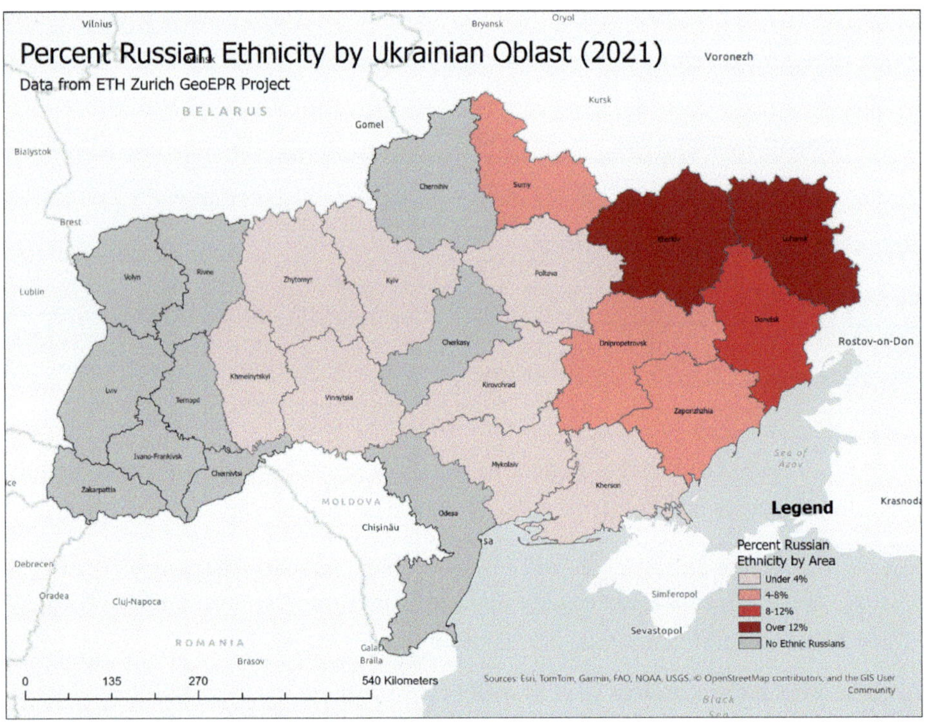

Create a map of your second cultural variable

1. Create a new layout using the **ANSI – Landscape Letter** size template.
2. Add a new **Map Frame** to your layout using the **Ukraine** bookmark and fill the layout using the map.
3. In the **Contents** pane, turn off all layers except the following:
   - Your selected second cultural variable—for example, **Most Common Native Language (Dec. 2015)**
   - **Ukrainian Oblasts**
4. Add the following elements to your map. Consider your intended audience in each decision you make.
   - An appropriate title relating to the cultural variable you selected
   - Data credit

   > **Tip:** Check the metadata of your layer to find the data source.

   - A legend
   - Scale bar (optional)
   - North arrow (optional)
   - Change the base layer (optional)
   - Other elements and modifications as you see fit
5. Export your layout as a high-quality JPEG file. Give it an appropriate name.

Create a map of Russian attack data

1. Create a new layout using the **ANSI – Landscape Letter** size template.
2. Add a new **Map Frame** to your layout using the **Ukraine** bookmark and fill the layout using the map.
3. In the **Contents** pane, turn off all layers except:
   - **Russian Attacks Summarized (Feb 22 2022–Jun 30 2023)**
   - **Ukrainian Oblasts**
4. Add the following elements to your map. Once again, consider your intended audience in each decision you make.
   - **Title**: Number of Russian Attacks on Ukrainian Civilians (Feb 22, 2022–June 30, 2023)
   - **Data credit**: Data from ACLED
   - A legend
   - Scale bar (optional)
   - North arrow (optional)
   - Change the base layer (optional)
   - Other elements and modifications as you see fit

Interpretation of analysis

After completing this exercise, answer the following questions in detail. The answers to these questions, as well as the analysis results from the exercise workflow, will inform the intelligence brief.

1. Based on this analysis, what patterns are present between Russian attacks on Ukrainian civilians and Ukrainian cultural variables? Does it appear that Russian attacks are targeting certain cultural groups more than others?
2. Based on your analysis, do you think that Russia has been truthful about its reasons for invading Ukraine (to protect and reassimilate Russian Ukrainians)? If not, what goals might their actions suggest?
3. What other variables could explain Russian attack patterns (for example, topography, climate, hydrology)? How could you collect data to analyze other potential variables of importance?
4. What actionable information is derived from this analysis? What efforts can this information support? Who should this information be disseminated to?
5. What other non-GEOINT analyses could support this research?
6. You analyzed ethnicity based on area, not population. What effect could this have on the accuracy and confidence level of your findings?

Intelligence brief

Based on the analysis from the exercise workflow, produce a three- to five-page intelligence brief by responding to the following question prompts. Incorporate your geospatial outputs from the exercise to enrich and illustrate your conclusions.

- **BLUF:** List three to five primary observations from your analysis in bulleted-list format that summarize findings of this work. Keep your intended audience in mind (for example, humanitarian organizations, US military, international news media).
- **Top-level analysis:** Report the primary data points supporting your BLUF.
- **Secondary analysis:** Provide the data sources and deeper analytic products from this analysis.
- **Impact:** Describe the impacts of your findings on the department or organization.
- **Confidence:** Report the level of confidence of your assessment based on available data, accuracy, and so on.
- **Bias:** Explain any background or information you have that could have affected your analysis.
- **Conclusions:** Determine what type of analytic conclusions there may be and how the data requirements, analysis, and brief writing may change based on what analytic framework you are following (descriptive, predictive, prescriptive).

Data confidence

The data provided in this chapter was precleaned and modified to suit the purposes of this tutorial. For a deeper dive into data confidence and collection methodologies, visit the links provided in the chapter references to examine the original data provided and how and when it was collected.

Applying and extending your skills

The following suggestions provide more information and enhance your data analysis and visualization skills.

Data analysis

- Explore the two variables you chose not to use in your initial analysis. What new findings and patterns emerge? Do the new findings support, refute, or enhance your initial findings?
- Analyze more than two cultural variables at once. How can you find ways of visualizing more complex spatial patterns in an intuitive and reader-friendly way?

> **Tip:** The use of bivariate color schemes is one way to show multiple variables at once.

- Your initial analysis focused on a large temporal range of attack data. Narrow the analysis to specific months or seasons. Do attack patterns shift over time from the start of the war through June 2023?
- How else might you analyze the given attack data besides counting the number of attacks? Perhaps you look at attacks or fatalities per capita, total fatality counts, or the average number of fatalities per attack.
- Your initial analysis looked only at Russian ethnicity. How can you bring other ethnicities into your analysis? What might these additions show you?

Data visualization

- Create a bivariate map of two cultural variables. What are the benefits of using a bivariate map vs. comparing two maps side by side? What are the challenges?

The following applications offer options for post-GIS editing work, which is commonly used by cartographers in their workflow. Using one or more, explore how you can use these graphics editing tools to modify your map designs:

- ArcGIS Maps for Adobe Creative Cloud (subscription based)
- Adobe Illustrator (subscription based)
- Inkscape (open source)

Additional resources

- "Effects of the Current Russo-Ukrainian War on Ethnolinguistic Groups in Ukraine," by Madeline Rouse and Daniel Simonds (April 21, 2023): https://storymaps.arcgis.com/stories/4632a586faee407680f00e666b711430

Current war

- "The UN and the War in Ukraine: Key Information," by United Nations Regional Information Centre for Western Europe (Sept. 3, 2022): https://unric.org/en/the-un-and-the-war-in-ukraine-key-information
- "Ukraine in Maps: Tracking the War with Russia," by BBC: www.bbc.com/news/world-europe-60506682
- "Russian Invasion of Ukraine," by CNN: www.cnn.com/world/europe/ukraine
- "War in Ukraine," by BBC: www.bbc.com/news/war-in-ukraine

Ukrainian history

- "Ukraine Has Seen Centuries of Conflict," by Jesse Greenspan (March 18, 2025): www.history.com/news/ukraine-timeline-invasions
- "Euromaidan: Rebirth of a Nation" by Euromaidan Press: https://euromaidanpress.com/euromaidan
- "Understanding Ukraine's Euromaidan Protests," by Open Society Foundations (May 2019): www.opensocietyfoundations.org/explainers/understanding-ukraines-euromaidan-protests
- "How Modern Ukraine Was Made on Maidan," by Peter Dickinson (August 21, 2021): www.atlanticcouncil.org/blogs/ukrainealert/how-modern-ukraine-was-made-on-maidan

Cultural variables

- "An Introduction to Linguistic Anthropology, Ethnolinguistics, and Their Connections to Language," by Shantal Garces (May 31, 2023): www.babbel.com/en/magazine/linguistic-anthropology

Data sources

- Conflict data comes from the ACLED Project: https://acleddata.com
- Ethnicity data comes from the ETHZürich Department of Humanities, Social and Political Sciences and Center for Comparative and International Studies, GeoEPR project: https://icr.ethz.ch/data/epr/geoepr/
- Linguistic and political data comes from the Harvard University MAPA Digital Atlas of Ukraine: https://gis.huri.harvard.edu

This work was originally created as a capstone for the Human Security and Geospatial Intelligence MS program at the University of Southern California, Spatial Sciences Institute. Special thanks to Daniel Simonds and Dr. Darren Martin Ruddell for their help with this project.

About the author

Madeline Rouse is a graduate of the University of Southern California, Spatial Sciences Institute and has an MS in human security and geospatial intelligence. She specializes in psychologically informed cartography and data visualization.

Chapter 13

Monitoring human rights violations using satellite remote sensing

Rebecca Bosworth, *United States Space Force*
Yi Qi, *University of Southern California, Spatial Sciences Institute*

Overview

In this chapter, you will learn how to use GIS tools and satellite remote sensing data to assess infrastructure damages indicating potential human rights violations as a result of Russia's 2022 incursion into Ukraine. The workflows and the final deliverables provide actionable information to help guide human rights research to promote international justice in the context of HSGI.

Learning objectives

- Perform exploratory data visualization of satellite remote sensing data.
- Apply geospatial filters to sort and analyze patterns of synthetic aperture radar (SAR) pixels that may indicate potential war-induced infrastructure damages.
- Calculate total SAR damage pixels within an area of interest.
- Analyze SAR damaged areas over time.
- Fuse geospatial results with open-source reporting.

Technical components

- Software: ArcGIS Pro 3.4

- Datasets:
  - European Space Agency Sentinel-1 RTC
  - ArcGIS Living Atlas of the World
  - Alaska Satellite Facility (ASF) Vertex
  - United Nations Satellite Centre (UNOSAT)

Prerequisites

- Proficiency with ArcGIS Pro is required.
- Technical skill level: Intermediate GIS.
- Time to complete chapter: One to two hours without intelligence brief; three to five hours with intelligence brief.

Background and situational context

Human rights law and international humanitarian law oversee universal rights upholding human dignity and protection of war victims. Since Russia's 2022 invasion of Ukraine, reports of violations of international human rights and international humanitarian law have accompanied devastating civilian casualties. The process of gathering evidence to prove violations is extremely challenging in unsafe areas of ongoing war. Satellite remote sensing can address this challenge by providing a means to detect, monitor, and document potential violations in active war zones. A wide range of satellite missions has been successfully deployed in the past few decades. The selection of suitable satellite data sources is determined by application conditions and requirements. For example, optical imagery, commonly recognized as traditional satellite photos, is affected by time of day, weather, and clouds. Very high-resolution (< 1 meter) optical imagery is also often expensive, limiting access by human rights practitioners. For some human rights researchers, availability and affordability may be prioritized over high-resolution imagery.

Thus, this chapter explores the use of freely available, medium-resolution synthetic aperture radar (SAR) imagery from the European Space Agency's Sentinel-1 satellite. SAR is a powerful active remote sensing technology that provides day or night, all-weather-capable imagery that can penetrate clouds and smoke. These capabilities make SAR suitable for monitoring unpredictable activity in war zones where optical imagery may not be available. Changes in Earth's surface, such as war-induced infrastructure damage, can be detected by differences in radar backscatter between before and after images. These changes in backscatter intensity between images taken at different times can indicate potential war-induced damages where potential human rights violations have taken place. Geospatial analysis using this imagery intelligence can corroborate additional evidence of violations in international courts.

The following workflow explores SAR damage detection trends in the Ukrainian city of Mariupol to investigate potential human rights violations and promote justice for civilians in Ukraine. Results are then compared with UNOSAT data to validate the performance of the SAR analysis.

Exercise workflow 13

This exercise consists of three parts. In the first part, you will import data provided for this chapter into ArcGIS Pro in preparation for analysis. In the second part, you will conduct backscatter intensity change analysis between pairs of SAR images. In the third part, you will evaluate detected damages.

1 Import imagery and data into ArcGIS Pro

1. Open **ArcGIS Pro**. Under **New Project**, click **Map** to create a new map.
2. In the **New Project** window, apply the following settings:
    - **Name**: Chapter_13
    - **Location:** C:\SecurityFirst

    Note: This tutorial will use the folder C:\SecurityFirst.

    The first step in the analysis is to import data provided for this chapter into ArcGIS Pro.
3. In the **Catalog** pane, click the **Computer** tab and navigate to the location of the data folder for this chapter, **SecurityFirst_ExerciseData/Chapter13.** Right-click the folder and click **Add To Project.**
4. Click the **Project** tab and expand **Folders.** Expand **Chapter13.**
5. In the data folder, right-click **UNOSAT_Mariupol_Livoberezhyny_AOI_12May2022** and click **Add To Current Map.**

    This layer represents the area of interest (AOI) and consists of two residential districts within Mariupol. The Zhovtnevyi District is on the left and the Livoberezhnyi District is on the right.

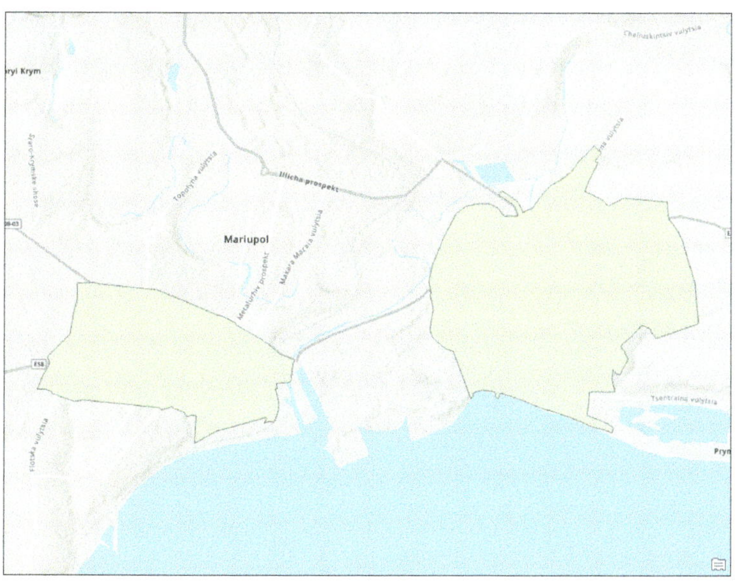

6. In the **Contents** pane, right-click the layer and click **Symbology**.
7. In the **Symbology** pane, click the **Symbol** and change it to **Black Outline (2 pts)**.
8. From the **Chapter13** folder, add the **UNOSAT_Mariupol_Livoberezhyny_DA_12May2022** shapefile to the map.

   The point data represents damages within the AOI reported by UNOSAT. Later, you will validate the UNOSAT data through a SAR analysis.
9. In the **Catalog** pane, click the **Portal** tab. Click the **Living Atlas** button to open ArcGIS Living Atlas of the World and search for Sentinel-1 RTC.
10. Right-click **Sentinel-1 RTC** and click **Add To Current Map**.

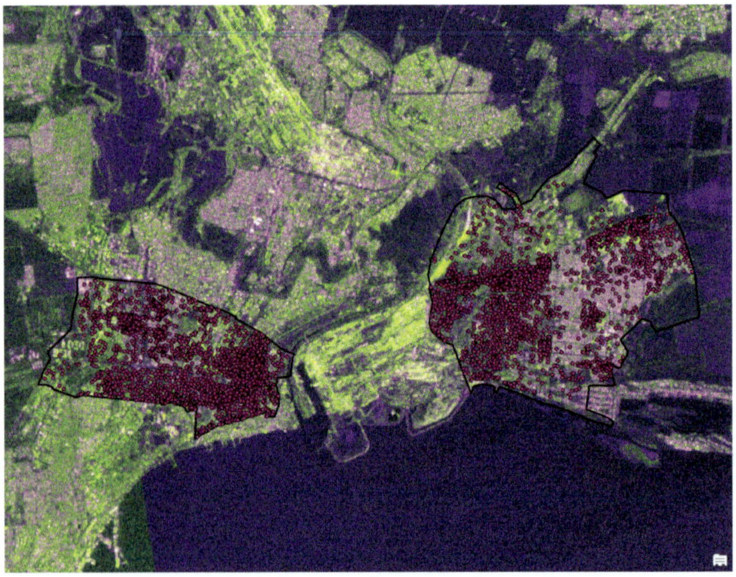

11. With the **Sentinel-1 RTC** layer selected in the **Contents** pane, click the **Data** tab on the ribbon. In the **Selection** group, click **Explore Raster Items**.
12. In **Raster Item Explorer**, apply the following settings:
    - **By Area Of Interest:** Display Extent
    - **By Attribute:** Field > acquisitiondate
    - **Start Date:** 2/16/2022 3:20:20 PM
    - **End Date:** 2/28/2022 3:20:20 AM

Tip: Select the Start Date and End Date from the lists.

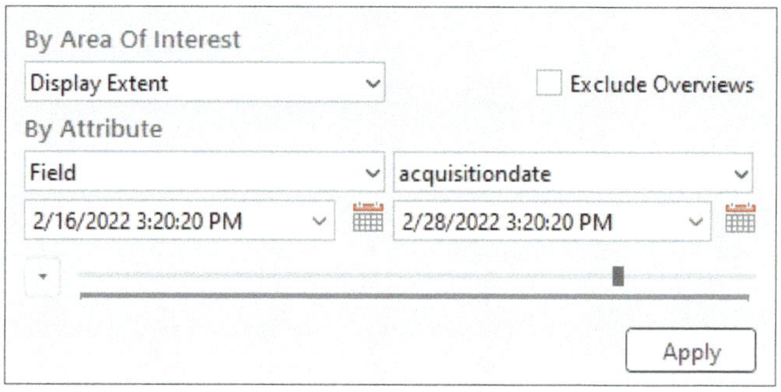

13. Click **Apply**.

    The query will produce five results; you will select the image acquired February 16, 2022.

14. Select the first image with an **acquisitiondate** of **2/16/2022**. On the corner of the image query, click **Add To Current Map**.
15. In the **Contents** pane, rename the image Feb 16 2022.
16. Repeat the preceding steps to select and add the image from February 28, 2022, to the map. Rename the image Feb 28 2022.
17. Select the **Feb 16 2022** imagery layer and click the **Data** tab on the ribbon. In the **Processing** group, open the **Processing Templates** from the list of options and select **VV Amplitude with Despeckle**.
18. Repeat this process for **Feb 28 2022**.

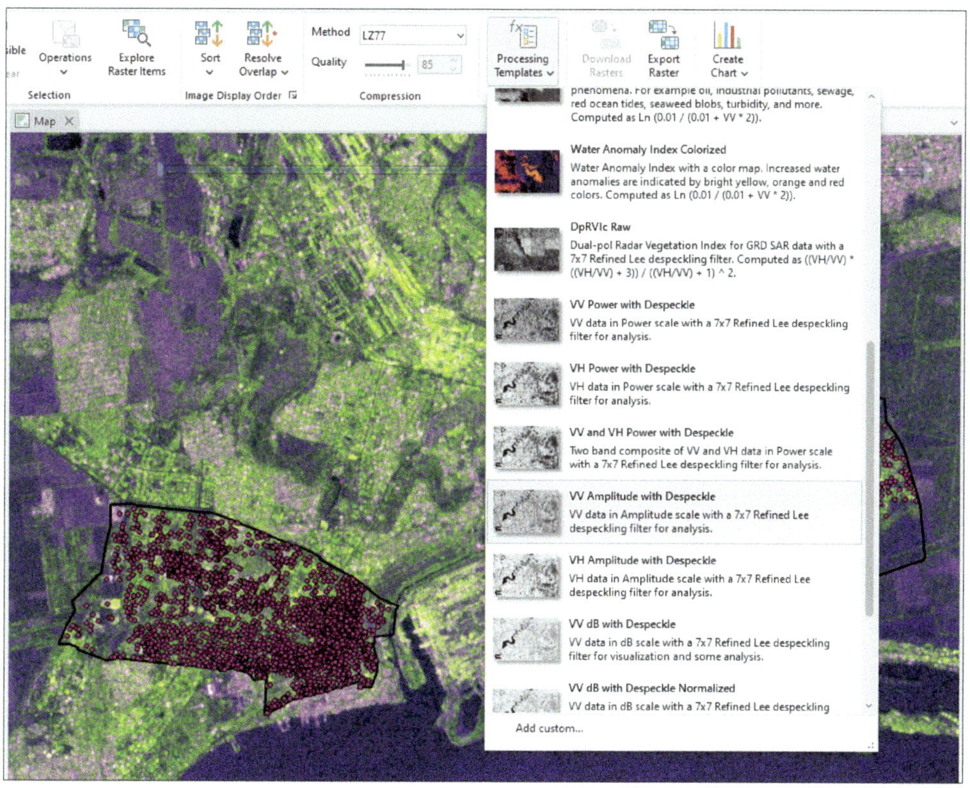

Tip: If your raster doesn't load after applying the Processing Template, it appears once you export the raster in the next steps.

19. In the **Contents** pane, right-click **Feb 16 2022** and select **Data** > **Export Raster**.
20. In the **Export Raster** window, add the following parameters:
    - **Output Raster Dataset:** Feb 16 2022 AOI.tif
    - **Geographic Transformation:** None
    - **Clipping Geometry:** UNOSAT_Mariupol_Livoberezhnyn_AOI_12May2022
    - **Use input features for clipping geometry:** <check the box>
    - **Clipping Type:** Outside
    - **Cell Size:** X = 10; Y = 10
    - **Pixel Type:** 32 Bit float
21. Click **Export**.
22. Repeat this process for **Feb 28 2022** and export the raster as Feb 28 2022 AOI.tif.

2 Calculate backscatter intensity change

Calculate backscatter differences

The next step will be to calculate the log difference between two Sentinel-1 RTC images to highlight areas of significant changes in backscatter over time.

1. In the **Catalog** pane, from the chapter data folder, expand the **ASF_Tools.pyt** toolbox.

2. Double-click the **Calculate Log Difference** tool to calculate the log ratio of pixel values from two images over the same area of interest.

> **Tip:** A detailed step-by-step process can be viewed at links.esri.com/SF-CalcLogDiffTutorial.

3. In the **Calculate Log Difference** pane, enter the following settings:
   - **Comparison raster**: Feb 28 2022 AOI.tif
   - **Reference raster**: Feb 16 2022 AOI.tif
   - **Destination folder for output file**: Chapter_13
   - **File name for output log difference file**: Feb 16 2022 AOI_Feb 28 2022 AOI_LogDiff.tif
4. Click **Run**.
5. Open the **Symbology** pane for the newly created log difference TIFF and set the symbology as follows:
   - **Primary symbology**: Classify
   - **Color scheme**: Red-Yellow-Blue (Continuous)

   Positive values (blue) indicate an increase in backscatter and negative values (red and orange) indicate a decrease in backscatter. Yellow represents little to no change.

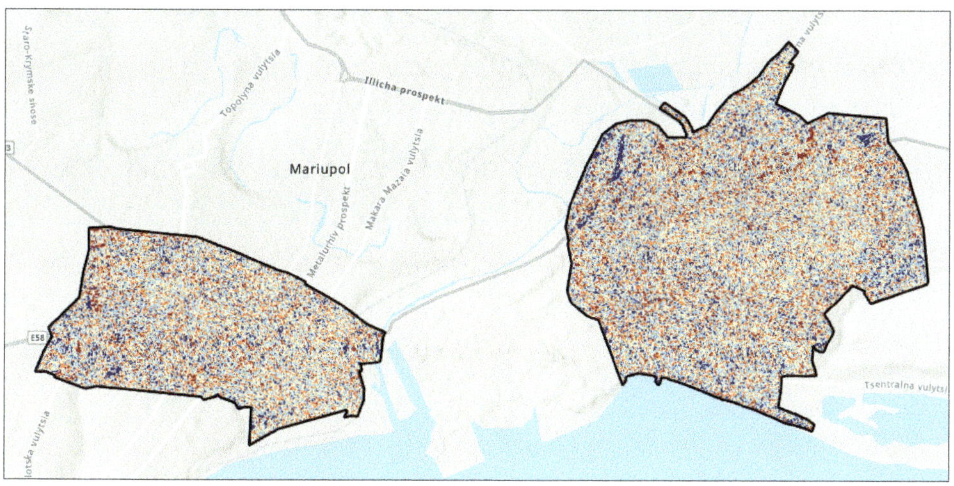

Apply map algebra and identify damage

In its current state, the log difference raster is not easy to interpret. The next step is to select a binary classification to view damaged and undamaged areas more easily. Based on the literature review on the interpretation of log difference radar data (Kristenson 2020a; Braun 2018; Van Heyningen 2018), you will use the standard deviation as a simplified threshold for determining changes.

For simplicity, you will not attribute significant positive or negative backscatter intensity change values to certain types of damage because of their dependence on building geometry and nature of collapse (Van Heyningen 2018). Any change in backscatter greater than one standard deviation from the mean, positive or negative, will be classified as potential damage.

1. On the ribbon, click the **Analysis** tab. In the **Geoprocessing** group, click **Tools**.
2. In the **Geoprocessing** pane, search for the Raster Calculator (Image Analyst Tools) tool. Open it.
3. In the **Raster Calculator** tool, create the following map algebra expression by double-clicking **Rasters** or **Tools** from the list and typing values:

```
("Feb 16 2022 AOI_Feb 28 2022 AOI_LogDiff" >= 0.11) | ("Feb 16 2022 AOI_Feb
28 2022 AOI_LogDiff" <= -0.083)
```

   In the current file (**Feb 16 2022_Feb 28 2022_LogDiff**), the mean is 0.013 and the standard deviation is 0.096. The expression is referred to either one standard deviation above (0.11 = 0.013 + 0.096) or below the mean (0.083 = 0.013 − 0.096).

4. Name the **Output raster** Feb162022_Feb282022_change.
5. Click **Run**.

> **Tip:** Detailed steps for finding the standard deviation and changing the raster color scheme can be viewed at links.esri.com/SF-CalcLogDiffTutorial.

Examine and interpret the damage raster

1. Select two colors to represent damaged and undamaged areas.

   The following example shows undamaged areas in gray and damaged areas in red. The damaged area is represented as 1 in pixel values, and the undamaged area as 0.

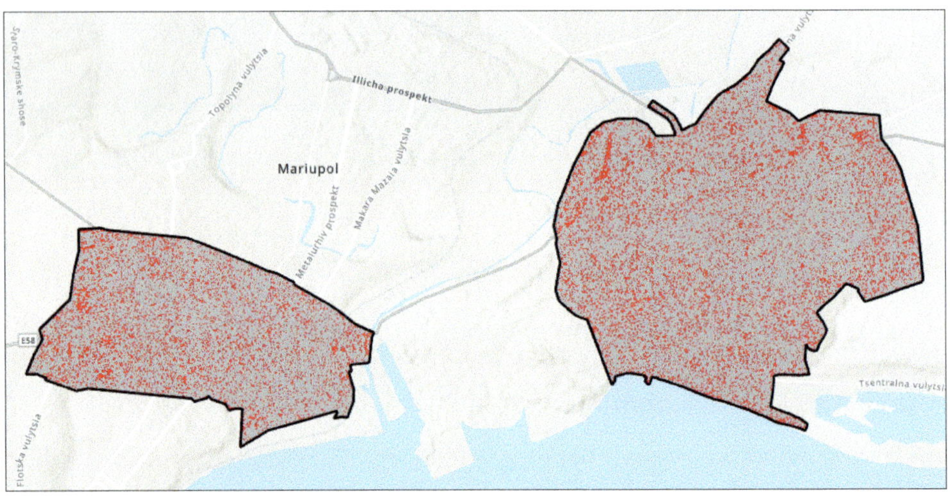

2. Zoom to the study area to examine damages in the study area.
3. In the **Contents** pane, turn on the **UNOSAT_Mariupol_Livoberezhyny_DA_12May2022** shapefile. Compare the damages between the UNOSAT point data vs. the results of the SAR analysis.

On your own

Examine images from April 29, 2022, and May 11, 2022, to determine the damage that occurred during this period. Results will show the progression of damage as the war continued.

3 Evaluate damages

In the second part, you visualized damaged areas by calculating and reclassifying backscatter intensity change. In the third part, you will quantify the mean percentage of damage within the UNOSAT AOI.

You will use the Summarize Categorical Raster tool to calculate the number of damaged areas and undamaged areas, which can be used to calculate the percentage of damaged pixels using the following equation:

$$\text{Damaged percentage (\%)} = \frac{\text{Damaged area}}{\text{Damaged area} + \text{Undamaged area}} \times 100$$

1. In the **Geoprocessing** pane, search for the Summarize Categorical Raster tool. Open it and apply the following settings:
   - **Input Categorical Raster:** Feb162022_Feb282022_change
   - **Output Summary Table:** FebChange_SummarizeCatRaster
2. Click **Run.**
3. In the **Contents** pane, scroll down to **Standalone Tables**. Right-click the new table and click **Open.**
4. Analyze the number of damaged and undamaged pixels.
5. On the ribbon, click the **Share** tab. In the **Output** group, click **Export Map.** Configure your preferred file type and specifications.

Interpretation of analysis

After completing this exercise, answer the following questions in detail. The answers to these questions, as well as the analysis results from the exercise workflow, will inform the intelligence brief. Answers are provided to questions 2, 4, and 6 to promote critical thinking.

1. Based on this analysis, what patterns do you observe when viewing the SAR analysis damages within the study area?
2. Did the detected damage levels increase, decrease, or stay the same between the two time periods examined in this chapter?
3. How do the results between the SAR analysis and the UNOSAT data compare? Do the datasets offer any insight to confirm or contradict damages within the study area?
4. What might you attribute as possible causes of the assessed damage levels?

5. What actionable information can be derived from this analysis? How might the UN, Ukraine, or human rights groups use this information?
6. What other intelligence sources might you use to enhance this geospatial analysis? Are there additional data sources that may be available to help gain a deeper understanding of this research?

Answers to select questions

2. Increase. Between February 16 and February 28, 2022, assessed damage was 27.2 percent. Between April 29 and May 11, 2022, assessed damage was 29.6 percent.

4. Analysis of SAR backscatter intensity change shows increased damage during the period of April 19 to May 11, 2022, compared with damages during the period of February 16 to February 28, 2022.

6. Open-source reports of increasing levels of conflict after Russia's invasion of Ukraine on February 24, 2022. Other intelligence sources that support this analysis include media (www.wsj.com/articles/russia-presses-offensive-as-ukrainians-try-to-evacuate-11646819525), social media (https://twitter.com/AP/status/1655675564743598112), and gray literature (https://crsreports.congress.gov/product/pdf/LSB/LSB10710). These sources can help build a timeline of events and complement imagery intelligence and geospatial analysis to better understand conflict in Ukraine. Such multidisciplinary intelligence fusion can help human rights researchers detect and monitor possible human rights violations to hold aggressors accountable in the International Criminal Court.

Intelligence brief

Based on the analysis from the exercise workflow, produce a three- to five-page intelligence brief by responding to the following question prompts. Incorporate your geospatial outputs from the exercise to enrich and illustrate your conclusions.

- **BLUF:** List three to five primary observations from your analysis in bulleted-list format that summarize findings of this work.
- **Top-level analysis:** Report the primary data points supporting your BLUF.
- **Secondary analysis:** Provide the data sources and deeper analytic products from this analysis.
- **Impact:** Describe the impacts of your findings on the department or organization.
- **Confidence:** Report the level of confidence of your assessment based on available data, accuracy, and so on.
- **Bias:** Explain any background or information you have that could have affected your analysis.
- **Conclusions:** Determine what type of analytic conclusions there may be and how the data requirements, analysis, and brief writing may change based on what analytic framework you are following (descriptive, predictive, prescriptive).

Applying and extending your skills

Although the field of SAR imagery is growing, more research is needed to refine techniques to help the human rights community. This chapter introduced basic SAR imagery analysis concepts using ArcGIS Pro with the goal of engaging human rights researchers and organizations that may not have advanced GIS skills. Professionals can refer to the ArcGIS Pro Image Analyst Synthetic Aperture Radar Toolset for more advanced analysis.

UNOSAT maintains a map library with public data produced in response to humanitarian emergencies, including natural disasters and conflict, at https://unosat.org/products.

This data is derived from very high-resolution (< 1 meter) optical imagery. You can download data from an event of interest and view it in ArcGIS Pro. Advance your skills by comparing UNOSAT data derived from optical imagery with SAR imagery. Keep in mind that although very high-resolution optical imagery offers several benefits, it is not always available and therefore may be unsuitable for some researchers during ongoing conflict.

Optional steps for downloading SAR imagery directly

Tip: This approach can be taken for regions of interest that are larger than 4,000 × 4,000 pixels. The ArcGIS Living Atlas image service is limited to AOIs of 4,000 × 4,000 pixels.

- Download the ASF Vertex Python toolbox for ArcGIS Pro. You can use the ASF Data Search Vertex to download Sentinel-1 SAR imagery.
- Create a free account and access ASF Vertex: https://search.asf.alaska.edu/#/.
- SAR imagery may contain distortions owing to the satellite's side-looking sensor. A process called radiometric terrain correction (RTC) stabilizes backscatter values to reduce geometric distortions that can lead to geolocation errors. This can be a time-consuming and expertise-demanding process, and the ASF provides downloading RTC imagery ready for GIS. In ASF Vertex, draw an AOI over the city of Mariupol, Ukraine.
- Sentinel-1 imagery is collected on a 12-day revisit rate corresponding to the temporal resolution of the satellite. Ensure that you select the same satellite footprint over the AOI to maintain consistency.
- Select the following dates for RTC: Feb. 16, 2022; Feb. 28, 2022; April 29, 2022; May 11, 2022.
- For detailed instructions on drawing the AOI box and downloading RTC imagery, see links.esri.com/RTCOnDemand.
- Select the following processing options:
  - Radiometry: gamma0
  - Scale: amplitude
  - Pixel Spacing: 30 meters
  - DEM Name: Copernicus DEM
- The RTC process and download procedure might take more than two hours, depending on the internet connection and computing environment. Download the images once RTC is complete.

Additional resources

- For an introduction to SAR, see ArcGIS Pro online Help (www.esri.com/en-us/arcgis/products/arcgis-pro/overview) and search Resources > Help > Analysis and Geoprocessing > Image Analyst > Synthetic Aperture Radar.
- To use the NASA Alaska Satellite Facility (ASF) Data Search Vertex, see https://search.asf.alaska.edu/#/.
- For an introduction to the ASF Data Search Vertex, see https://docs.asf.alaska.edu/vertex/manual/.
- For more information about the NASA ASF ArcGIS Python toolbox, see https://asf.alaska.edu/how-to/data-tools/gis-tools/.

References

- Amnesty International. 2022. “Key Recommendations for EU Leaders.” www.amnesty.org/en/wp-content/uploads/2022/03/TIGO_IOR_10_2022_2787_Ukraine-and-Russia_Amnesty-Internationals-Key-Recommendations-for-EU-leaders_22032022.pdf.
- Bradley, Bethany. 2021. “Summarizing Categorical Raster in Polygons.” YouTube video on zonal statistics. www.youtube.com/watch?v=_QOi6YN2dLU.
- Braun, Andreas. 2018. “Assessment of Building Damage in Raqqa During the Syrian Civil War Using Time-Series of Radar Satellite Imagery.” *GI_Forum* 6 (1): 228–42. doi.org/10.1553/giscience2018_01_s228.
- Human Rights Council. 2023. *Report of the Independent International Commission of Inquiry on Ukraine*. www.ohchr.org/sites/default/files/documents/hrbodies/hrcouncil/coiukraine/A_HRC_52_62_AUV_EN.pdf.
- Kristenson, Heidi. 2020a. “Log Difference Tool.” Alaska Satellite Facility. https://storymaps.arcgis.com/stories/743d1126594b46e0b43be36473dba0e3.
- Kristenson, Heidi. 2020b. “RTC on Demand! Working with Radiometrically Terrain Corrected Sentinel-1 SAR Datasets from the Alaska Satellite Facility.” ASF. https://storymaps.arcgis.com/stories/2ead3222d2294d1fae1d11d3f98d7c35.
- Masters, Jonathan. 2023. “Ukraine: Conflict at the Crossroads of Europe and Asia.” Council on Foreign Relations. www.cfr.org/backgrounder/ukraine-conflict-crossroads-europe-and-russia.
- Mulligan, Stephen P. 2022. “The Law of War and the Russian Invasion of Ukraine.” Legal sidebar of the Congressional Research Service. https://crsreports.congress.gov/product/pdf/LSB/LSB10710.
- Van Heyningen, J. C. 2018. “Rapid Building Damage Detection Through SAR Timeseries Analysis in the Google Earth Engine.” links.esri.com/SARDamageDetection.

About the authors

Rebecca Bosworth is an officer in the United States Space Force. She has a bachelor's degree in atmospheric science from the University of California, Berkeley, and a master's degree in geographic information science and technology from the University of Southern California.

Dr. Yi Qi is an associate professor in the Spatial Sciences Institute at the University of Southern California. His teaching and research are focused on remote sensing and spatial data science. He has a PhD in geography from the University of Utah.

Part 4

Safer seas

Chapter 14

Assessing illegal fishing and monitoring maritime activities with dashboards

Diana Ter-Ghazaryan, *University of Southern California*
Bruce Vitor, *Florida International University*

Overview

In this chapter, you will learn how to use ArcGIS Online to assess illegal, unreported, and unregulated (IUU) fishing in the Eastern Tropical Pacific Marine Corridor (CMAR) in the context of HSGI. You will also learn how to build a dashboard to monitor suspected IUU fishing activity and other maritime activities. The workflows introduced and the final deliverable will provide actionable information to help guide efforts against IUU fishing activities.

Learning objectives

- Add and publish spatial and nonspatial CSV datasets to ArcGIS Online.
- Perform exploratory data visualization of maritime activities to identify potential locations of IUU fishing activity in the CMAR.
- Apply geospatial filters to sort and analyze patterns of maritime activities that may constitute IUU fishing.
- Build a dashboard in ArcGIS Dashboards to monitor potential IUU fishing and other maritime activities near exclusive economic zones (EEZs) of countries.

Technical components

- Software:
  - ArcGIS Online (at least Publisher level access)
  - ArcGIS Dashboards
- Datasets:
  - ArcGIS Living Atlas of the World basemap
  - World Exclusive Economic Zone (EEZ) boundaries
  - Three CSV datasets from Global Fishing Watch (loitering events, encounter events, and a list of vessels involved in transshipment)

Prerequisites

- Basic knowledge of or experience with ArcGIS Online Map Viewer
- Technical skill level: Introductory GIS
- Time to complete chapter: One to two hours without intelligence brief, three to five hours with intelligence brief

Background and situational context

IUU fishing threatens marine ecosystems, exacerbates food insecurity and regional instability, is linked to major human rights violations, and is a global issue with national security implications. Some 90 percent of the fisheries in the world are overfished or fully fished. IUU fishing is the sixth most common illicit activity, estimated to be worth between $15 billion and $36 billion annually. It is the third most lucrative natural resource crime, following timber and mining, and an estimated one in five fish eaten by Americans is illegally caught (Ian Urbina Subcommittee on Water, Oceans, and Wildlife 2019). The US Southern Command commander stated in her 2023 posture statement that IUU fishing is one of the top security concerns by partner nations, rivaling transnational organized crime and cybersecurity threats (US Southern Command 2023).

IUU fishing is tremendously complicated. According to INTERPOL, any IUU fishing case potentially has nine different countries with jurisdiction: (1) vessel's flag state, (2) coastal state, (3) company owner, (4) captain's nationality, (5) crew nationalities, (6) corruption or money-laundering individuals or organizations, (7) import states, (8) export states, and (9) port states (INTERPOL 2018). The Food and Agriculture Organization (FAO) International Plan of Action to Prevent, Deter, and Eliminate Illegal, Unreported, and Unregulated Fishing, passed in 2001, provides a framework to define IUU fishing, which includes activities conducted by individuals, vessels, or companies that violate national or international fishing regulations or operate outside the framework of applicable laws.

Illegal fishing refers to fishing activities that violate national or international laws, including fishing without a valid license or permit, fishing in prohibited areas, fishing with prohibited gear, exceeding catch quotas or size limits, and targeting protected or endangered species. Unreported fishing involves fishing activities that are not reported or are underreported to relevant authorities, making it difficult to monitor and manage fish stocks effectively, including failing to report, discarding or hiding catches, or misreporting catch locations or species. Unregulated fishing refers to fishing activities conducted outside the regulatory framework or in areas where no regulations or conservation measures exist. It often occurs in international waters or in regions with weak governance, where vessels can operate without restrictions and exploit fish stocks. All three practices can lead to overfishing and depletion of fish populations, disrupt marine ecosystems, and pose a threat to food security and the livelihoods of coastal communities, especially in places experiencing food insecurity (Food and Agriculture Organization of the United Nations 2001). Combating IUU fishing requires international cooperation, effective monitoring and surveillance systems, and stronger legislation and enforcement, as well as efforts to promote sustainable fishing practices and responsible fisheries management.

Transshipment refers to the direct transfer of any quantity of fish onboard from one vessel to another vessel, regardless of the location of the event, without the fish being recorded as landed. It is a key indicator of IUU fishing and other criminal activities. Transshipments largely happen out of sight of flag, coastal, and port states and regional fishery management offices (RFMOs). Aggregating fish makes it difficult to know which ones are legal or illegal and contributes to unreported fishing (Miller et al. 2018).

Determining if a vessel is loitering can help us understand if a vessel is involved in transshipment (Food and Agriculture Organization of the United Nations 2023). Global Fishing Watch (GFW) defines loitering as an occurrence “when a single vessel exhibits behavior indicative of a potential encounter event.” Other activities, such as maintenance or waiting for permission to return to port, could result in a vessel loitering, but loitering is considered one of the best indicators of transshipment. For investigators, using the vessels’ automatic identification system (AIS) data, such as vessel speed, length of time at a location, and proximity to shore, adds important context to what may be occurring (Global Fishing Watch, n.d.).

This exercise will explore loitering and encounter events near EEZs of several countries in the CMAR. The final few steps will consist of building a dashboard in ArcGIS Dashboards to explore data and patterns dynamically.

Exercise workflow 14

1 Import and join the datasets

GFW is a collaborative website that enables users to monitor global fishing activity using data received from AIS trackers on vessels (more details on AIS in this chapter). The website became publicly available in 2016. You will import datasets downloaded from GFW (https://globalfishingwatch.org/datasets-and-code) and move the data into ArcGIS Online to prepare for analysis. The three GFW datasets are loitering, encounters, and transshipment vessels. GFW created these datasets through processing with machine learning more than 32 billion AIS signals broadcast from ships at sea.

1. In your browser, navigate to www.arcgis.com. Sign in to your ArcGIS account.
2. On the navigation bar, click **Content**.
3. On the left, click **New item**.
4. In the **New item** window, from **SecurityFirst_ExerciseData/Chapter14**, add the **loitering_events.csv** file.
5. Confirm that the option to **Add loitering_events.csv and create hosted feature layer or table** is selected and click **Next**.

   The **loitering_events** and **encounter_events** datasets contain geographic coordinates, whereas the transshipment vessel dataset does not. You will complete an attribute join to add the transshipment vessel dataset later in this workflow.
6. In the next window, review the field data types. For the **transshipment_mmsi** field, change the **Type** to **String**. Keep the remaining default selections and click **Next**.

   Setting this field to use a string data type will allow that field to later be joined to another dataset's field, which also uses a string data type.

   Next, you will specify the location settings for the new feature layer.
7. In the next window, click the down arrow for **None**. From the list, select **Latitude and longitude**.
8. For **Latitude**, select **starting_latitude**. For **Longitude**, select **starting longitude**. Click **Next**.

   Finally, you will name the dataset and publish it.
9. In the next window, add the following settings and click Publish:
   - **Title**: Loitering_<your initials>
   - Folder: SecurityFirst
   - **Tags**: loitering, Security First
   - **Summary**: This is a dataset summarizing loitering events.

   You will now repeat the steps to add the **encounter_events** dataset.
10. Create a new item and add the **encounter_events.csv** file. Specify the field data types as follows and click **Next**:
    - **fishing_vessel_mmsi**: String
    - **transshipment_vessel_mmsi**: String
    - **start_time**: Date
    - **end_time**: Date
    - **mean_latitude**: Double
11. In the **Location settings** window, add the following settings and click **Next**:
    - **Latitude**: mean_latitutde
    - **Longitude**: mean_longitude

12. In the next window, add the following settings and click **Publish**:
    - **Title**: Encounters_<your initials>
    - **Folder**: SecurityFirst
    - **Tags**: encounter events, Security First
    - **Summary**: This is a dataset summarizing encounter events.

    You will now repeat the steps to add the **transshipment_vessels** dataset.

13. Create a new item and add the **transshipment_vessels.csv** file. Specify the field data types as shown and click **Next**:
    - **mmsi**: String
    - **shipname**: String
    - **callsign**: String
    - **flag**: String
    - **imo**: IMO, string

14. In the **Location settings** window, keep the option as **None** (this CSV contains no location data and will be added as a table). Click **Next** to upload the dataset following the same naming conventions as the previous items.

    You now have three new datasets: **Encounters**, **Loitering**, and **Transshipment Vessels**.

15. On the navigation bar, click **Map** to open a new map.
16. In the **Layers** pane, click **Add**.
17. In the **Add layer** pane, locate and add the **Loitering** and **Encounters** datasets.

    These are geospatial layers, and the **Transshipment Vessels** dataset is a table.

18. On the **Contents** toolbar, click **Tables**. Click **Add table**.
19. Locate the **Transshipment Vessels** dataset and click **Add**.

    Your map shows the points that represent the **Loitering** and **Encounters** datasets. It is also connected to the **Transshipment Vessels** table.

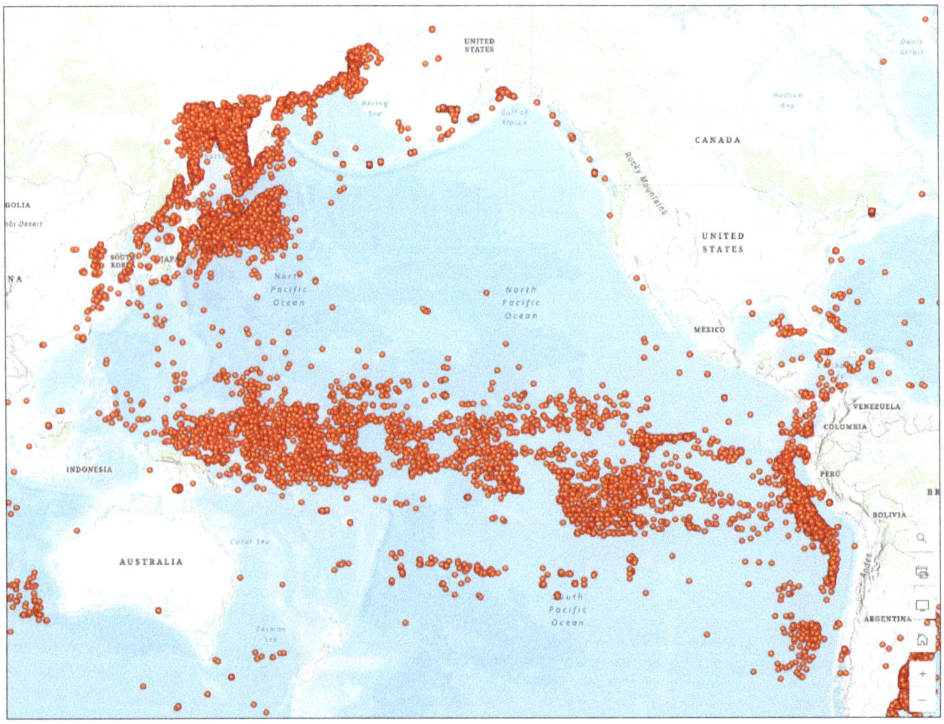

20. On the **Contents** toolbar, click **Save and open**. Save your map with the following settings:
    - **Title**: Global Transshipment_<your initials>
    - **Folder**: SecurityFirst
    - **Tags**: transshipment, Security First
    - **Summary**: This is a dataset summarizing global transshipment events.
21. Click **Save.**
22. Review both layers and the table.

    The **Loitering** dataset is large, with more than 17,000 individual points. The **Transshipment** table lists vessel flags under the **Flag** field. To visualize the loitering events according to the flag state of each vessel, you will complete an attribute join based on the **mmsi** field.

    > **Tip:** The Maritime Mobile Service Identities (MMSIs) are nine-digit numbers used by the AIS and other equipment to uniquely identify a ship or a coast radio station. The International Telecommunications Union in Geneva, Switzerland, regulates and manages MMSIs, much like radio call signs.

23. On the **Settings** toolbar, click **Analysis.**
24. In the **Analysis** pane, click **Tools.** Under the **Summarize data section**, select **Join Features.**

25. Configure the settings for the **Join Features** tool as follows:

    - **Target layer:** Loitering
    - **Join layer:** Transshipment Vessels
    - **Target field:** transshipment_mmsi
    - **Join field:** mmsi
    - **Output name:** LoiteringJoin_<your initials>

    Before you run the tool, you will estimate the amount of credits that this analysis operation will cost.

26. At the bottom of the tool, click **Estimate credits**.

    Note: At full extent, this join operation will consume about 18 credits.

    > **Tip:** You can reduce credit usage by limiting the processing extent or the dataset extent. To reduce the processing extent, zoom to a desired smaller area. In the tool, open Environment Settings, and under Processing Extent, select Display Extent. If that number is too high still, you can zoom in more.

27. Click **Run**.

    The new layer is added to your map automatically.

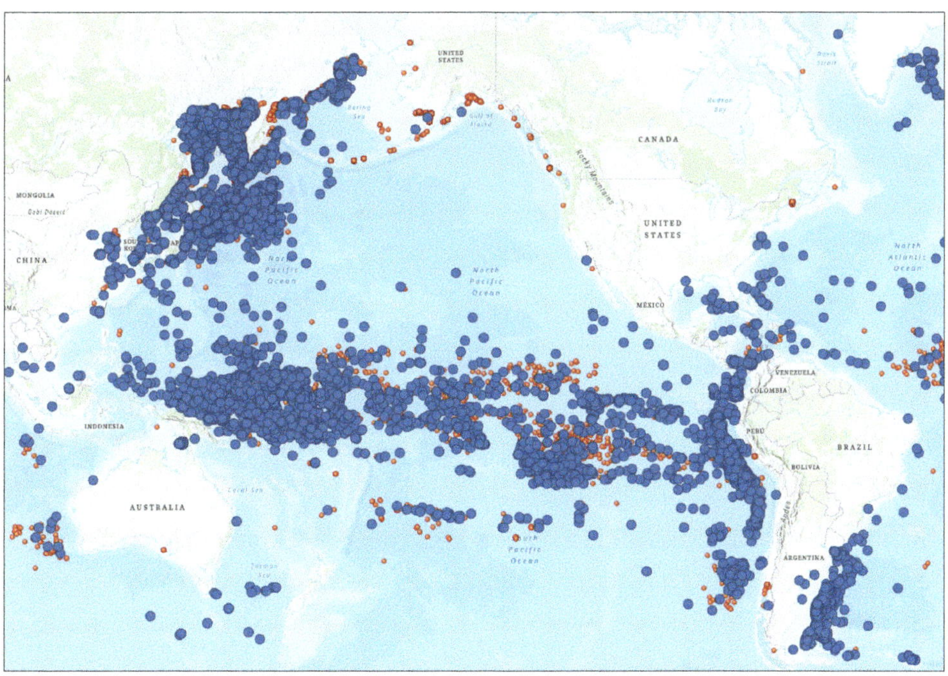

2 Visualize and quantify patterns

The newly joined dataset now contains information about the vessels, including the flag.

1. In the **Layers** pane, click the **Visibility** button to turn off every layer other than **LoiteringJoin**.

   You will visualize the dataset based on a few attributes.

2. On the **Settings** toolbar, click **Styles**.
3. In the **Styles** pane, under **Choose attributes**, click **Fields**.
4. From the **Select fields** list, select **flag** and click **Add**. Click **Done**.

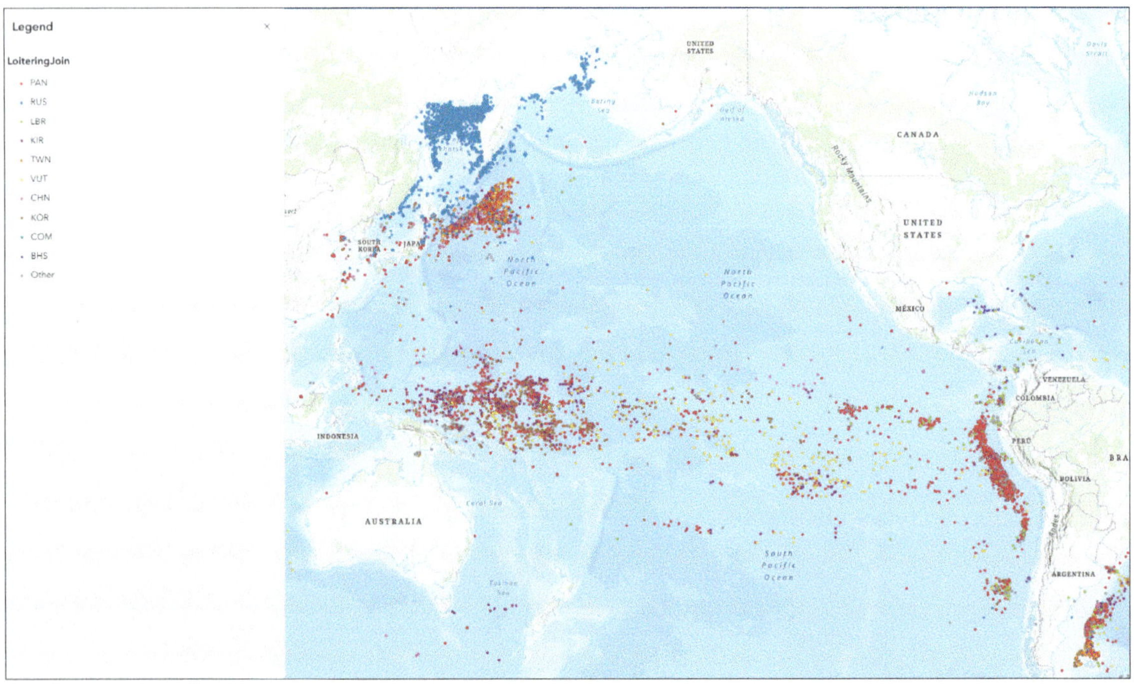

5. In the **Layers** pane, click the **Options** button for **LoiteringJoin** and click **Save**.

   This will save the new flag symbology. Patterns of loitering events are emerging based on country flag. There is a clear pattern of vessels bearing the Russian flag near Russia's shores, but there isn't another clear pattern at first glance. Next, you will generate a pie chart that shows vessel registration by country.

6. In the **Layers** pane, select **LoiteringJoin**. In the **Settings** pane, click **Configure charts**.
7. In the **Charts** pane, click **Add chart**. From the list, select the **Pie chart**.
8. For **Category**, select **flag**.

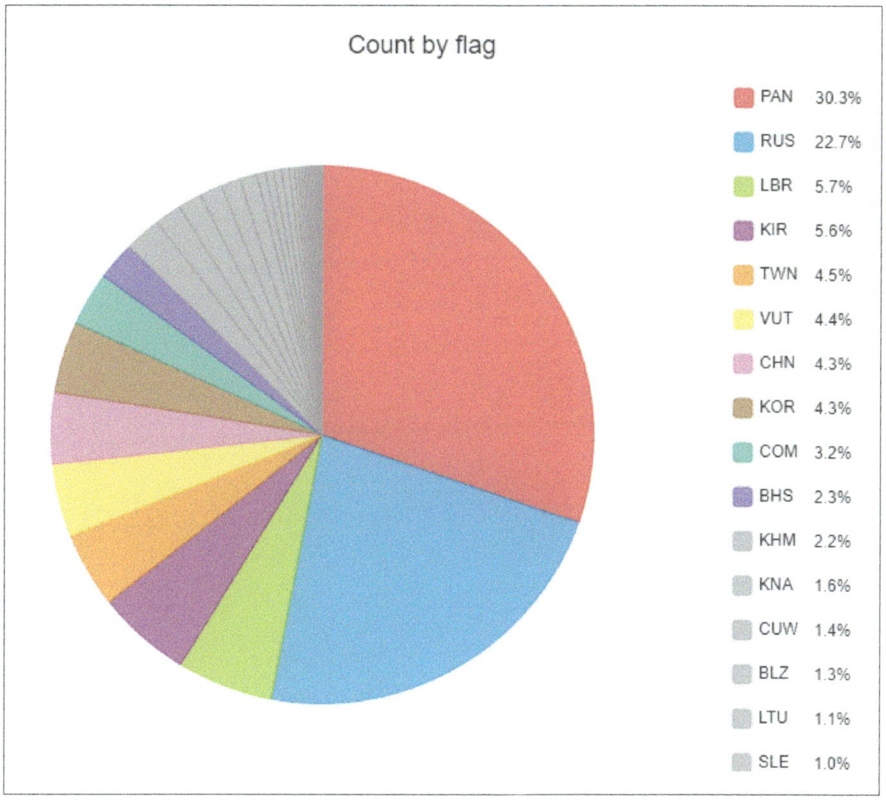

More than 25 percent of vessels in this dataset bear the Panamanian flag. Some refer to this as a "flag of convenience," for a flag other than the country of ownership. To learn more about this phenomenon, see links.esri.com/SF-Flags.

Next, you will investigate patterns around South America in the CMAR territory. The **Encounters** dataset documents occurrences of transshipment-capable vessels drifting slowly enough and long enough to receive a transfer of catch from one or more fishing vessels that are actively transmitting an AIS signal.

The **Encounters** dataset uses the **duration_hr** field to specify the amount of time that the encounter lasted.

9. Adjust layer visibility so that **Encounters** is the only visible layer and select it.

10. In the **Styles** pane, click **Field.** In the **Select fields** pane, select **duration_hr** and click **Add.**

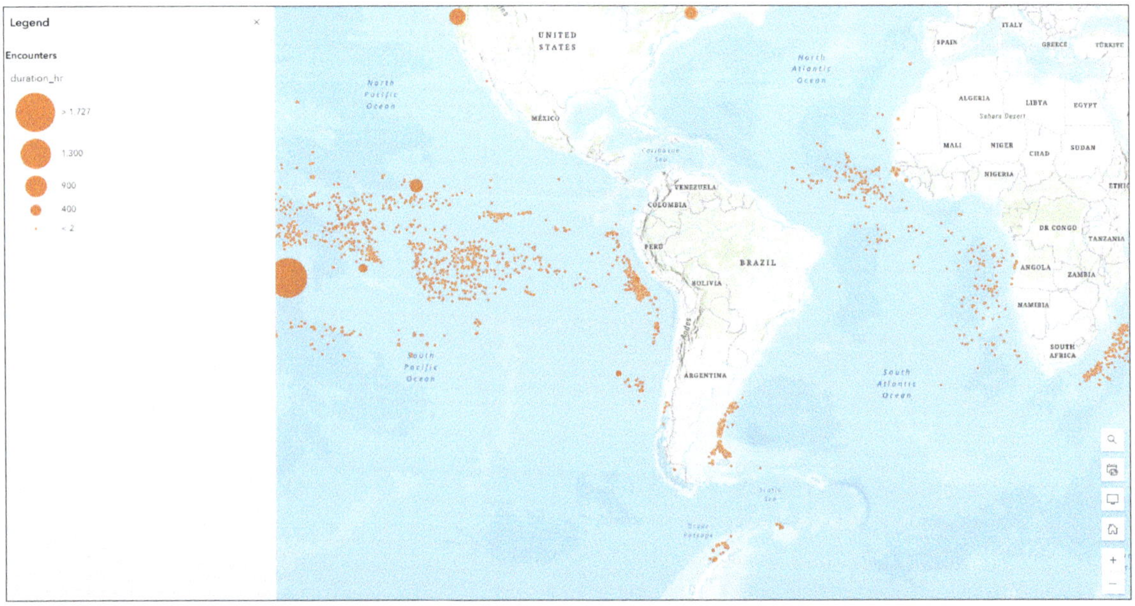

Patterns are starting to emerge—clusters of encounter activity to the west of Peru and to the east of Argentina. Curiously, the length of most encounters in this part of the world is relatively short (you may want to generate a chart to visualize this).

The **Encounters** dataset contains potential transshipment events that are deemed legitimate because both the transshipment and the fishing vessel are transmitting AIS signals. The **Loitering** dataset, on the other hand, includes occurrences of transshipment-capable vessels drifting slowly enough and long enough to receive a transfer of catch but during which time no fishing vessels are transmitting AIS signals. Some of these loitering events are suspected to be instances of transshipment from fishing vessels that have been engaged in illegal fishing.

11. Adjust visibility to see the **Encounters** and **Loitering** datasets on the map. Visually investigate the patterns that emerge while looking at both layers.

    Next, you will turn your attention to the **Loitering** dataset and add countries' EEZs to investigate these clusters of encounters further.

12. Turn off visibility for the **Encounters** and **Loitering** layers and turn on the **LoiteringJoin** layer.
13. In the **Layers** pane, click **Add.**
14. In the **Add layer** pane, next to **My Content**, click the list and change the selection to **Living Atlas.**
15. In the search bar, type World Exclusive Economic Zone Boundaries. Locate the layer in the results and click **Add.**

16. Click the back arrow to return to the **Layers** pane. Expand the **World Exclusive Economic Zone Boundaries** group layer and turn off visibility for the **Sovereign Country** layer.

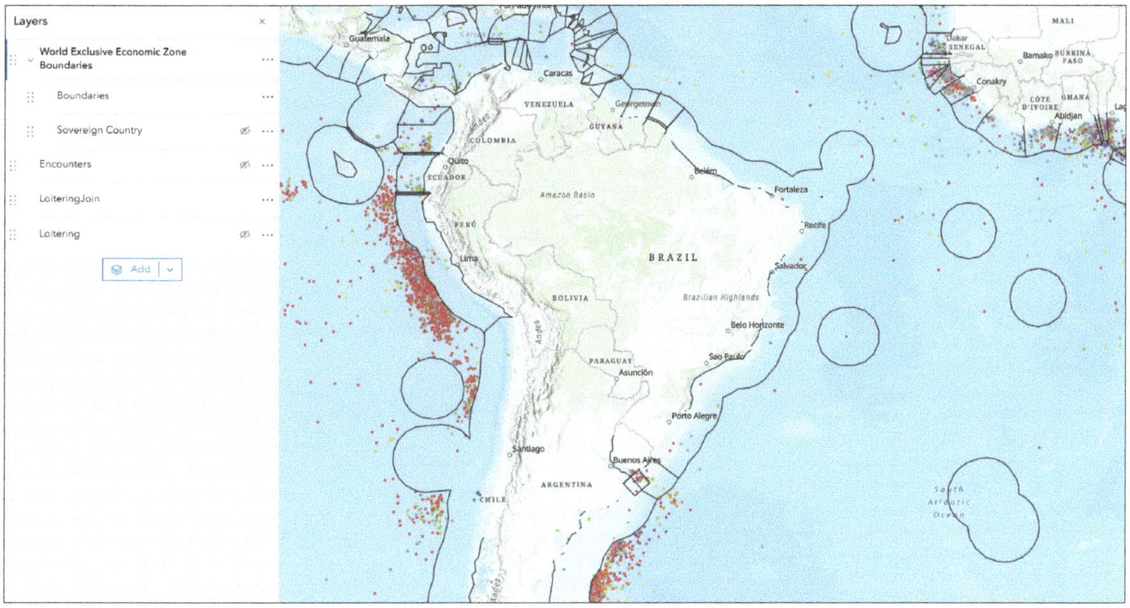

You can see that many loitering events happen just outside of countries' EEZs, although some instances of loitering also occur within EEZs. Much has been written about various countries' distant fishing fleets overfishing the high seas and coming dangerously close to other countries' EEZs (see the references and additional resources for more on this).

Next, you will perform some analysis to quantify the scale of this challenge. While running the tool, you will request useful summary statistics (majority and minority values and percentages per field) to get an idea of which country flags have loitered in specific country EEZs.

17. In the **Analysis** pane, open the **Summarize Within** tool and configure it as follows:

- **Input features**: LoiteringJoin
- **Summary polygon layer**: World Exclusive Economic Zone Boundaries–Sovereign Country
- **Group by field**: flag
- **Include minority and majority values:** <check the box>
- **Include percentage of summarized feature:** <check the box>
- **Output name**: Loitering_EEZ_Stats
- **Save in folder**: SecurityFirst

18. Estimate the credits and click **Run** to execute the tool.

19. Adjust layer visibility so that the **Loitering_Join** layer is visible. Investigate the resulting layer.

    When you click individual summary results on the map, a pop-up includes a data table and a pie chart that breaks down the flags of vessels loitering in a specific EEZ.

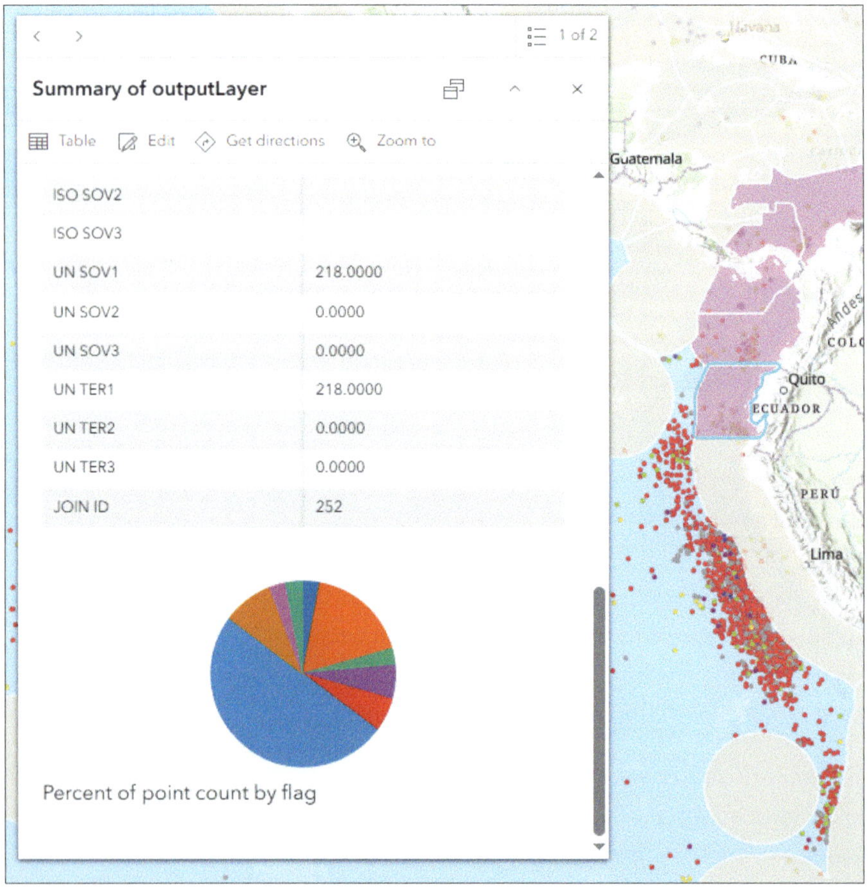

The pie chart is an interactive object, and labels for country flags pop up with the cursor. The majority flag within the Peruvian EEZ is Liberia, with Panama not far behind. Both of these are considered flags of convenience. This dataset is a great candidate for a dashboard to investigate these patterns interactively, especially when dynamically updated data is available.

Tip: You could extend this analysis by using the Summarize Nearby tool as well.

20. **Save** the map.

3 Create a dashboard

You will now create a dashboard in ArcGIS Dashboards with some of the datasets that you generated in the previous steps.

1. Create a new map called Loitering by EEZ for Dashboard. From **My content**, add the **Loitering_EEZ_Stats** layer and the **LoiteringJoin** layer. Save this new map.

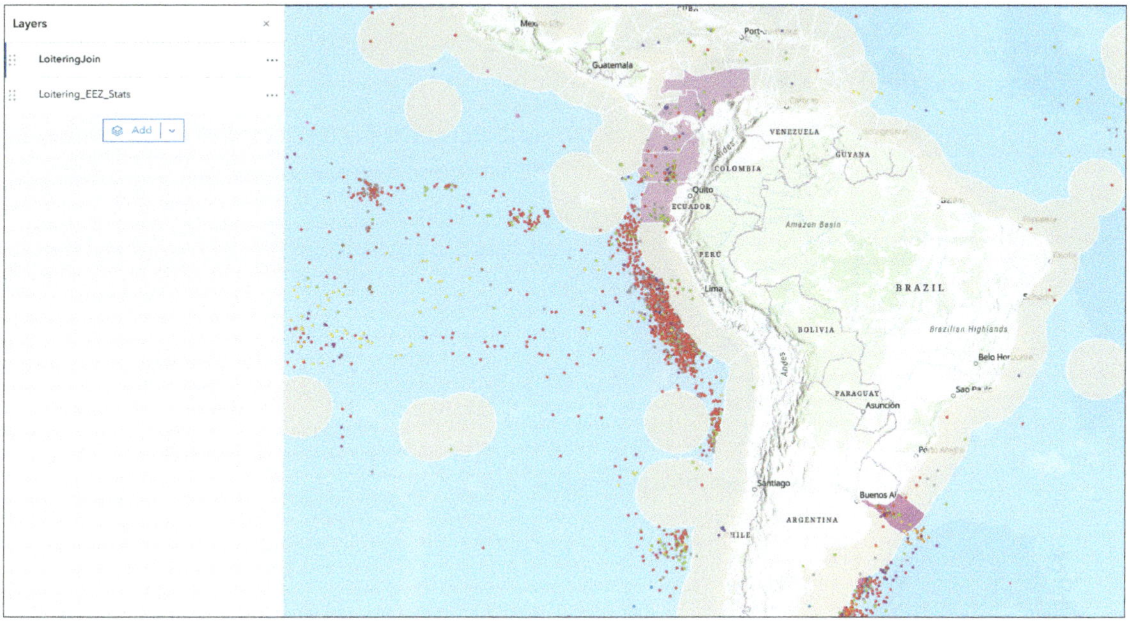

2. On the top navigation bar, next to your profile picture, click the **App Launcher** button and click **Dashboards**.

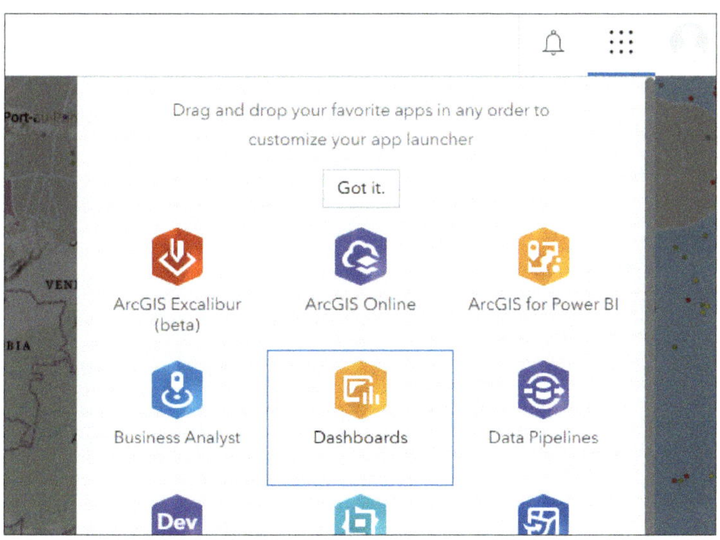

A new window opens.

3. Click **Create dashboard.** Complete the dashboard details as follows:
   - **Title**: Loitering by EEZ
   - **Tags**: Security First
   - **Summary**: This is a dashboard to explore loitering events near countries' EEZs.
   - **Folder**: SecurityFirst
4. Click **Create dashboard.**

   You now have a new dashboard. First, you will add the web map that you created to serve as the main map in the dashboard.
5. In the center of the new dashboard, click the **Add elements** button. From the list, select **Map.**

   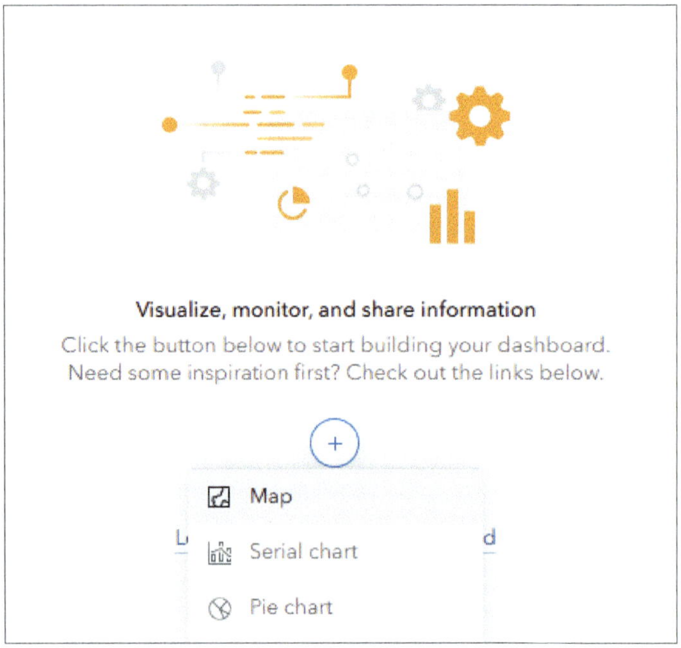

   You are redirected to your ArcGIS Online content, where you can add any of your web maps. Other web maps should be used with caution because you never know when the owner of the web map may delete it or stop updating the layers. It is always safest to rely on your own web maps for dashboards and other ArcGIS Online apps.
6. Locate your **Loitering by EEZ for Dashboard** map and select it.

> **Tip:** Under Settings, various choices for customizing your dashboard are provided. Experiment with these choices. You do not need to make your final choices at this moment—you can always go back and edit the dashboard to modify the theme, add elements, and so on.

7. Accept the default settings for now and click **Done**.

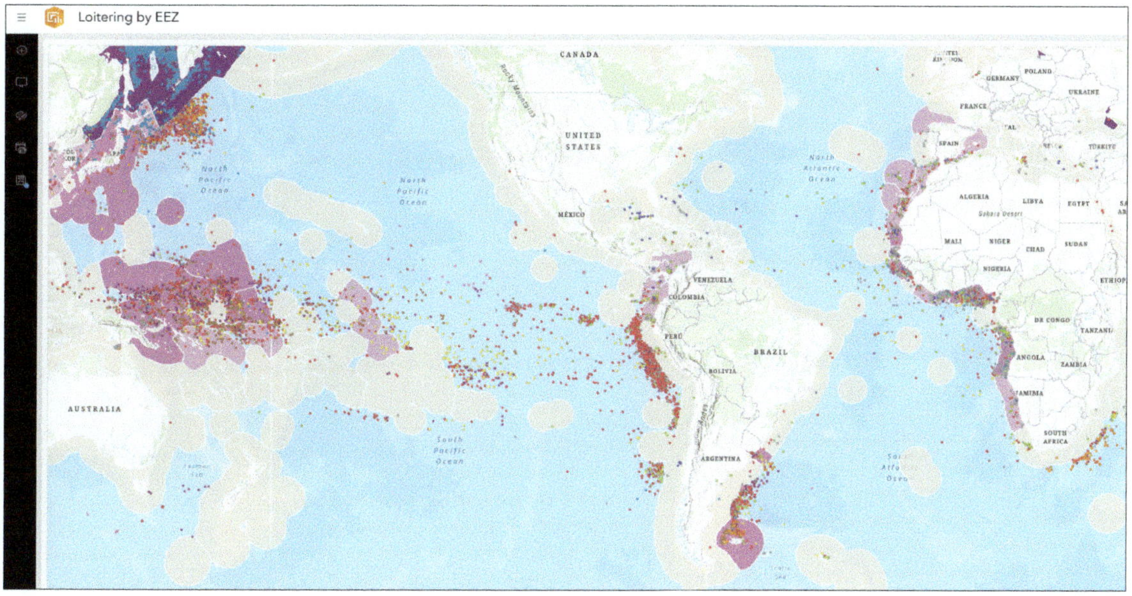

Next, you will add a few elements.

8. On the left toolbar, click the **Add element** button.

   Several plus signs appear on the dashboard.

9. Click the plus sign on the right side of the dashboard to add a new element in that location. Select the **Pie chart** element.

   Earlier, you generated a pie chart to represent the vessel flags in the **LoiteringJoin** dataset. Now you will create a similar pie chart with the benefit of interactivity and dynamism.

10. In the **Select a layer** window, select **LoiteringJoin**.

11. Configure the pie chart as follows:

    - **Category field**: flag
    - **Statistic**: Count

> **Tip:** On the left of the Pie chart window are many customization options on the tabs. You can try these if you want.

12. Click **Done**.

    The new chart is added on the right side of your dashboard as an element.

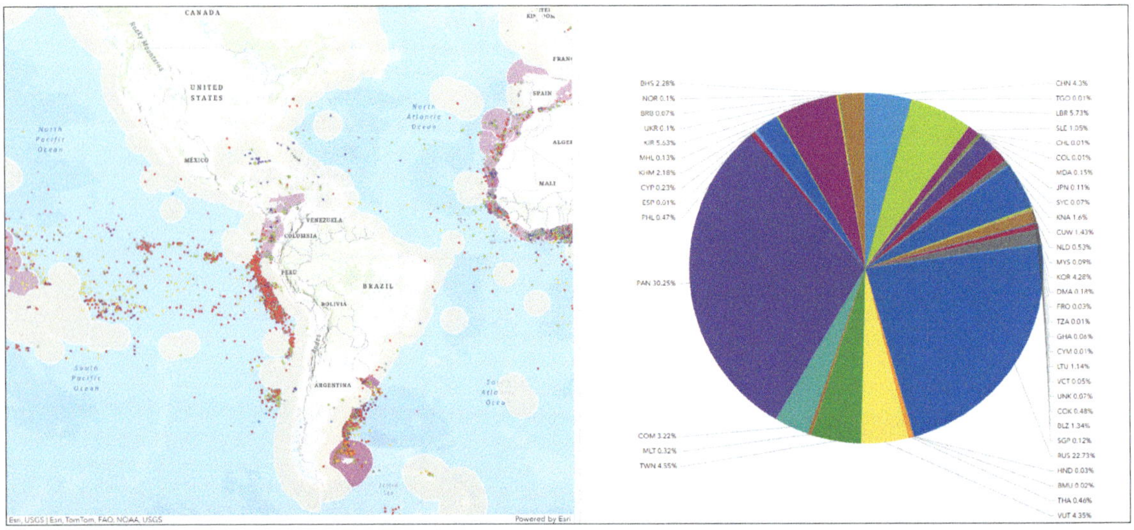

13. Save the dashboard.

    You can resize the elements, as well as move them around on the screen and dock them in different places. Next, you will customize the map.

14. Point to the top left of the map. In the options that appear, click the **Configure** button.

15. On the **Settings** tab, adjust the settings so that only **Legend** and **Zoom in/out** are turned on.

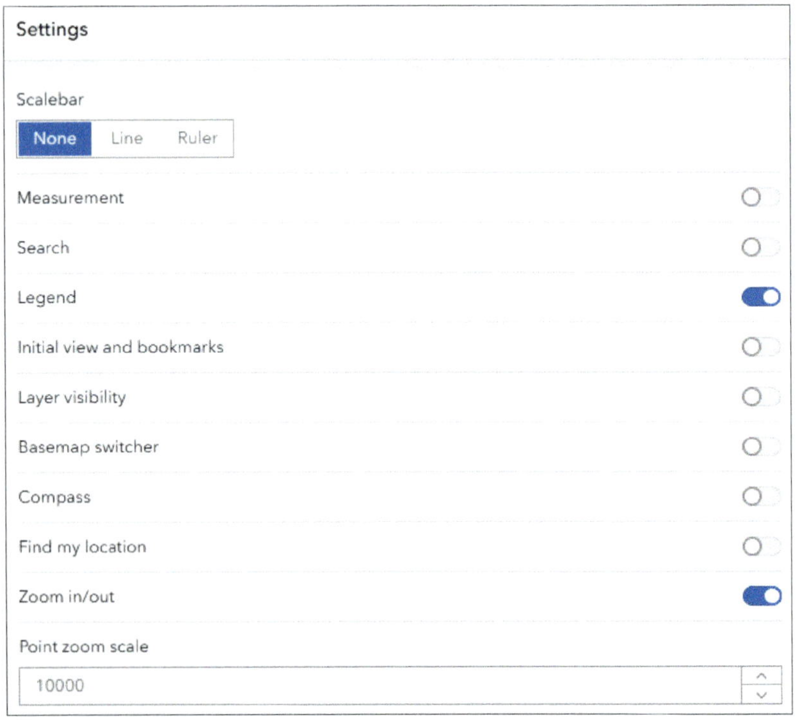

16. Click **Done**.

    A **Legend** button is added to the top right of the map. When you click the button, the legend appears as a panel.

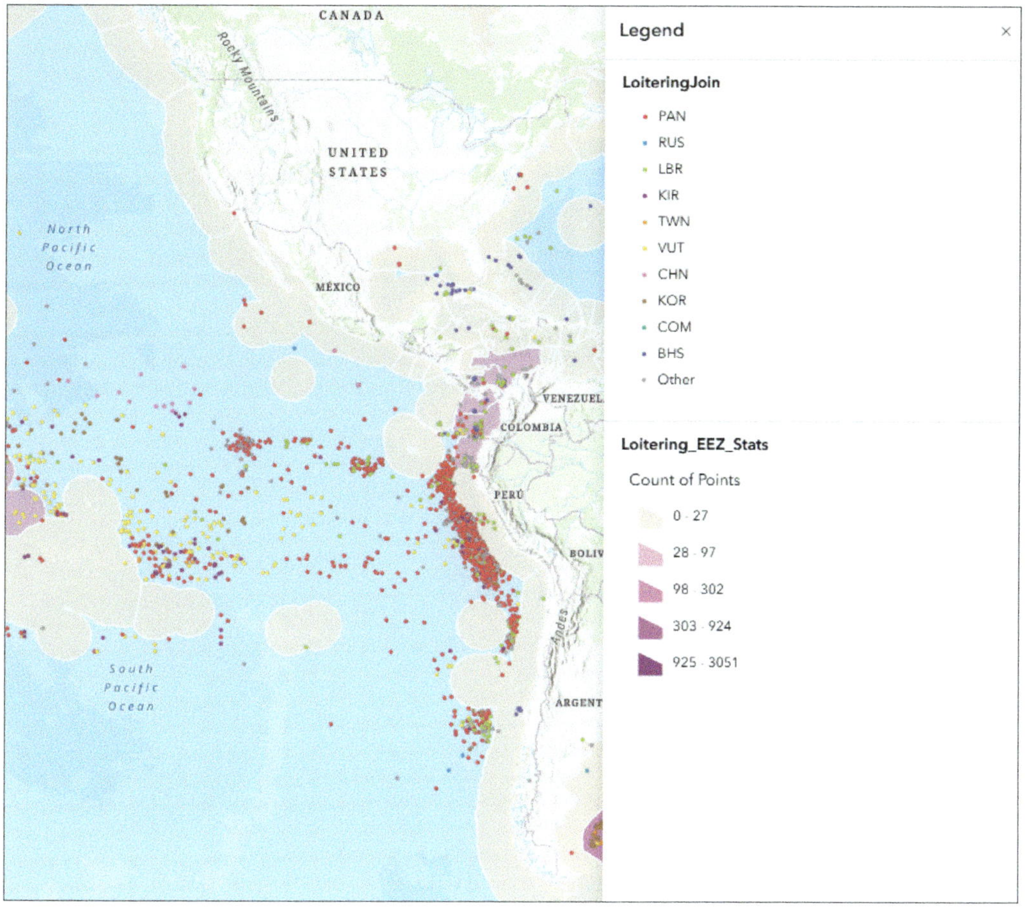

> **Tip:** If you want to modify the layer names and otherwise customize the legend, you must modify the web map that powers the dashboard map.

17. Save the dashboard.

    You have several ways to access and modify your dashboard later. You can select it from the list of items on your **My content** page. When you select it there, you will be redirected to the item page, where you can choose to view, edit, or share the dashboard. Alternatively, to see all your dashboards on one page, click the **App Launcher** in ArcGIS Online, and then click **Dashboards**. You will see a page with all your dashboards, from which you can click through to view or edit each one.

18. Return to ArcGIS Online and open the item page for the **Loitering by EEZ** dashboard.

19. Click **Share** to share the dashboard.

You may share it within your ArcGIS organization. Here, you can share it within your group (groups must be set up ahead of time for group sharing to be possible). You can also share the dashboard publicly. In this case, anyone with a link will be able to view and interact with it. Simply sharing the dashboard publicly is not sufficient for someone to view it. Every layer and component that makes up the dashboard must be shared publicly as well.

20. To share all items, navigate to the **My content** tab on ArcGIS Online and share the appropriate items:

- **Loitering by EEZ** dashboard
- **Loitering by EEZ for Dashboard** web map
- **Loitering_EEZ_Stats** hosted feature layer
- **Loitering** hosted feature layer
- **Encounters** hosted feature layer

Now that everything has been shared properly, your new dashboard works as intended.

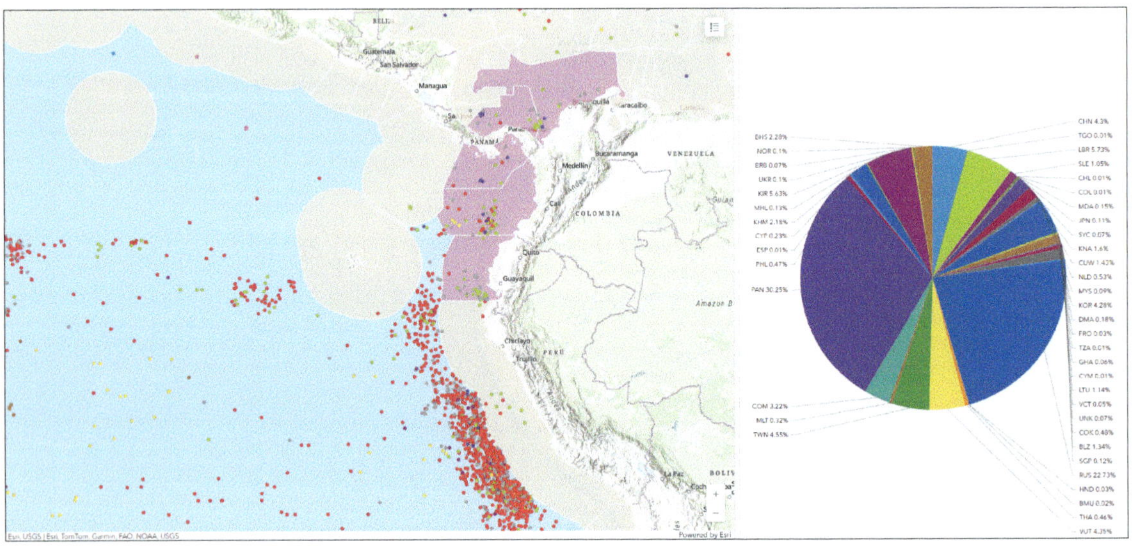

Such a dashboard might be especially useful when using a dynamically updating dataset for the web map and pie chart.

Interpretation of analysis

After completing this exercise, answer the following questions in detail. The answers to these questions, as well as the analysis results from the exercise workflow, will inform the intelligence brief. Including the dashboard as a component of the brief will be an innovative and useful way to share your data and analysis results.

1. What is the most important takeaway from the analysis workflow you completed in this chapter? What actionable information can border authorities use to monitor and patrol countries' maritime borders?
2. What are some security challenges that the flag of convenience phenomenon presents?
3. What time of year (season) do most loitering and transshipment events occur in various regions? What type of vessels are these (likely tied to the type of species being fished, which is seasonal)? What are the flag states for these vessels? What patterns begin to emerge?

Intelligence brief

Based on the analysis from the exercise workflow, produce a three- to five-page intelligence brief by responding to the following question prompts. Incorporate your geospatial outputs from the exercise to enrich and illustrate your conclusions.

- **BLUF:** List three to five primary observations from your analysis in bulleted-list format that summarize findings of this work.
- **Top-level analysis:** Report the primary data points supporting your BLUF.
- **Secondary analysis:** Provide the data sources and deeper analytic products from this analysis.
- **Impact:** Describe the impacts of your findings on the department or organization.
- **Confidence:** Report the level of confidence of your assessment based on available data, accuracy, and so on.
- **Bias:** Explain any background or information you have that could have affected your analysis.
- **Conclusions:** Determine what type of analytic conclusions there may be and how the data requirements, analysis, and brief writing may change based on what analytic framework you are following (descriptive, predictive, prescriptive).

Applying and extending your skills

Accessing AIS data

AIS is an automatic tracking system that uses transceivers on ships and is used by vessel traffic services (VTS). AIS information supplements marine radar, which continues to be the primary method of collision avoidance for water transport. When vessels purposely turn off their AIS transceivers, this activity can be an indication of illicit activity on the water.

AIS datasets can be enormous and require heavy computational resources. Global AIS transceiver data collected from satellite and internet-connected shore-based stations is aggregated and made available on the internet through a number of service providers. Data aggregated this way can be viewed on any internet-capable device to provide near-global, real-time position data from anywhere in the world. Typical data includes vessel name, details, location, speed, and direction on a map; it is searchable, has potentially unlimited, global range, and its history is archived. Most of this data is free of charge, but satellite data and special services such as searching the archives are usually supplied at a cost.

Sources of AIS data

- https://insights.spire.com/maritime
- https://marinecadastre.gov/ais
- https://www.aishub.net
- https://www.orbcomm.com
- https://www.skylight.global
- https://windward.ai
- https://starboard.nz
- https://seavision.volpe.dot.gov
- Esri Vessel Traffic Hub: https://vessel-traffic-esrioceans.hub.arcgis.com

Additional resources

The following online resources provide information about human security challenges:

- "Chinese Fishing Fleet Encroaches on the Galapagos Islands" (HawkEye 360): www.he360.com/resource/potential-illegal-fishing-seen-from-space
- Collyns, Dan. "Cat and Mouse on the High Seas: On the Tail of China's Vast Squid Fleet." 2020. *The Guardian*. www.theguardian.com/environment/2020/oct/21/china-squid-fleet-peru-coastguard-fishing.
- Collyns, Dan. "Illegal Fishing Spurs Billions in Losses for Developing Countries, Study Says." 2022. *The Guardian*. www.theguardian.com/environment/2022/oct/26/illegal-fishing-billions-losses-developing-countries.
- Eradication of Illegal, Unreported and Unregulated Fishing (IUUF) and to Promote Fisheries Crimes as a Transnational Organized Crime (United Nations Department of Economic and Social Affairs): https://sdgs.un.org/partnerships/eradication-illegal-unreported-and-unregulated-fishing-iuuf-and-promote-fisheries
- Illegal, Unreported, and Unregulated Fishing Strategic Outlook (US Coast Guard): www.uscg.mil/iuufishing
- Illegal, Unreported and Unregulated (IUU) Fishing (Food and Agriculture Organization of the United Nations): www.fao.org/iuu-fishing/background/what-is-iuu-fishing/en
- Understanding Illegal, Unreported, and Unregulated Fishing (NOAA Fisheries): www.fisheries.noaa.gov/insight/understanding-illegal-unreported-and-unregulated-fishing
- Vessel Traffic (Marine Cadastre): https://hub.marinecadastre.gov/pages/vesseltraffic

References

- Food and Agriculture Organization of the United Nations. 2001. "International Plan of Action to Prevent, Deter, and Eliminate Illegal, Unreported, and Unregulated Fishing." Accessed 2001. www.fao.org/iuu-fishing/international-framework/ipoa-iuu/en/.
- Food and Agriculture Organization of the United Nations. 2023. "Voluntary Guidelines for Transshipment." Accessed 2023. https://openknowledge.fao.org/items/f7253f5f-1145-4aa7-9003-0b968abb8640.
- Global Fishing Watch. n.d. "What Is a Loitering Event?" Frequently Asked Questions. https://globalfishingwatch.org/faqs/what-is-loitering-event.
- Ian Urbina Subcommittee on Water, Oceans, and Wildlife. 2019. Written testimony, "Oversight of NOAA's Report at Illegal, Unreported, and Unregulated Fishing." Accessed November 14, 2019. www.congress.gov/116/meeting/house/110213/witnesses/HHRG-116-II13-Wstate-UrbinaI-20191114.pdf.
- INTERPOL. 2018. "International Law Enforcement Cooperation in the Fisheries Sector: A Guide for Law Enforcement Practitioners." Accessed February 2018. https://cites.org/sites/default/files/eng/prog/implementation/enf/International%20Law%20Enforcement%20Cooperation%20in%20the%20Fisheries%20Sector.pdf.
- Miller, Nathan A., Aaron Road, Timothy Hochberg, John Amos, and David A. Kroodsma. 2018. "Identifying Global Patterns of Transshipment Behavior." *Frontiers in Marine Science* 5 (July). doi.org/10.3389/fmars.2018.00240.
- US Southern Command. 2023. "Statement of General Laura J. Richardson, Commander, United States Southern Command Before the 118th Congress, House Armed Services Committee."

About the authors

Diana Ter-Ghazaryan is an associate professor of spatial sciences at the University of Southern California Spatial Sciences Institute, where her teaching portfolio includes courses on human security and geospatial intelligence. She codirects the institute's Human Security and Geospatial Intelligence lab. Her current research interests are in geospatial intelligence and security, as well as the intersection of geospatial analysis and spatial humanities. She has also held positions as a faculty research associate at Florida International University, GIS Center, as well as on the faculty of University of Miami, Department of Geography. She has a PhD in international relations and geography and an MS in environmental studies from Florida International University and a GISP certification from the GIS Certification Institute.

Bruce Vitor is the director of research and operations at Florida International University's Jack D. Gordon Institute for Public Policy. He oversees the university's Security Research Hub, an open-source research initiative, and he provides leadership in national security and strategy development with a wide variety of US and international partners. He retired as a colonel from the US Army after 27 years of service in multiple assignments throughout the world as an infantry officer and strategic planner. He graduated from the United States Military Academy at West Point. He has an MBA from Embry-Riddle Aeronautical University and a master's degree in military art and science from the Command and General Staff College.

Chapter 15

Detecting ship encounters at sea with GIS

Ana Catarina Nunes, *Portuguese Navy, NOVA IMS, Nova University, Lisbon*
Marco Painho, *NOVA IMS, Nova University, Lisbon*

Overview

This chapter presents a workflow for visualizing maritime traffic in a study area and detecting encounters between ships that may represent possible transshipment activities in the context of maritime security and GEOINT. The methodology focuses on using geospatial tools to process and analyze data shared by ships and integrating the results in a GIS.

Learning objectives

- Perform exploratory analysis with automatic identification system (AIS) data.
- Construct ship trajectories using ArcGIS Pro.
- Build and analyze maps of maritime traffic density using ArcGIS Pro.
- Detect encounters between ships using ArcGIS Pro.

Technical components

- Software:
  - ArcGIS Pro 3.4
- Datasets:
  - Study area (southernmost part of Portugal's Exclusive Economic Zone [EEZ])
  - AIS data from satellites (SAT-AIS)
  - AIS data from vessel traffic services (T-AIS)

- Esri Ocean basemap
- Land boundaries

Prerequisites

- Basic knowledge of Python and access for installing and using libraries
- Technical skill level: Introductory GIS
- Time to complete chapter: Two to three hours without intelligence brief, four to five hours with intelligence brief

Background and situational context

According to the United Nations, more than 80 percent of world trade is currently transported by sea (UNCTAD 2021). Areas with high maritime traffic are susceptible to a higher number of illicit activities. Examples of these are drug smuggling, human trafficking, illegal fishing, and pollution by illegal discharge, among others. Another activity of interest is illegal transshipment, in which two or more ships meet at sea to transfer any kind of load illicitly (Miller et al. 2018). This chapter will present a methodology that contributes to the control and consequent mitigation of illegal transshipment events by providing a workflow for detecting encounters between ships at sea.

The study area used in this exercise is in the southern area of the Portuguese EEZ. Tracking ships based on AIS is one of the few tools available to detect these transshipment events. However, some ships may purposely turn off signal emission or send incorrect information while carrying out illegal actions, which makes detecting them difficult. Despite these constraints, the use of AIS and its tracking present a useful way to detect encounters between vessels.

Exercise workflow 15

1 Explore and analyze data for a 24-hour period

Your first task is to perform exploratory analysis of the data for a 24-hour period.

1. Start ArcGIS Pro. Under **New Project**, click **Map**.
2. In the **New Project** window, apply the following settings:
    - **Name**: Chapter_15
    - **Location**: C:\SecurityFirst

    Note: This tutorial will use the C:\SecurityFirst folder.

3. Click **OK**.
4. On the ribbon, click the **Map** tab. In the **Layer** group, click **Add Data**.

5. From the chapter data folder (**SecurityFirst_ExerciseData/Chapter16**), add the following datasets:
   - **ais_points.shp**
   - **ais_ptstat_day.shp**
   - **study_area_PT.shp**
   - **land_boundaries.shp**
   - **monthly_trajectories.shp**
6. Save your project.

   You will now create a clone of the default environment so that you can add the necessary Python libraries.
7. On the ribbon, click the **Project** tab. Click the **Package Manager** tab.
8. On the **Package Manager** page, next to **Active Environment**, click the **Environment Manager** button.
9. In the **Environment Manager** window, click the **More options** button for **arcgispro-py3** and click **Clone**.

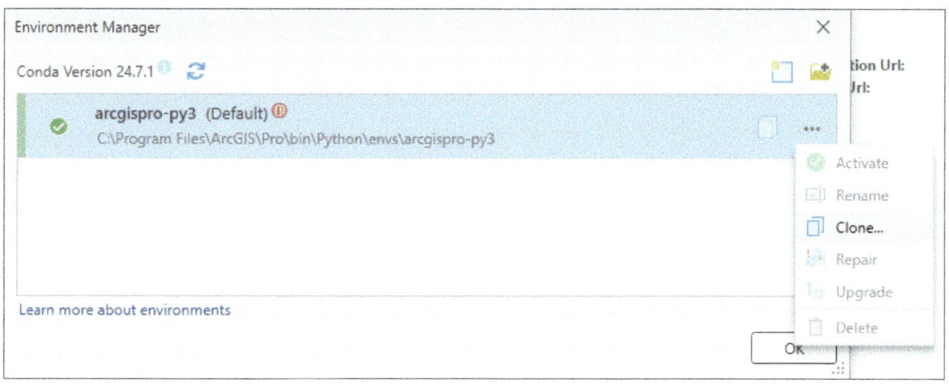

10. Accept the defaults of the clone environment and click **OK**.
11. Switch the **Active Environment** to **arcgispro-py3-clone**.
12. Restart ArcGIS Pro for the environment changes to take effect.
13. In the **Package Manager**, search for geopandas. Confirm that you have **geopandas** and **geopandas_base** installed. If not, click the **Add Packages** tab and install them. Refresh your environment to apply the changes.

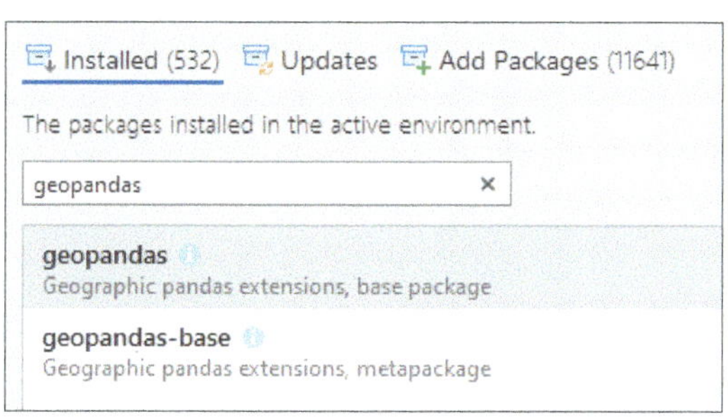

You will now use ArcGIS Notebooks in ArcGIS Pro to run blocks of Python code step by step.

14. Return to the **Chapter_15** project. On the ribbon, click the **Analysis** tab. In the **Geoprocessing** group, click the arrow to the right of **Python** and select **Python Notebook**.
15. In the next cell, type or copy the following blocks of code, cell by cell. Each comment represents a new code block.

```
#Code for Exploratory Data Analysis on AIS Data
# Import necessary Python libraries and abbreviate names for convenience
import numpy as np
import pandas as pd
import geopandas as gpd
from geopandas import GeoDataFrame, read_file
import shapely as shp
from shapely.geometry import Point, LineString, Polygon
from datetime import datetime, timedelta
import matplotlib.pyplot as plt

# Enable inline plotting for visualization within the notebook
%matplotlib inline

# Load geospatial data from a shapefile and convert it into a GeoDataFrame
# Replace '…/ais_ptsat_day.shp' with the actual path to your shapefile.
↪ Example: If your shapefile is in a folder named "data" on your desktop, you
↪ would replace '…/ais_ptsat_day.shp' with something like: 'C:/Users/YourName
↪ /Desktop/data/ais_ptsat_day.shp' (for Windows)
t_start = datetime.now()
gdf_orig = gpd.read_file('…/ais_ptsat_day.shp')
print("Finished reading {} rows in {}".format(len(gdf_orig),datetime.now() -
↪ t_start))

# Convert MultiPoint geometries to individual Point geometries
gdf_p=gdf_orig.explode(index_parts=True)

# Display the first row of data to inspect the structure
gdf_p.head(1)

# Remove records with MMSI (Maritime Mobile Service Identity) equal to 0
print("Original size: {} rows".format(len(gdf_p)))
gdf_p1 = gdf_p[(gdf_p.mmsi!=0)]
x=len(gdf_p)-len(gdf_p1)
print("Reduced to {} rows after removing {} mmsi records".format(len
↪ (gdf_p1),x))
```

```
# Filter out speed values: retain only records where speed is between 0 and 30
↪ knots
print("Original size: {} rows".format(len(gdf_p1)))
gdf = gdf_p1[(gdf_p.speed_over>=0)&(gdf_p1.speed_over<=30)]
x=len(gdf_p1)-len(gdf)
print("Reduced to {} rows after removing {} speed records".format(len(gdf),x))

# Plot a histogram to visualize the distribution of vessel speeds
gdf['speed_over'].hist(bins=100, figsize=(15,3))
plt.xlabel("Speed vessel (knots)",fontsize=12)
plt.ylabel("Number of records",fontsize=12)

# Plot a bar chart showing the distribution of vessel types
gdf['vessel_typ']. value_counts().plot (kind='bar', figsize=(15,3))
plt.xlabel("Vessel type",fontsize=12)
plt.ylabel("Number of records",fontsize=12)
```

Note: The arrow in the code block indicates a continuation of the code line. When you type the code, it will all be on one line.

After the notebook tasks have completed, you will use geoprocessing tools to build the ship trajectories and analyze the density of ship traffic in the study area. You may save the notebook for later use and close it.

16. In the **Contents** pane, right-click **ais_points** and select **Properties**. Click the **Time** tab.
17. Under **Filter using time**, select **Filter layer content based on attribute values** and apply the following settings:
    - **Layer Time**: Each feature has a single time field
    - **Time Field**: instant

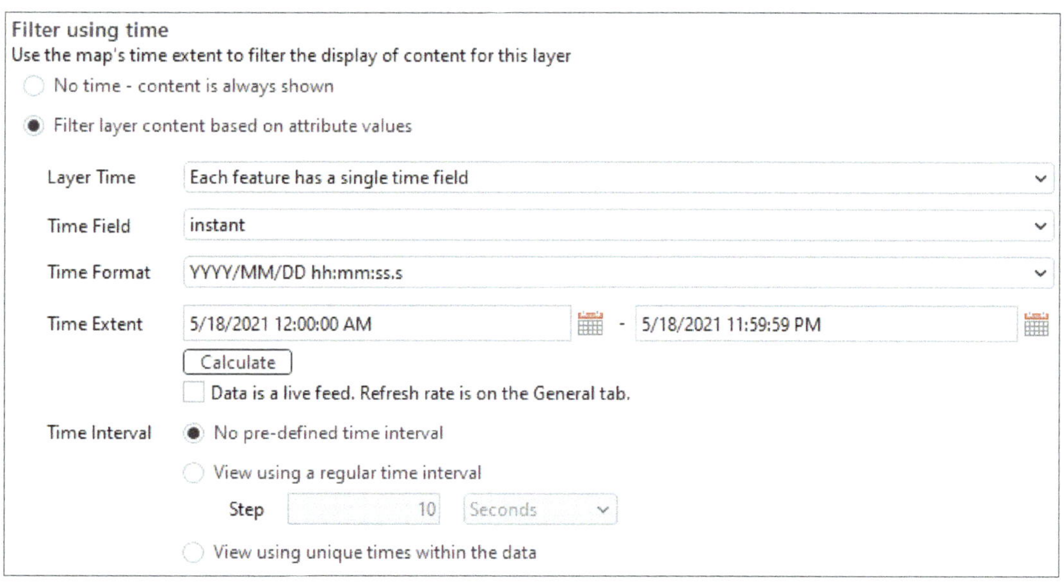

18. Click **OK.**

    Next, you will reconstruct line tracks from entered point data. These are ship trajectories.

19. On the ribbon, click the **Analysis** tab. In the **Geoprocessing** group, click **Tools.**

20. In the **Geoprocessing** pane, search for the Reconstruct Tracks tool. Open the tool and add the following parameters:

    - **Input Layer**: ais_points
    - **Output Feature Class**: Tracks
    - **Track Fields**: mmsi
    - **Method**: Geodesic

21. Click **Run.**

    Ship trajectories are now shown as lines rather than a collection of points.

22. In the **Contents** pane, uncheck the boxes for the **ais_pstat_day**, **ais_points**, and **monthly_trajectories** layers to hide them.

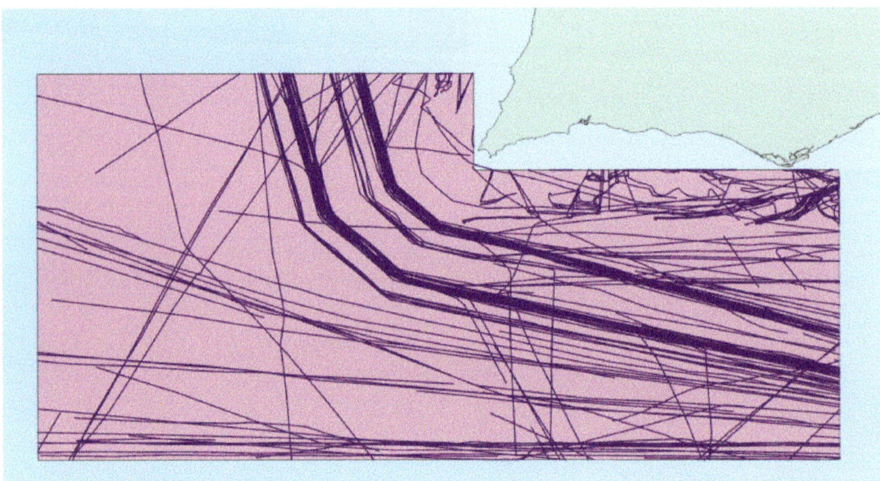

23. Save your project.

2 Create maritime traffic density maps

You will now create maps of maritime traffic density using ArcGIS Pro.

1. On the ribbon, click the **Insert** tab. In the **Project** group, click **New Map**. Double-click **Map1** in the **Contents** pane and rename the map Maritime Traffic Density.

2. From the chapter data folder, add the following datasets, and save the map:

    - **study_area_PT.shp**
    - **land_boundaries.shp**

    Before running the following geoprocessing tools, you will need to reproject all your datasets to the ETRS 1989 UTM Zone 29N projection.

3. In the **Geoprocessing** pane, search for the Project tool. Open the tool and apply the following settings:
    - **Input Dataset or Feature Class:** land_boundaries
    - **Output Dataset or Feature Class:** land_boundaries
    - **Output Coordinate System:** ETRS 1989 UTM Zone 29N
4. Click **Run**.
5. Rerun the tool for the **study_area_PT** layer. When the tool finishes running, remove the old layers from the **Contents** pane.

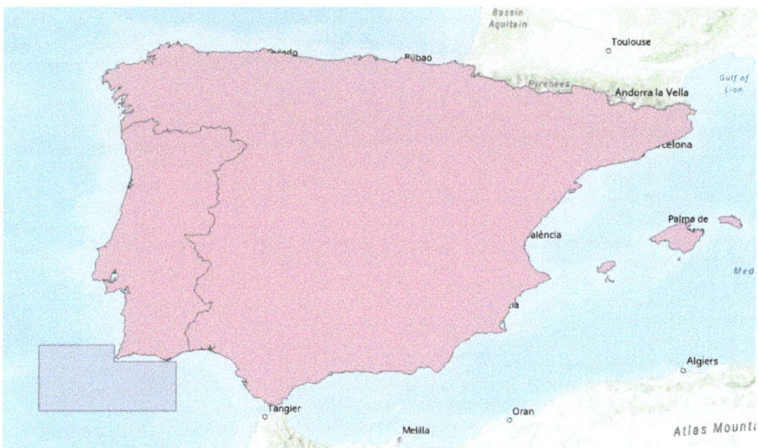

> **Tip:** It is important to use a projected coordinate system and planar coordinates. Otherwise, using a shapefile with units in degrees will make a nautical mile look larger than it is.

6. On the map, zoom in so that the **study_area_PT** layer fills up most of the screen.
7. In the **Geoprocessing** pane, search for the Generate Tessellation tool. Apply the following settings and click **Run**:
    - **Output Feature Class:** Generate_Tessellation
    - **Extent:** Extent of data in all layers > study_area_PT
    - **Shape Type:** Hexagon
    - **Size:** 0.0001 Unknown
    - **Spatial Reference:** ETRS_1989_UTM_Zone_29N

    Next, you will clip the tessellation to the study area boundaries.
8. In the **Geoprocessing** pane, search for the Clip Layer tool. Apply the following settings and click **Run**:
    - **Input Layer:** Generate_Tessellation
    - **Clip Layer:** study_area_PT
    - **Output Feature Class:** Clipped_Tessellation

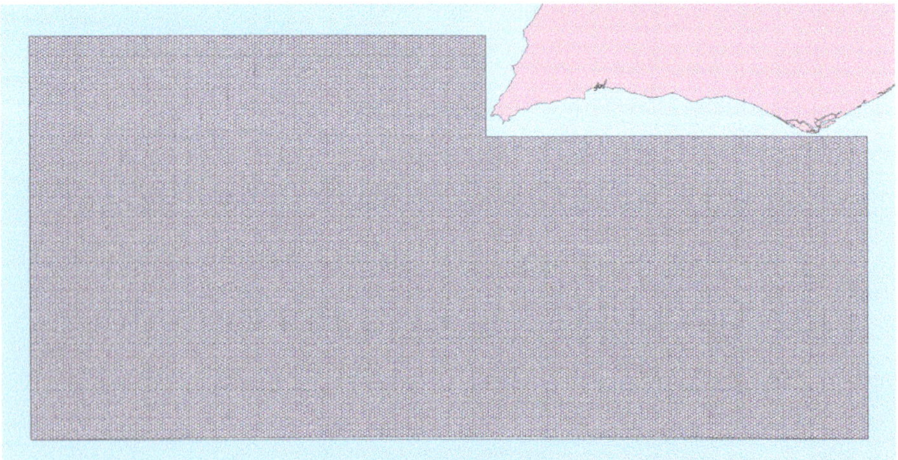

The tessellation is now clipped to the study area.

9. From the chapter folder, add the **monthly_trajectories** dataset.

   This dataset has been prepared in advance using the monthly AIS data. The **monthly_trajectories** and the **Clipped_Tessellation** datasets overlap.

   > **Tip:** In the monthly_trajectories folder, there are three other datasets besides the monthly_trajectories dataset: monthly_trajectories_fishing, monthly_trajectories_merchant, and monthly_trajectories_passenger. These datasets are subsets of the monthly_trajectories dataset, each containing one type of vessel trajectory. You can use these datasets to generate density maps by type of vessel within the study area.

10. In the **Geoprocessing** pane, search for the Spatial Join tool. Apply the following settings and click **Run**:

    - **Target features**: Clipped_Tessellation
    - **Join features:** monthly_trajectories
    - **Output Feature Class**: New_Tessellation
    - **Match Option**: Intersect

    You now have a dataset of the study area that shows the density of trajectories that cross each cell. You will symbolize the dataset accordingly.

11. In the **Contents** pane, right-click the **New_Tessellation** dataset and select **Symbology**.

12. In the **Symbology** pane, change the **Primary Symbology** to **Graduated Colors**.

    Graduated color ramps are used to show increases in color intensity. In this case, cells that are crossed by more trajectories will have a more intense color.

    Note: On the Advanced Symbology Options tab, increase the sample size to 100000 to ensure that all the data is being used.

13. On the top right of the **Symbology** pane, click the three lines and select **Import symbology**.
14. In the **Apply Symbology From Layer** tool, for **Symbology Layer**, navigate to the chapter's data folder, open the **symbology** folder, and select the **density_map.lyrx** file. Click **Run**.
15. Review the resulting density map.

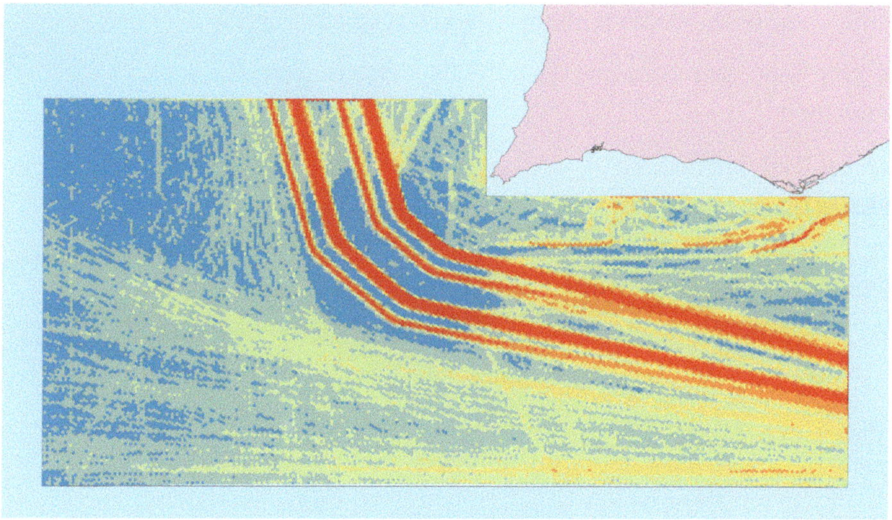

3 Create a ship encounter detection model

Finally, you will build a ship encounter detection model, using the ship trajectories and other datasets that you generated earlier.

1. Create a new map and name it Ship Encounter Detection. Add the following layers to the map:
   - **ais_points.shp**
   - **land_boundaries.shp**
   - **study_area_PT.shp**

   You will select vessel trajectories with speeds equal to or less than five knots.
2. In the **Contents** pane, right-click **ais_points** and click **Properties**.
3. On the **Definition Query** tab, click **New definition query**.
4. Add the following query: `Where speed_over is less than or equal to 5.`

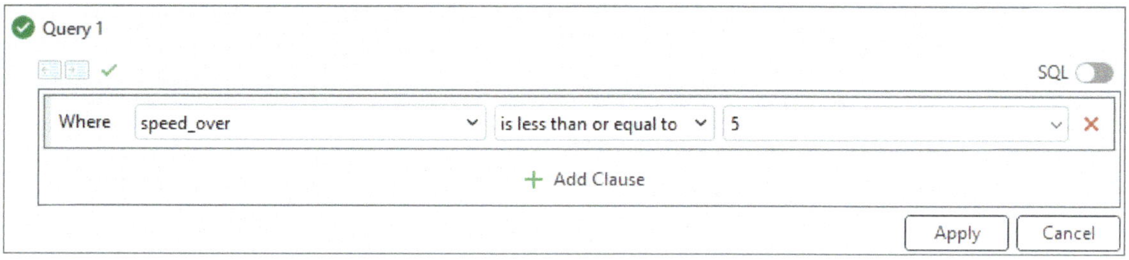

5. Click **Apply**.
6. Click **OK** to close the **Layer Properties** window.

   The number of points on the map decreases.

   Next, you will create a buffer that extends five nautical miles from the coast using the **Pairwise Buffer** tool, and then overlay the buffer with the selected vessel trajectories.

7. In the **Geoprocessing** pane, search for and open the **Pairwise Buffer** tool. Apply the following settings and run the tool:
   - **Input Features**: land_boundaries
   - **Output Feature Class**: land_Buffer
   - **Distance**: 5 International Nautical Miles
   - **Method**: Geodesic (shape preserving)
   - **Dissolve Type**: Dissolve all output features into a single feature.

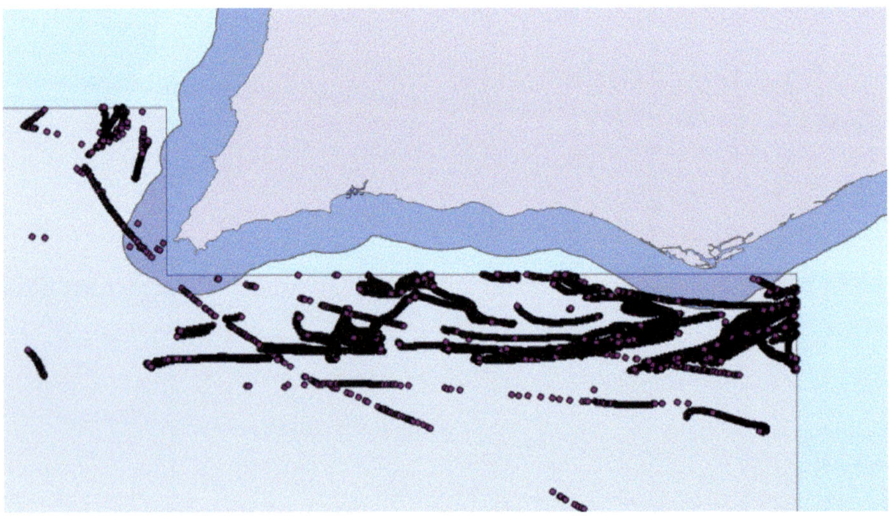

   Now you will overlay the five-nautical-mile buffer you created in the previous step with the ship trajectories that have a speed of less than or equal to five knots.

8. Search for and open the Pairwise Erase tool. Apply the following settings and run the tool:
   - **Input Features**: ais_points
   - **Erase Features**: land_Buffer
   - **Output Feature Class**: ais_points_sea

   Note: The tool will notify that you are using the filtered records that were defined using the definition query you created earlier. There should be 132,377 records.

   The resulting layer, **ais_points_sea**, is a new layer of daily AIS data in which ships, in addition to having a speed of five knots or less, are more than five nautical miles from the coast.

9. In the **Contents** pane, turn off the **ais_points** layer.

You will now use the **ais_points_sea** dataset to detect ships that are within 150 meters of another ship for at least 10 minutes.

10. Right-click the **ais_points_sea** layer and click **Properties**.
11. Click the **Time** tab. For **Filter using time**, select **Filter layer content based on attribute values.** Apply the following parameters:
    - **Layer Time**: Each feature has a single time field
    - **Time Field**: instant
12. Click **OK.**

You will now run a join operation to examine patterns of ships that are within 150 meters of another ship for at least 10 minutes.

13. Search for the Join Features tool and open it. Set the following parameters and run the tool:
    - **Target Layer**: ais_points_sea
    - **Join Layer**: ais_points_sea
    - **Output Dataset**: ais_ship_pairs
    - **Join Operation**: Join one to many
    - **Spatial Relationship**: Geodesic Near
    - **Spatial Near Distance:** 150 Meters
    - **Temporal Relationship:** Near
    - **Temporal Near Distance**: 10 minutes
    - **Include Distance**: <check the box>
    - **Distance Unit:** Meters
    - **Advance Options** > **Join Condition**: $target.mmsi != $join.mmsi

You now have a dataset that shows patterns of ships that are within 150 meters of another ship for at least 10 minutes, pointing out encounter events.

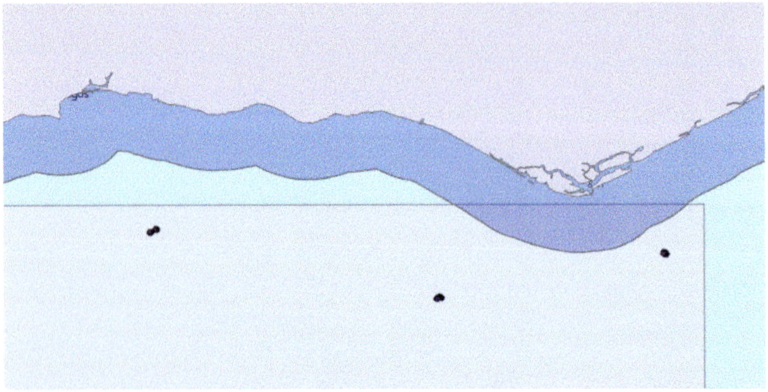

On your own

Try different values for the distance between ships and for the temporal relationship to examine different results. Overlay the results with the density maps built earlier to visualize the locations where ship-to-ship encounters were detected. Overlay the results with the daily trajectories built earlier to analyze the behavior of those ships throughout the day.

Interpretation of analysis

Takeaways

In the exploratory analysis of the data performed over a 24-hour period, speed outliers with values below 0 and above 30 knots were excluded. Figure 15-1 displays the distribution of vessel speeds, showing a peak between 10 and 15 knots, which may correspond to vessels in transit. Another peak between 0 and 5 knots is observed, which may correspond to vessels engaged in fishing activity or that have reduced speed for other reasons. This low-speed interval was used in the algorithm for detecting encounters between ships because illegal transhipment activities often occur at low speeds.

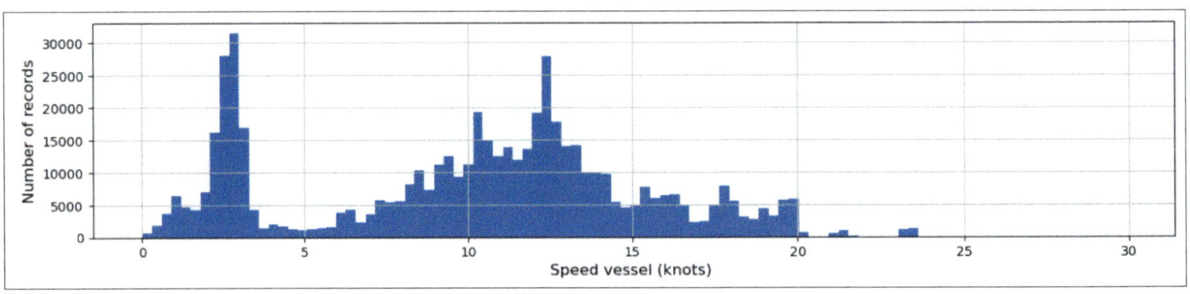

Figure 15-1. Histogram of vessel speed plotted according to number of records.

Figure 15-2 illustrates the distribution of ships by type in the study area. It indicates that types 70 and 30 are the most frequent. These types correspond to merchant and fishing vessels.

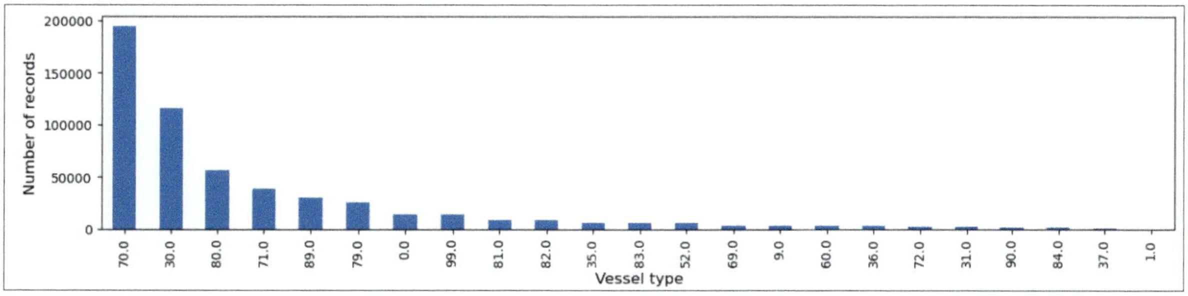

Figure 15-2. Histogram of vessel type plotted according to number of records.

Table 15-1 lists a selection of the AIS type codes (US Coast Guard, NOAA, BOEM 2018).

Table 15-1. Select AIS type codes

Vessel Type Code	Vessel Type
0	Not available, default
1–19	Reserved for future use
30	Fishing
31	Towing
35	Military operations
36	Sailing
37	Pleasure craft
52	Tug
60	Passenger, all ships of this type
69	Passenger, no additional information
70	Cargo, all ships of this type
71	Cargo, hazardous category A
72	Cargo, hazardous category B
79	Cargo, no additional information
80	Tanker, all ships of this type
81	Tanker, hazardous category A
82	Tanker, hazardous category B
83	Tanker, hazardous category C
84	Tanker, hazardous category D
89	Tanker, no additional information
90	Other type, all ships of this type
99	Other type, no additional information

After completing this exercise, answer the following questions in detail. The answers to these questions, as well as the analysis results from the exercise workflow, will inform the intelligence brief.

1. Examine the following maps of maritime density, broken out by vessel type. What patterns do you observe?

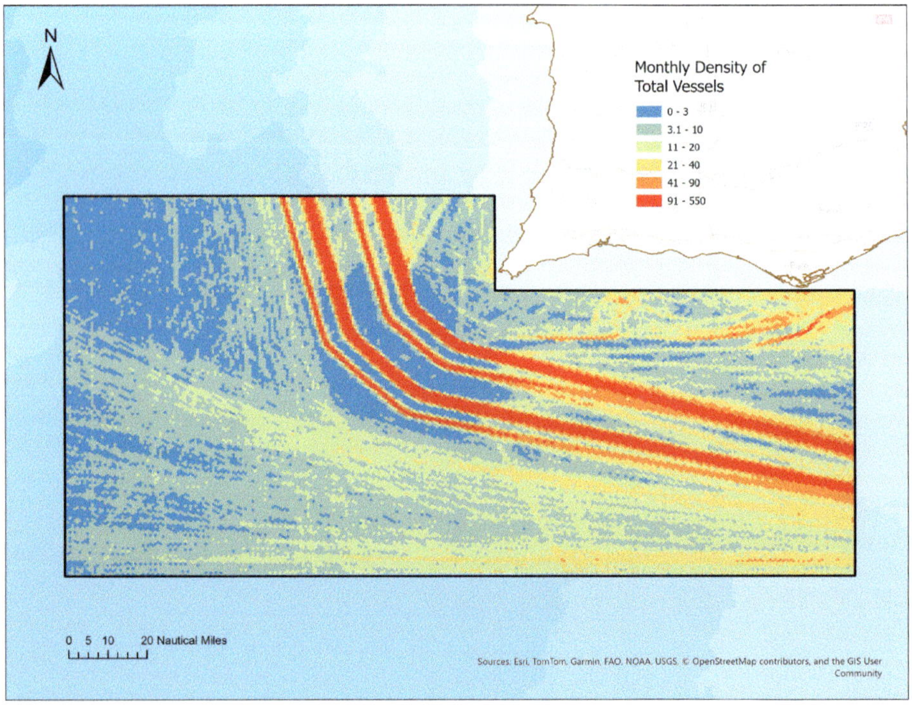

Figure 15-3. The monthly density of total vessels.

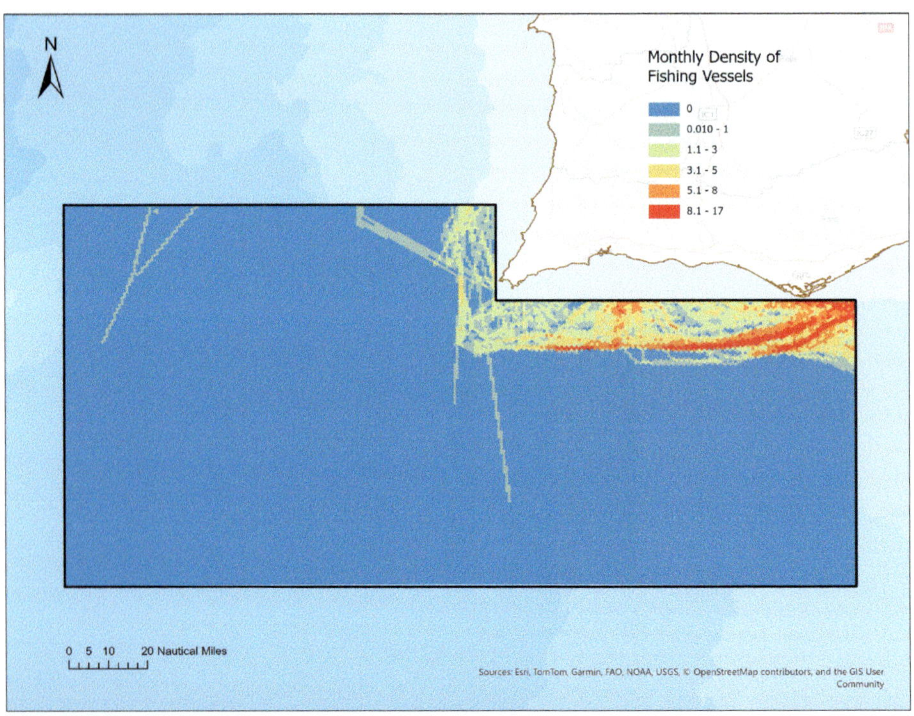

Figure 15-4. The monthly density of fishing vessels per cell.

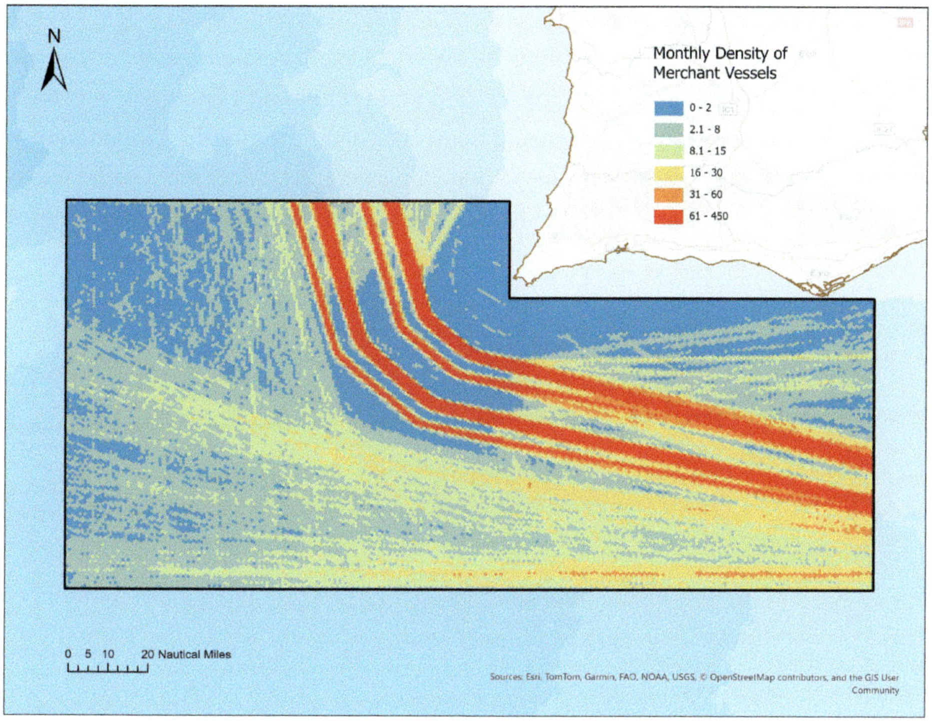

Figure 15-5. The monthly density of merchant vessels per cell.

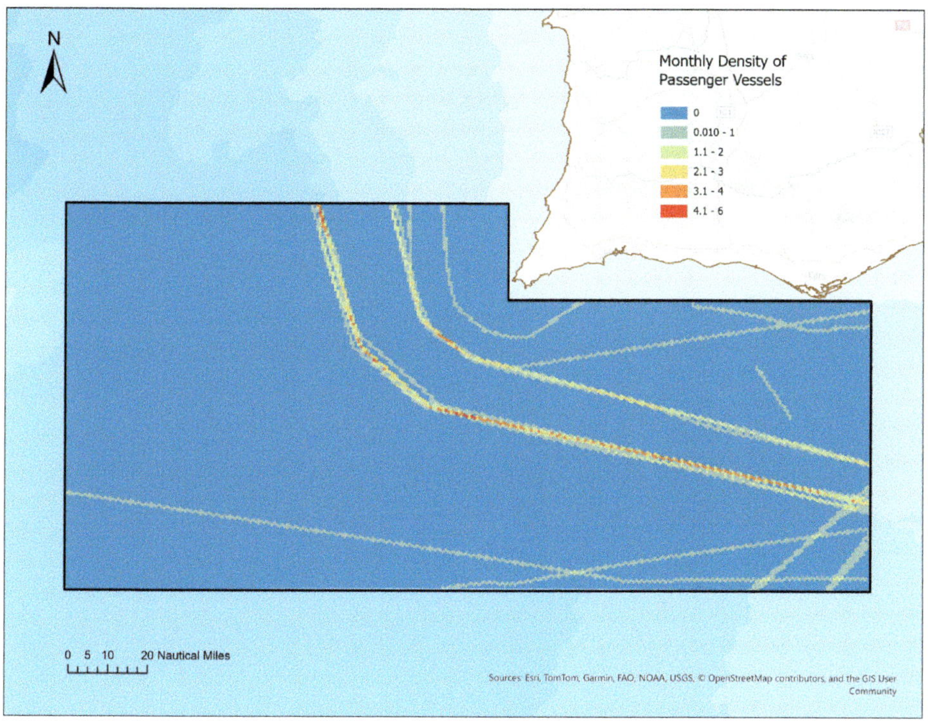

Figure 15-6. The monthly density of passenger vessels per cell.

2. The results of building the ship encounter detection model are visualized in figure 7, showing parts of trajectories that were identified as possible encounters between ships. If you overlay this map with different density maps (figures 8 and 9: merchant and fishing vessels), you will notice that possible encounters were detected in an area where the traffic of merchant vessels is lower and that of fishing vessels is higher. What can this tell you about potential encounters at sea? What additional datasets should you investigate next to help answer further questions about this finding?

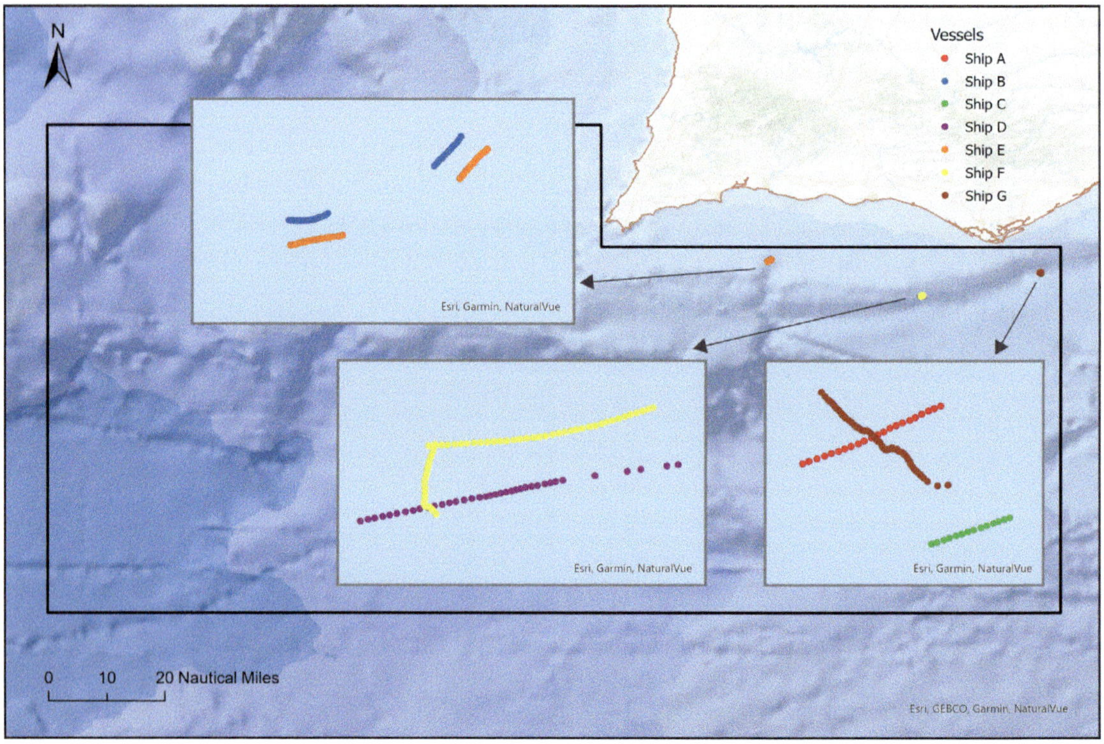

Figure 15-7. Possible encounters between ships mapped in three insets.

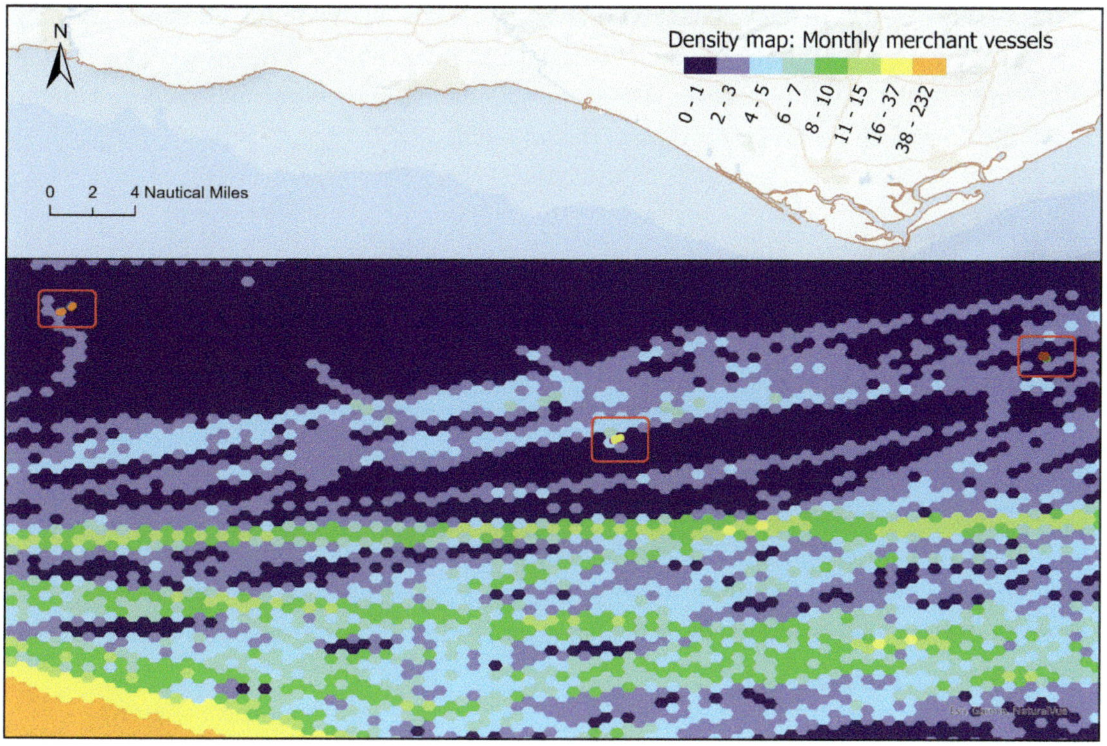

Figure 15-8. Density map of monthly merchant vessels.

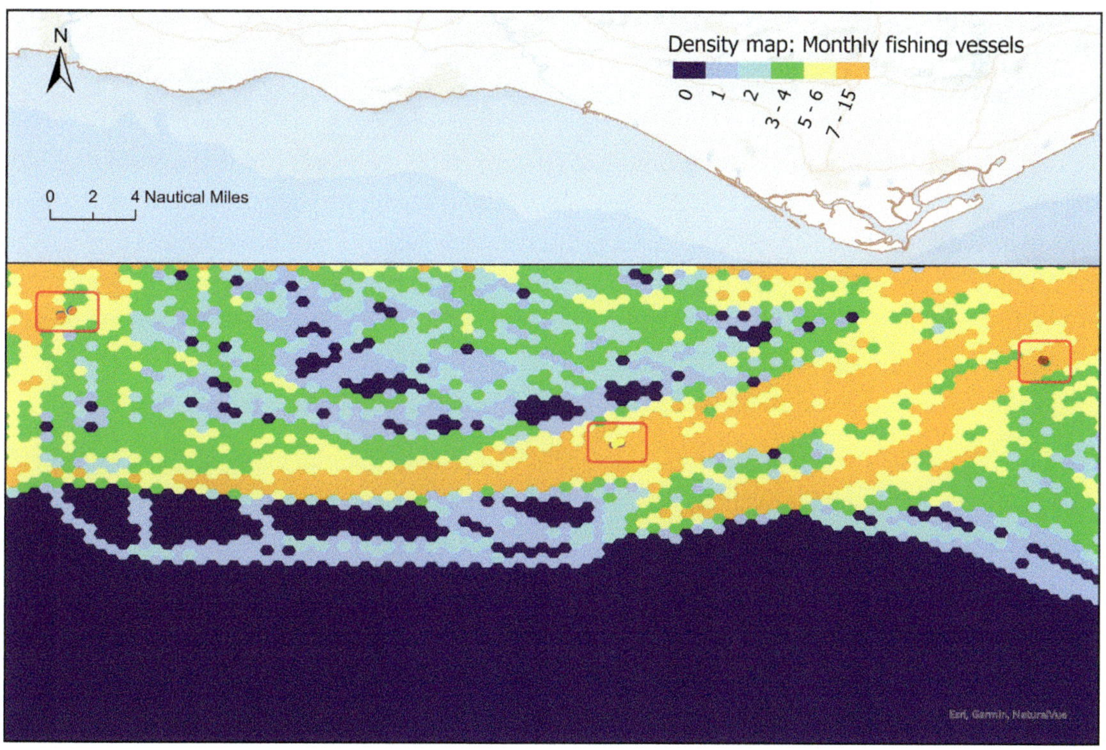

Figure 15-9. Density map of monthly fishing vessels.

3. Suppose your workflow flags a potential encounter between two vessels in an area known for illegal fishing activity. How would you prioritize this event for further investigation, and what additional datasets (satellite imagery, weather data, historical vessel behavior) would you incorporate to confirm the nature of the interaction?

Intelligence brief

Based on the analysis from the exercise workflow, produce a three- to five-page intelligence brief by responding to the following question prompts. Incorporate your geospatial outputs from the exercise to enrich and illustrate your conclusions.

- **BLUF:** List three to five primary observations from your analysis in bulleted-list format that summarize findings of this work.
- **Top-level analysis:** Report the primary data points supporting your BLUF.
- **Secondary analysis:** Provide the data sources and deeper analytic products from this analysis.
- **Impact:** Describe the impacts of your findings on the department or organization.
- **Confidence:** Report the level of confidence of your assessment based on available data, accuracy, and so on.
- **Bias:** Explain any background or information you have that could have affected your analysis.
- **Conclusions:** Determine what type of analytic conclusions there may be and how the data requirements, analysis, and brief writing may change based on what analytic framework you are following (descriptive, predictive, prescriptive).

Applying and extending your skills

The International Convention for the Safety of Life at Sea (SOLAS) of 1974 is the main regulation for the use of AIS on ships (IMO 2015). It states that its use is mandatory for ships of more than 300 gross tons in international traffic or 500 gross tons and carrying cargo in noninternational traffic, as well as by all ships carrying passengers. The European Union also requires all fishing vessels over 15 meters in length to use AIS. This requirement allows the automatic exchange of information between ships, coastal stations, and communication satellites. The use of AIS enhances safety, navigation efficiency, protection of marine environments, and tracking of vessels of interest.

Owing to the sheer volume of AIS data, transforming it into geospatial information requires effective and systematic spatiotemporal analysis and a high computational capacity for its processing, as well as databases that allow for its management.

Regarding the procedure presented in this workflow, the investigation of the identified vessels can be continued as follows:

- Select the full daily track of each ship from the trajectories layer and examine its behavior and maneuvers throughout the day, or extend the period of the analysis, depending on the availability of data.

- Integrate satellite images to identify the vessels in review to examine whether they were carrying out illegal transshipment activities.
- Use near-real-time systems, integrating AIS data and satellite imagery, to provide more efficient and effective decision support.

Other methodologies can be used to detect anomalies in the behavior of ships at sea:

- Clustering
- Neural networks
- Statistical analysis of patterns
- Density estimation and its outliers
- Velocity profiles

These methodologies can be applied in ArcGIS Pro using several available tools, such as Optimized Hot Spot Analysis, statistical analysis of patterns (for example, Spatial Autocorrelation or Nearest Neighbor) and Kernel Density. External deep learning tools can also be integrated, such as TensorFlow and PyTorch.

References

- European Maritime Safety Agency (EMSA). 2019. Traffic Density Mapping Service Methodology. SCRIBD. www.scribd.com/document/621195289/TDMS-methodology-V-1-0.
- Graser, Anita. 2019. "MovingPandas: Efficient Structures for Movement Data in Python." *GI_Forum* 7 (1): 54–68. doi.org/10.1553/GISCIENCE2019_01_S54.
- Hsu, Hua-Zhi. 2014. "Safety Domain Measurement for Vessels in an Overtaking Situation." *International Journal of E-Navigation and Maritime Economy* 1: 29–38. doi.org/10.1016/J.ENAVI.2014.12.004.
- IMO (International Maritime Organization). 2015. "Revised Guidelines for the Onboard Operational Use of Shipborne Automatic Identification Systems (AIS)." Resolution A.1106(29). https://wwwcdn.imo.org/localresources/en/OurWork/Safety/Documents/AIS/Resolution%20A.1106(29).pdf.
- Jordahl, Kelsey, Joris Van den Bossche, Martin Fleischmann, et al. 2020. "geopandas/geopandas: v0.8.1." *Zenodo*. doi.org/10.5281/zenodo.3946761.
- Kluyver, Thomas, Benjamin Ragan-Kelley, Fernando Pérez, et al. 2016. "Jupyter Notebooks: A Publishing Format for Reproducible Computational Workflows." In *Positioning and Power in Academic Publishing: Players, Agents, and Agendas*. IOS Press Ebooks: 87–90. doi.org/10.3233/978-1-61499-649-1-87.
- Miller, Nathan A., Aaron Roan, Timothy Hochberg, John Amos, and David A. Kroodsma. 2018. "Identifying Global Patterns of Transshipment Behavior." *Frontiers in Marine Science* 5. doi.org/10.3389/fmars.2018.00240.
- PostGIS. n.d. About PostGIS. https://postgis.net.
- PostgreSQL Global Development Group. n.d. "What Is PostgreSQL?" Online documentation for current and supported versions. www.postgresql.org/docs/14/intro-whatis.html.

- UNCTAD (UN Trade and Development). 2021. "Review of Maritime Transport 2021." https://unctad.org/webflyer/review-maritime-transport-2021.
- US Coast Guard, NOAA, and BOEM. 2018. "AIS Vessel Type and Group Codes Used by the Marine Cadastre Project 2018-05-23." https://coast.noaa.gov/data/marinecadastre/ais/VesselTypeCodes2018.pdf.
- Xiao, Fangliang, Han Ligteringen, Coen Van Gulijk, and Ben Ale. 2015. "Comparison Study on AIS Data of Ship Traffic Behavior. *Ocean Engineering* 95: 84–93. doi.org/10.1016/J.OCEANENG. 2014.11.020.

About the authors

Ana Catarina Nunes is a strategic data analyst in the Portuguese Navy, with more than seven years' experience onboard scientific vessels, as well as three years working in GEOINT. She has a master's degree in naval military sciences, specializing in hydrography, and a master's degree in GISS from NOVA IMS, Nova University, where she also received the USGIF-accredited certificate in geospatial intelligence.

Marco Painho is a full professor of GISS at NOVA IMS, Nova University of Lisbon. He has a bachelor's degree in environmental engineering from Nova University, a master's degree in regional planning from the University of Massachusetts, and a PhD in geography from the University of California, Santa Barbara. He coordinates several programs at Nova University: the Master of Science in Geospatial Technologies, the Master of Science in Geographic Information Systems and Science, and the Certificate in Geospatial Intelligence.

Part 5

Global security

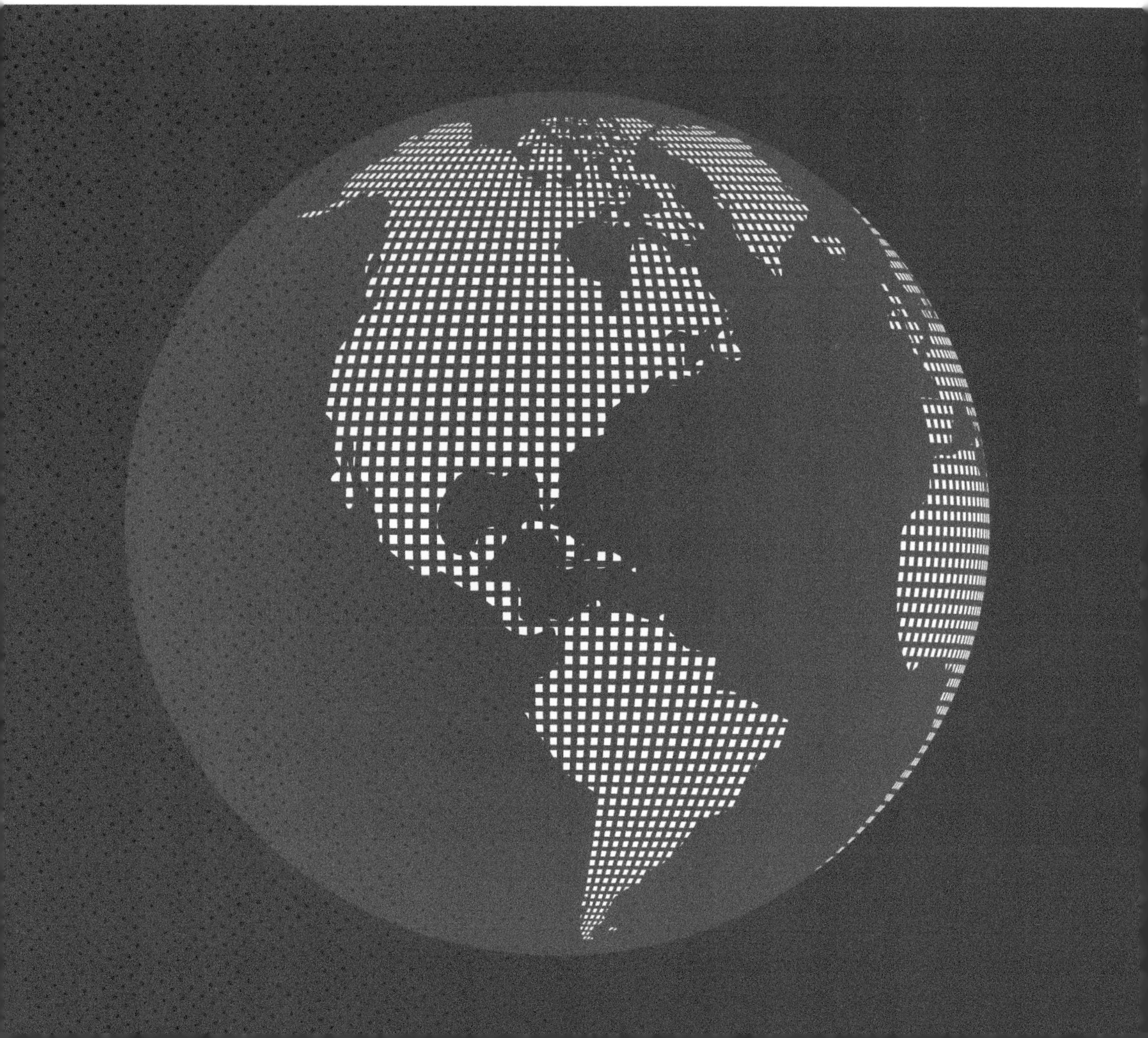

Chapter 16

Aggression or defense? Assessing Russian intentions in the Arctic

Michael R. Pfonner, *University of Southern California, Spatial Sciences Institute*
Darren Martin Ruddell, *University of Southern California, Spatial Sciences Institute*

Overview

In this chapter, you will learn how to apply GIST tools and methods—as well as open-source intelligence (OSINT) sources—to identify potential HSGI threats pertaining to Russian military movements in the Arctic Sea region.

By analyzing changes in Arctic sea ice over time, you will develop and build on your spatial and spatiotemporal analytic skills and proficiency with ArcGIS Pro. You will also become familiar with Russian and North Atlantic Treaty Organization (NATO) military forces to predict future Russian movements and gauge Russian intentions toward this region as sea ice continues to recede. By doing so, you will be able to identify human security threats to communities that rely on the Arctic—particularly those that reside in the region—and may be affected by militarization, based on your perception of Russian priorities.

We use "Arctic Sea region" to describe the geographic area adjacent to the body of water, whereas "Arctic sea ice" denotes sea ice—frozen seawater atop the ocean's surface—in the Arctic Sea.

Learning objectives

- Perform exploratory data visualization of changes in Arctic sea ice distribution across time.
- Conduct geospatial data analysis of changes in sea ice patterns.
- Read and understand commonly used graphical schematics of force structures.
- Produce and interpret static and dynamic mapping products for intelligence purposes.
- Analyze military intentions using a human security lens.
- Identify locations vulnerable to potential human security threats resulting from movements of military forces.

Technical components

- Software: ArcGIS Pro 3.4
- Datasets:
  - Arctic sea ice, National Snow and Ice Data Center
  - Arctic regions, ecological and biological significance areas
  - Shipping and natural resources, National Geospatial-Intelligence Agency
  - Esri portal data, ArcGIS Living Atlas of the World and ArcGIS Online
  - Land cover and vegetation, Arctic Data Center
  - Reference material for NATO and Russian forces

Prerequisites

- Prior knowledge of or experience using GIS software or spatial data is recommended.
- Technical skill level: Intermediate GIS.
- Time to complete chapter: One to two hours without intelligence brief, three to five hours with intelligence brief.

Background and situational context

Since 2005, the Russian state has refurbished, modernized, and reactivated Soviet-era military bases in the Arctic. Russian military forces have been deployed and reinforced across the northern extent of Russia, from the Kola Peninsula in the west to the Chukchi Peninsula in the east. The study area is displayed in figure 16-1.

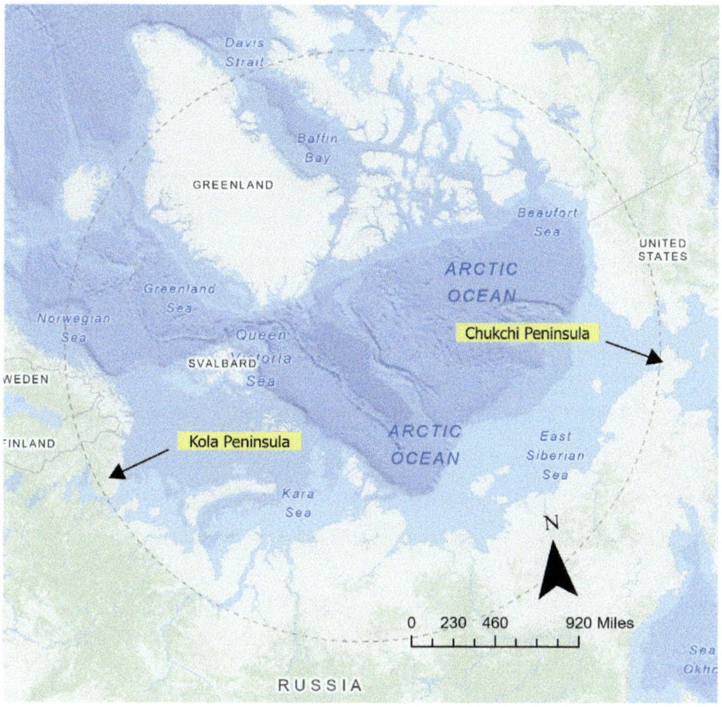

Figure 16-1. The Arctic Sea region.

As the yearly maximum extent of Arctic sea ice declines because of climate change, Russia has expanded its military operational footing in the region, even as NATO members struggle to maintain their current infrastructure while modestly increasing their members' contributions of forces to multilateral contingents. Misinterpretations of intentions and failure to reach mutually acceptable agreements are frequent causes of war, so the current imbalance in the Arctic poses potential threats to international and human security.

There are two dominant sets of interpretations of Russian behavior. Although these are not mutually exclusive (there is certainly room for overlap), analysts tend to group Russian behavior into two broad camps. For the purposes of this exercise, they will form two competing assumptions that you will consider:

- **Aggressive intentions**: Russia primarily seeks to deploy forces to the Arctic to improve its nuclear weapons capabilities and set up a launch pad for conflict with NATO in the event of a war, however unlikely.
- **Defensive intentions**: Russia primarily seeks to deploy forces to protect the Northern Shipping Route (also called the Northern Sea Route) and natural resource exploration activities from sabotage.

Keep these assumptions in mind as you conduct an analysis of sea ice. Also, consider the types of military units that have been deployed in the Arctic and think about their purpose. For example, would highly mobile armored units be more suitable for aggressive or defensive use? What about surface-to-air missile batteries?

This exercise focuses on human security impacts, so think critically about what kinds of effects aggressive and defensive deployments of military forces may have on the communities that rely on the Arctic for their way of life. Using the preceding example, think about the effects that a large number of personnel and vehicles would have on a community and its environment compared with installations that rely on automated systems. These considerations will help you gain insight into Russian actions and intentions in the region, which can then be used to help formulate a set of potential response scenarios.

Exercise workflow 16

The spatiotemporal analytic portion of this exercise provides users with multiple learning experiences. It introduces key context and major datasets used in the analysis of climate change and human security, including static and live datasets. By learning and using spatiotemporal analytic techniques, users will gain confidence in their technical skills in preparation for increasingly advanced applications.

Users will also develop a working familiarity with several GIS functions that they are likely to encounter in their professional lives:

- Build out a basic analytic map by configuring the study area, projection, and display settings.
- Add content to the map and confirm that it loads and displays correctly.
- Explore datasets by examining data tables and the relationships between the data and its objects (in this case, polygons).
- Apply spatiotemporal analysis tools to examine trends in data over time.
  - Statistically and visually compute changes in the spatial extent of sea ice over time.
  - Distinguish between seasonal variation and long-term trends.
- Assess the visualized change in sea ice across time in conjunction with revealed natural resources and shipping routes to identify regions of interest for Russian expansion or Russia–NATO contention.
- Consider human security concerns in user-specified areas.

In this exercise, you will first analyze changes in Arctic sea ice over time to help identify where Russian military expansion is most likely. Following your spatiotemporal analysis, you will integrate and apply OSINT sources of NATO and Russian military buildups to conduct an intelligence analysis and then produce an intelligence brief.

1 Set up your map

1. Open ArcGIS Pro. Under **New Project**, click **Map** to create a new map.
2. In the **New Project dialog box**, apply the following settings:
   - **Name:** Chapter_16
   - **Location:** C:\SecurityFirst

3. Navigate to the **Catalog** pane on the right. Click the **Portal** tab and then click the **Living Atlas** button. In the search bar, type Polar Basemap and press **Enter**.
4. Right-click **Arctic Ocean Base** and then click **Add To Current Map**.

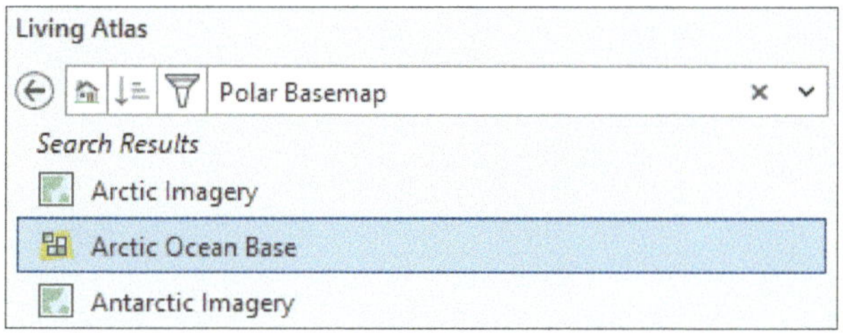

5. In the **Catalog** pane, click the **Computer** tab. Locate the location in which you saved the datasets provided with this exercise.
6. Navigate to the **NSIDC_arctic_ice** folder.

   There should be 20 individual shapefiles: four data layers per decade—each decade should have 1 (January); 4 (April); 7 (July); and 11 (November), corresponding to the calendar month of each recording.

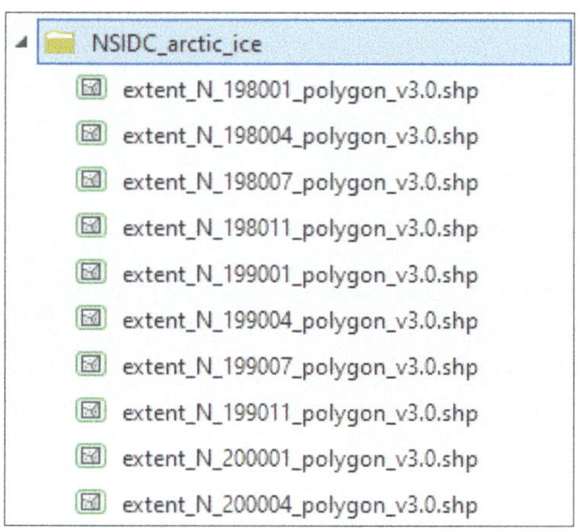

> **Tip:** Each layer has a number that correlates with the year and month of the recording. For example, in the shapefile extent_N_198001_polygon_v3.0, the number 198001 stands for January (01) 1980.

7. Right-click **NSIDC_arctic_ice** and click **Add to Project**.
8. Click the **Project** tab and expand **Folders**. Find the data folder you just added.
9. Select all the shapefiles, right-click, and click **Add To Current Map**.
10. In the **Contents** pane, confirm that the layers are selected and visible on your map.

    Your map should show a bird's-eye perspective of the Arctic. The colors used will likely vary by user.

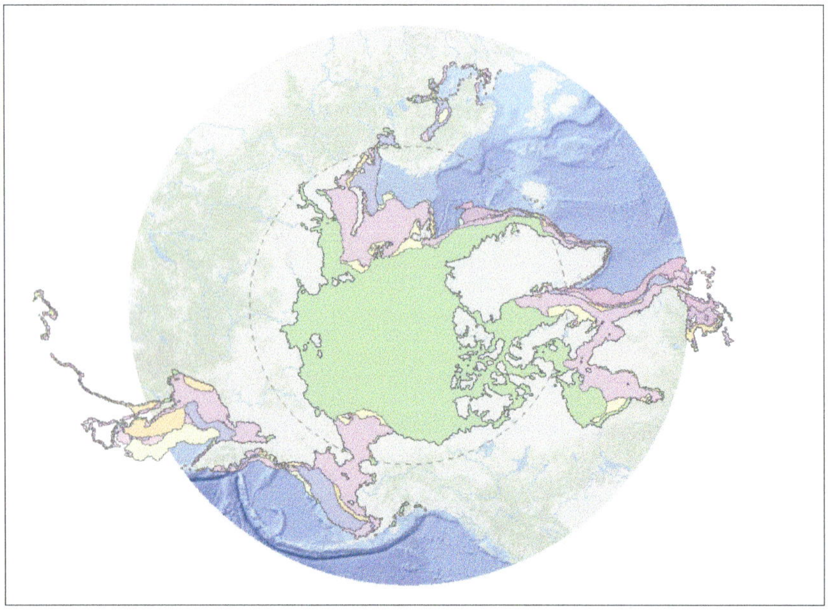

11. Uncheck the **box** for all data layers except **extent_N_198001_polygon_v3.0** to hide them.

> **Tip:** This can be done for all layers (except for January 1980) by using Shift + Select, which will save time in rendering each change. The remaining data layer represents sea ice extent from January 1980.

12. Examine the January 1980 (**extent_N_198001_polygon_v3.0**) data layer.

    What do you observe about the sea ice extent? Which countries are in contact with the sea ice extent?

> **Tip:** If needed, turn on labels with country names. Navigate to the Map tab in the ribbon at the top of your screen, click Basemap, and select Terrain with Labels. This layer can be turned on or off as needed throughout the exercise.

2 Explore the study area

You will now orient your map by rotating it. This will prominently display the Russian Arctic.

1. In the **Contents** pane, right-click **Map** and click **Properties** to open the **Map Properties** window.
2. On the **General** tab, set the **Rotation** to 110.00 to orient the view toward the study area.

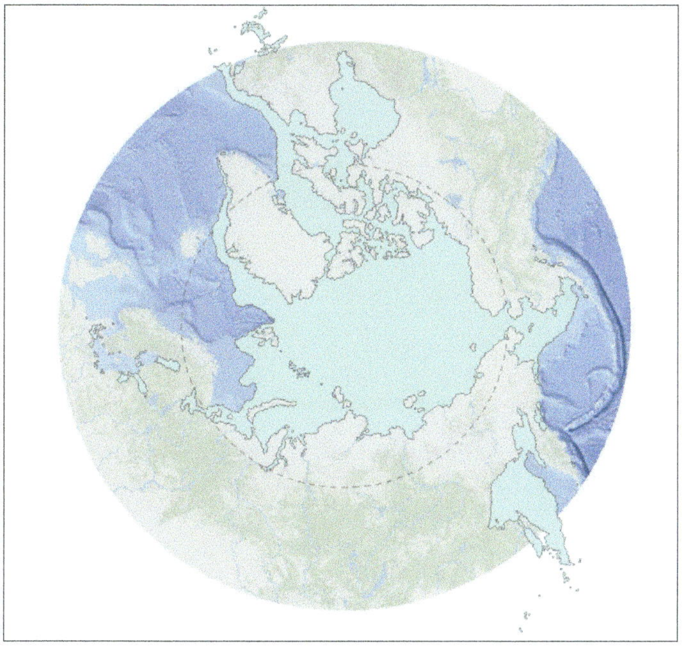

Next, you will select the appropriate projected coordinate system so that you can conduct spatiotemporal analyses.

3. In the **Map Properties** window, click the **Coordinate Systems** tab.
4. Look at the **Current XY** section to identify the current coordinate system.

   The default is probably **WGS 1984 EPSG Alaska Polar Stereographic**. But it may instead be set to **NSIDC_Sea_Ice_Polar_Stereographic_North**, depending on which stage you are at in the analysis.

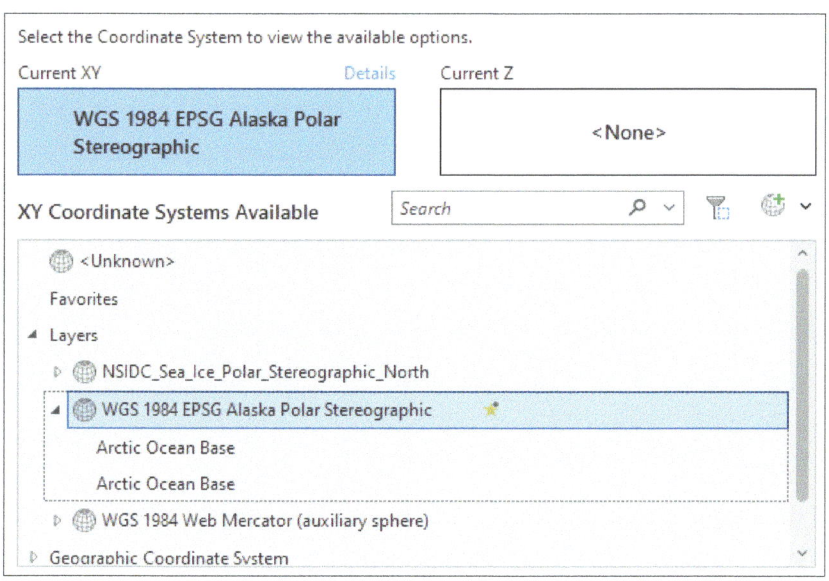

Your task is to assess Arctic sea ice extent, so you will use a projection that is conducive to spatial analysis and comparing areas.

5. On the **Coordinate Systems** tab, search for North Pole Lambert and press **Enter**.
6. Look at the **XY Coordinate Systems Available** list. Click the arrow next to **Projected Coordinate System**, followed by **Polar**, and then click **North Pole Lambert Azimuthal Equal-Area**. At the bottom of the window, click **OK**.
7. After your map redraws, set the **Rotation** to 250.00 using the procedure described earlier.

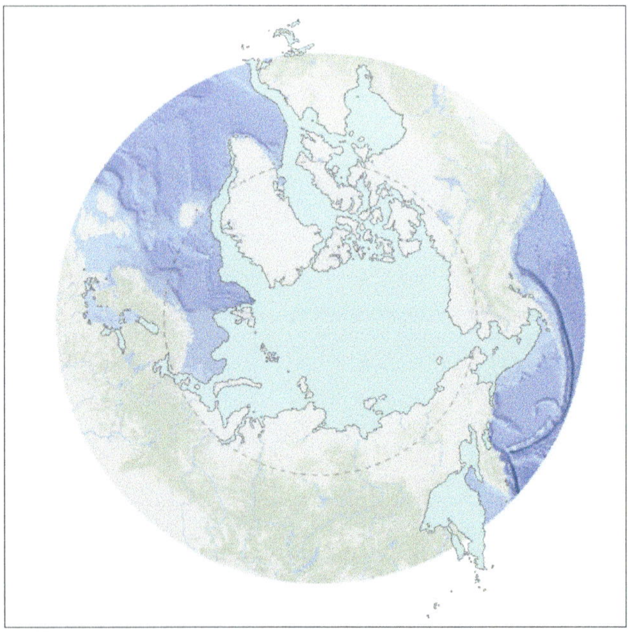

8. Keeping the January 1980 data visible, turn on the January 2020 shapefile (**extent_N_202001_polygon_v3.0**).

   What do you notice about the extent of sea ice compared with the older data? Which areas may be more accessible?
9. Turn off both datasets for January. Turn on the July data for both 1980 and 2020.

   What do you observe compared with the January data?

3 Perform an exploratory analysis of the sea ice extent

In this section, you will assess the variation of Arctic sea ice within a given year (seasonal change), as well as variation throughout the entire time series. First, you will examine the changes in sea ice extent visually on the map. You will create a chart to visualize the change.

1. Examine the four data layers within a given year—beginning with 1980—and note seasonal changes from January (01) to April (04) to July (07) to November (11).

   What seasonal changes do you observe? Which months possess the greatest sea ice extent and where?

2. Repeat this process with at least two additional years for the decade 1980–1990.
3. Repeat the process with at least three years in each of the other decades.

   Next, you will create a chart to visualize changes in sea ice extent for the full 40-year time series.

4. In the **Catalog** pane, click the **Computer** tab. Find the CSV file on sea ice extent, **Sea_Ice_Extent_Monthly_Data_by_Year_G02135_v3.0.csv**. Expand the file to show **'NH-Extent$'**. Right-click it and click **Add To Current Map.**

   These numbers report monthly average sea ice extent in millions of square kilometers.

5. In the **Contents** pane, under **Standalone Tables**, right-click the **'NH-Extent$'** file and click **Open** to open the **Attribute Table.**

   What years are reported in this dataset? Generally, which months see the greatest sea ice extent, and which months have the least sea ice extent?

6. Close the **Attribute Table.**
7. Right-click the **'NH-Extent$'** file and click **Create Chart** > **Line Chart** to open the **Chart Properties** pane on the right.
8. In the **Chart Properties** pane, enter the following details:
   - **Date or Number**: Year
   - **Aggregation:** <none>
   - **Numeric field(s)**: Annual

   Your chart appears under the map.

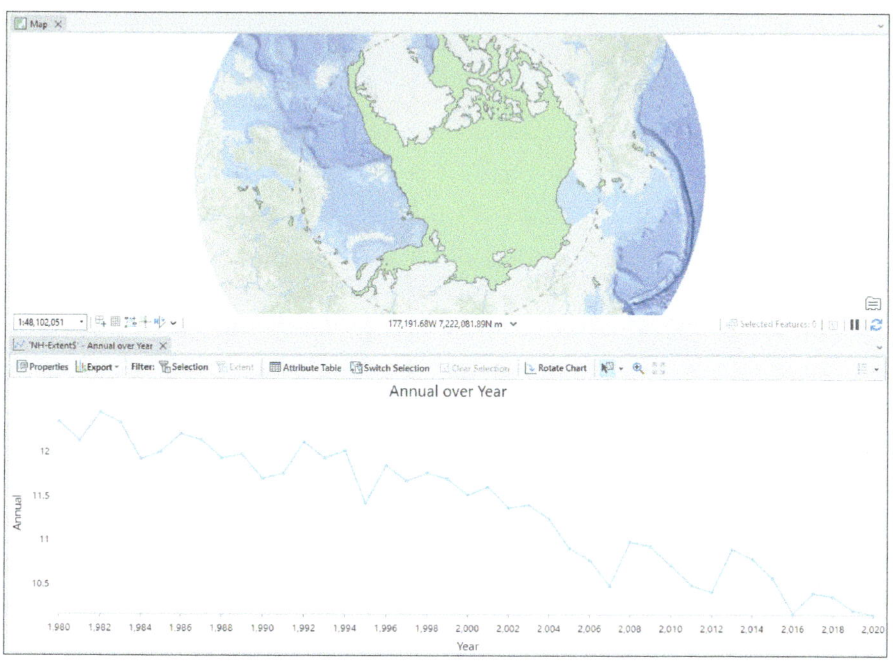

Now you can examine the time series expressed monthly over time to see whether there are trends in the data.

9. In the **Chart Properties** pane, enter the following details:
    - **Date or Number**: Year
    - **Aggregation**: Sum
    - **Numeric field(s)**: <Check the box for each month>

What results do you observe in the change in sea ice extent throughout the time series? What does each month exhibit as a general trend year over year?

> **Tip:** If you want, you can export the chart as a graphic or table by selecting Export > Export as Graphic (or Export as Table) in the chart window below the map.

10. Close the chart.

4 Map potential Russian interests in the Arctic with live data

In this section, you will map and analyze possible Russian interests in the region. You will begin by importing live data directly from an ArcGIS portal. This data is similar to the shapefile data that you used earlier but with a few features that lend themselves to different forms of analysis.

You will import the Arctic sea ice live feed data to compare datasets.

1. In the **Catalog** pane, navigate to the **Portal** tab and click the **Living Atlas** button.
2. Search for Arctic Sea Ice Extent and press **Enter**. Right-click the dataset and click **Add To Current Map**.

> **Tip:** Make a copy of the data to ensure correct use.

3. In the **Contents** pane, right-click **Arctic Sea Ice Extent** and click **Data Design** > **Fields**.

    The **Fields** table will open.
4. Double-click the **Visible** heading, which will check every field.
5. On the top ribbon, navigate to the **Fields** tab. In the **Manage Edits** group, click **Save**.

6. Close the **Fields** table.

> **Tip:** Reproject the data to make it compatible with the current map.

7. On the ribbon, click the **View** tab. In the **Windows** group, click **Geoprocessing.**

   The **Geoprocessing** pane will open at right.

8. Search for Copy Features and click the tool when it appears.

9. In the **Copy Features** tool, apply the following settings:

   - **Input Features:** Arctic Sea Ice Extent
   - **Output Feature Class:** ArcticSeaIceChangeSeptember

10. At the top of the tool, click the **Environments** tab. For **Output Coordinate System**, select **Current Map []**. Your map name appears within the brackets.

11. Click **Run.**

12. Open the **Attribute Table** of the newly created layer (**ArcticSeaIceChangeSeptember**) to become acquainted with the data.

    What similarities and differences do you notice between this data and the dataset used to examine sea ice extent earlier? What are some strengths of each data structure?

    Next, you will compare charts for sea ice over time.

13. Create a **Line Chart** for **ArcticSeaIceChangeSeptember.**

    The **Chart Properties** pane appears.

14. Apply the following settings:

    - **Date or Number:** Year
    - **Aggregation:** Sum
    - **Numeric field(s):** Extent (million km^2)

    Note: The last year and month will depend on the recency of your data. The dataset you are working with is a live feed, so it may vary depending on when it is retrieved.

    How does this chart compare with the chart generated by the static data you used earlier?

15. Change the **Aggregation** to **Mean and Median.**

    What do you observe in the trends?

    Finally, you will split the data to examine the time series by month over the 45-year time series.

16. In the **Chart Properties** pane, set **Aggregation** to **Sum** again. For **Split by (optional)**, click **Month.** If you get a warning, click **Yes.**

    What do you observe about monthly trends? Which months consistently exhibit the greatest extents of sea ice? Which months have the least extents?

Tip: If you want, you can export the chart using the same method described earlier.

The next steps show you how to illustrate changes in sea ice extent. You will start by creating a static map to illustrate changes over time. Thinking about your chart, you can see that February and March typically have the greatest extent of sea ice, whereas September has the least extent.

Why do you think this is the case?

Next, you will map the extent of September across the time series.

17. In the **Contents** pane, right-click **ArcticSeaIceChangeSeptember** and click **Properties**.
18. Click the **Time** tab, and for **Filter using time**, select **No time–content is always shown**.

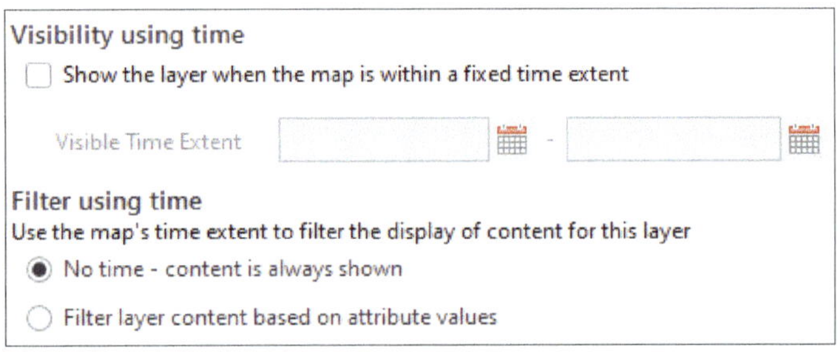

19. Click the **Definition Query** tab and click **New definition query**.
20. In the **Definition Queries** settings in the layer properties window, create the following query:
    - **First menu**: Month
    - **Second menu:** is equal to
    - **Third menu:** 9

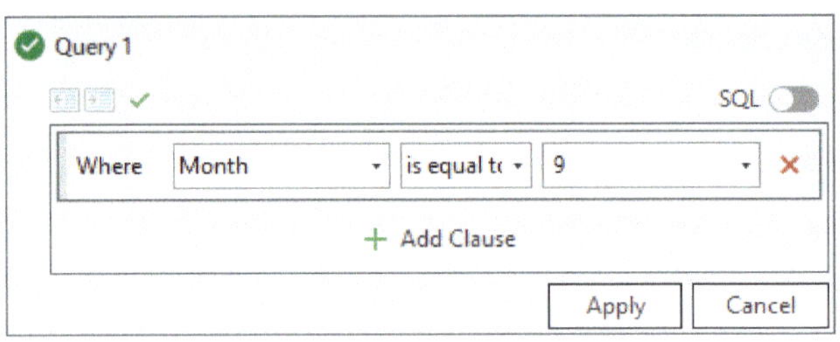

21. Click **Apply** and then click **OK**.

    What does the chart show now? Why was 9 selected for the definition query?

22. Close both the chart and the **Chart Properties** pane.

    Note: The chart appears in the Contents pane so you can return to it when needed.

    You have also applied a filter to the layer, which will make future, intensive analyses more manageable. The filtered number of records will depend on the recency of the data.

    How many records does your **ArcticSeaIceChangeSeptember** or equivalent dataset have? How many records per year will we see as the live data develops?

5 Map sea ice change over time

Next, you will create a map using the preparatory steps you have conducted so far. This map will form the basis of a comparative analysis between visualization techniques.

1. Right-click **ArcticSeaIceChangeSeptember** and click **Symbology**.
2. Navigate to the **Symbology** pane on the right.
3. Under **Primary Symbology**, select **Graduated Colors**.

   Note: The default may be a yellow-red color scheme. Do not manually select a new color scheme; the map display will automatically update with a new graduated color scheme once the following step is completed.

   Without additional preparation, the result will contain far too many features to make a meaningful analysis. You need to find all overlapping features.

4. Open the **Geoprocessing** pane. Search for the Count Overlapping Features tool and open it.
5. In the **Count Overlapping Features** tool, apply the following settings:
   - **Input Features:** ArcticSeaIceChangeSeptember
   - **Output Feature Class:** ArcticIceChangeSeptemberOverlap
6. Click the **Environment** tab. Set the **Output Coordinate System** to **Current Map []**.
7. Click **Run**.

   This will likely take some time to complete.

8. To view the result, confirm that the only visible layer is **ArcticIceChangeSeptemberOverlap**.

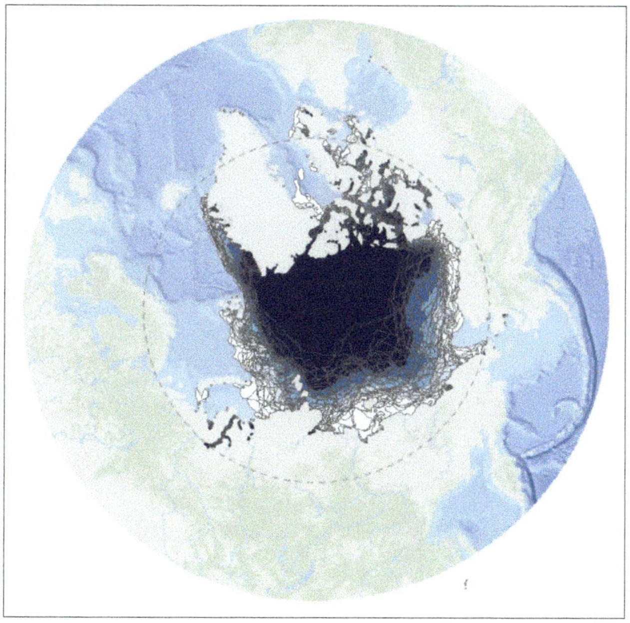

What do you observe about the map compared with the layers you examined in the first two parts of the exercise? How do you interpret the colors used for the map?

Next, clean up the map's appearance by continuing to tweak the symbology.

9. Go to the **Symbology** pane for **ArcticIceChangeSeptemberOverlap**.
10. Under **Primary Symbology**, change the setting to **Unclassed Colors**.
11. Click the **Field** menu and set it to **COUNT_**.
12. Change the **Color scheme** to **Blues (Continuous)**. If you can't see the names of the color schemes, open the **Color scheme** menu and check the box for **Show names**.

    You should see a warning about the maximum sample size at the top of the pane. The number of bins (categories for colors) and the number of records is mismatched. You will probably have 45 or more records and only six bins.
13. Navigate to the top of the **Symbology** pane and click the **Advanced symbology options** button. Go to **Sample size** and change the **Maximum sample size** to 100,000.

    This will be more than sufficient for your layer.
14. Press **Enter** to apply the change.

    You can now make the color scheme more intuitive by reversing the order used by the default color scheme.
15. Return to **Primary symbology**.

    The **Symbology** pane should display a histogram.
16. On the right of the histogram, click **More** and select **Reverse color scheme**.

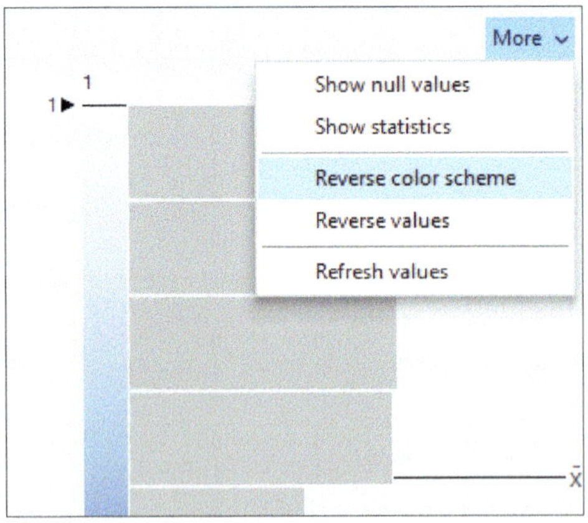

    Now you can reduce the optical clutter present from the many outlines.
17. In the **Symbology** pane, click the **Template** box and then click the **Properties** tab that appears. Under **Appearance**, change **Outline width** to 0 pt. Click **Apply**.

    Your map should show a smooth transition between colors.

How do you interpret the map? What do you observe about the retreat of Arctic sea ice? How does this data compare with the static data for the last few years?

6 Examine Russian movement or intentions in the region

The next part of the analysis will focus on examining Russian movement within the region to help understand the country's strategic intentions.

You will use the static map for one more application before moving on to a dynamic map. You will add some additional data to illustrate potential Russian interests in the region as sea ice retreats.

You will import natural resources data to illustrate potential sources of resource extraction.

1. From the **Energy_Resources_Potential** folder, add **Probability_of_Undiscovered_Oil_or_Gas_Fields.shp** to your map.
2. Change the symbology of the new layer to use **Single Symbol** as its **Primary Symbology**. Click the **Symbol** box.
3. Change the **Color** to **Gray 50%**. Change the **Outline width** to 0 pt.

   Note: You can use any color that makes sense for your analysis.

4. Click **Apply**.
5. On the ribbon, click the **Feature Layer** tab. In the **Effects** group, change the **Transparency** to 70.0%.

   Note: You can set the transparency to any value that enables you to view this data layer and the underlying sea ice layer.

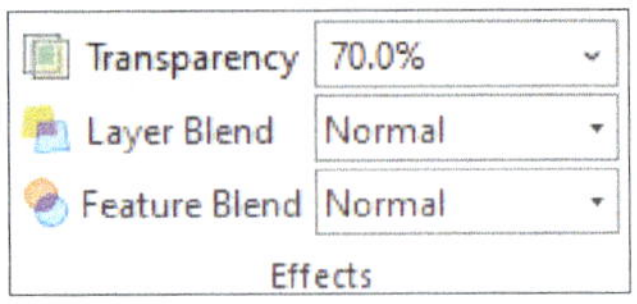

Next, you will add a dataset that illustrates shipping routes in the region.

6. From the **Shipping_and_Hydrography** folder, add **Arctic_Sea_Routes.shp** to your map.
7. Set the **Symbology** to anything that makes the routes easily visible.

   This layer can be labeled without obstructing the underlying map.
8. In the **Contents** pane, right-click **Arctic_Sea_Routes** and click **Label**.

   Note: This layer should be at the top of the map drawing order.

   Your map should now show labels describing Arctic sea routes.

What do you observe about the map regarding retreating sea ice? What countries and regions are seeing the most drastic changes?

Tip: You can turn the visibility of the reference layer on and off if needed.

How do you foresee the future of the hypothetical transpolar route? Where are natural resources becoming more accessible? Do the changes affect Russia or NATO states more? Does this favor either Russia or NATO? How so or why not?

Now you can set up a dynamic map—one that changes over time to illustrate the retreat of Arctic sea ice.

First, you need a duplicate layer of the live data.

9. Turn off the visibility for the static map layer.

   The visibility can be turned on and off later for reference when producing your intelligence brief.

   Next, you should use the skills you gained earlier to complete the following.

10. Import **Arctic Sea Ice Extent** from **ArcGIS Living Atlas** to add it to your map and make a copy of the data.
11. Reproject the layer in your map's current coordinate reference system. Give the layer a slightly different name, such as ArcticIceChangeAllMonths. Do not re-create the charts produced earlier.
12. Do not repeat the steps for limiting the data to September. Right-click the layer, go to **Properties**, and click the **Time** tab. Ensure that **Filter using time** is set to **Filter layer content based on attribute values.** Click **OK.**

    Now you can set up the increments, or time steps, used to illustrate change across the time series.

13. On the ribbon, click the **Time** tab.
14. Match the **Start** and **End** periods in the **Current Time** group to the **Start** and **End** periods in the **Full Extent** group.
15. In the **Step** group, click **Step Interval** and set the interval to 3 **Months.**

    Note: Your end date may vary based on the recency of the live data.

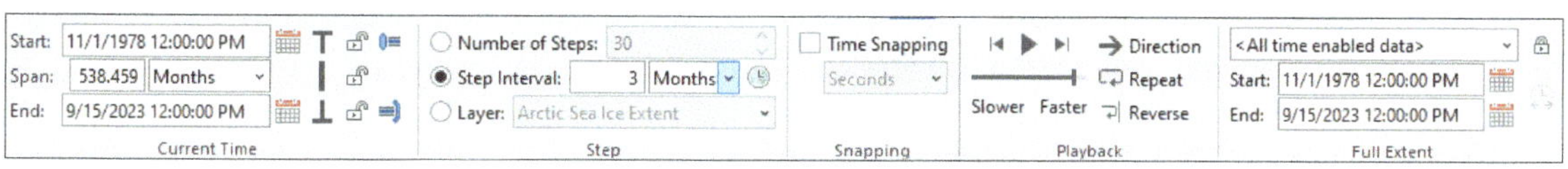

Next, you will select symbology that will contribute to a meaningful analysis. For this stage of the exercise, you will use area rather than extent.

16. Set the **Primary Symbology** of the layer to **Unclassed Colors** and set the **Field** to **Shape_Area.** Change the **Color scheme** to **Blues (Continuous).**
17. Using the steps you performed earlier when you mapped sea ice over time, set the **Outline width** to 0 pt.

    Next, you will fix the drawing order of the map so that you can visualize the changes in the time series while examining effects of the changes on sea routes and energy resources.

18. In the **Contents** pane, drag the **Probability_of_Undiscovered_Oil_or_Gas_Fields** layer above **ArcticIceChangeAllMonths.** Wait for the map to redraw.

19. Drag the **Arctic_Sea_Routes** layer above **Probability_of_Undiscovered_Oil_or_Gas_Fields**. Wait for the map to redraw.

    Before you begin examining the dynamic map, consider the following points:

    - The map will show change over three months for each time step it goes through.
    - The colors will begin as darker blues, and some portions will turn lighter over time. This is intentional because we are studying change. How do you interpret this implementation of color?
    - What time periods exhibit the most dramatic long-term changes (not seasonal change) in sea ice? (You can pause the time steps at any point.)
    - What regions in the Russian Arctic change the most over the time series?

20. From the **EBSA_0511** folder, import **EBSA_0511_JC_area_sort.shp** to your map.
21. Turn on the **Label** feature. Right-click the layer and click **Labeling Properties**.
22. In the **Label Class** pane, delete the line in the **Expression** field. In the **Fields** list, double-click **AREANAME**.

    The Expression field should now show the following:

    ```
    $feature.AREANAME
    ```

23. Click **Apply**.
24. Analyze the layer. When you are done, turn off its visibility.

    This layer can be turned off. When an area of interest is revealed during the time step progression, it can be turned on to make note of the region's name and then turned off again.

    Remember to turn the reference layer on and off as you need. Zooming in and out will label different scales of administrative zones.

    You will now begin the time series.

25. On your map, hover over the **Time slider** and click the **Play** button. Watch the changes take place over time.

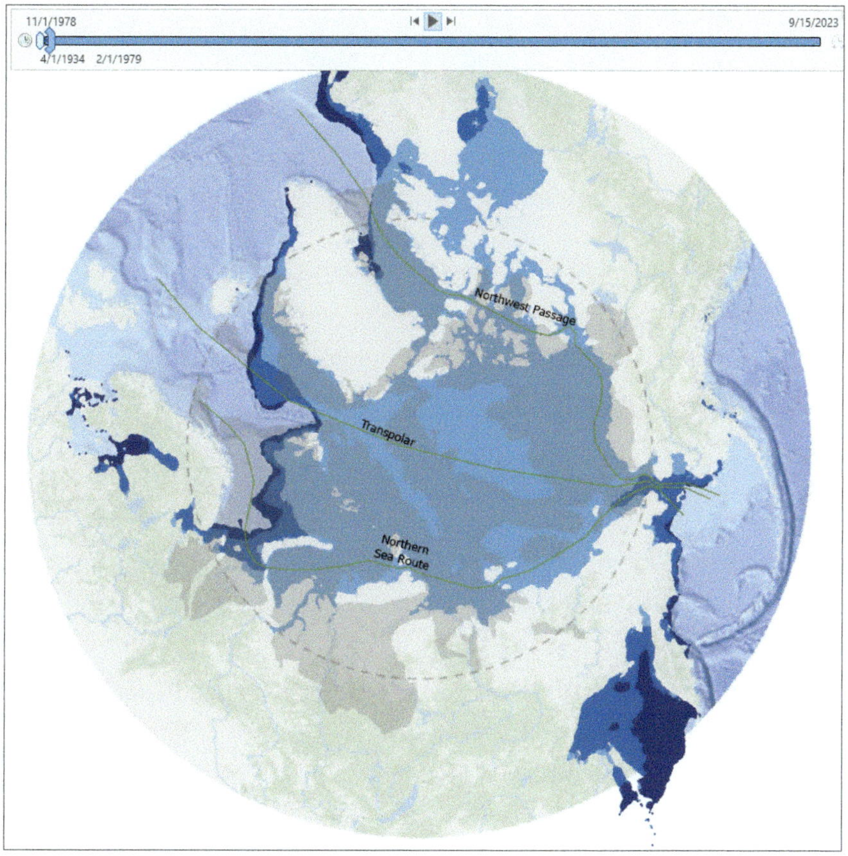

Do you prefer the static or the dynamic display for making intelligence recommendations? Why? What are the strengths and weaknesses of each?

Interpretation of analysis

After completing this exercise, answer the following questions in detail. The answers to these questions, as well as the analysis results from the exercise workflow, will inform the intelligence brief.

1. Based on your analysis, what observations can you identify regarding Arctic sea ice extent and NATO and Russian military deployments?
2. Do you see Russian intentions in the Arctic as being more aggressive or defensive in nature? What interests do you believe Russia is pursuing?
3. Based on your interpretation of Russian military intentions in the Arctic, where, geographically speaking, might there be future militarization? What kinds of military units might be deployed to these locations, and what would their purposes be?
4. What communities might be affected by the type of aggressive or defensive deployment predicted in the preceding question? What types of impacts would one expect to see on those communities?

Note: You can use the ArcGIS Online portal feature to search for "Arctic peoples" and select the Arctic_peoples_subdivided_according_to_language_families dataset to find the names of potentially affected peoples.

5. Are all effects in the region negative, or are there potentially positive impacts that could result? How can the human security element be addressed to mitigate negative impacts on Arctic communities?
6. For state security concerns, how should NATO member states respond to Russian militarization depending on your interpretation of Russian intentions?
7. What additional sources or types of sources might you employ to enhance support for your conclusions and recommendations for action?

Intelligence brief

Based on the analysis from the exercise workflow, produce a three- to five-page intelligence brief by responding to the following question prompts. Incorporate your geospatial outputs from the exercise to enrich and illustrate your conclusions.

- **BLUF:** List three to five primary observations from your analysis in bulleted-list format that summarize findings of this work.
- **Top-level analysis:** Report the primary data points supporting your BLUF.
- **Secondary analysis:** Provide the data sources and deeper analytic products from this analysis.
- **Impact:** Describe the impacts of your findings on the department or organization.
- **Confidence:** Report the level of confidence of your assessment based on available data, accuracy, and so on.
- **Bias:** Explain any background or information you have that could have affected your analysis.
- **Conclusions:** Determine what type of analytic conclusions there may be and how the data requirements, analysis, and brief writing may change based on what analytic framework you are following (descriptive, predictive, prescriptive).

Applying and extending your skills

Consider how you would further explore the topic of Russian militarization of the Arctic in ways that address state and human security concerns. This section provides multiple sources for enhancing your understanding of the topic and using additional data and OSINT sources to add more support or nuance to your conclusions and recommendations.

Data sources and analyses

- How would you structure a dataset to directly analyze Russian military forces? Describe the types of data you would need to collect (unit type, year deployed to the Arctic, and so on) and briefly discuss how you would operationalize it in an analysis.

- What other varieties of data would you seek to further test Russian intentions in the Arctic? Are there existing datasets that you could apply, or would you need to develop your own? What are some strengths and weaknesses associated with off-the-shelf data? What are the strengths and weaknesses of purpose-built data?
- Apart from Arctic sea ice, NATO activity, shipping routes, and natural resources, what other spatial variables can be included to further test Russian intentions? What nonspatial variables could be integrated into a comprehensive quantitative or mixed-methods analysis?
- Conduct a suitability analysis for Russian forces or commercial actors to land equipment in one area that you believe may be a potential target for development:

  a. Identify the area you want to assess. (*Tip*: The smaller the area, the more feasible the analysis.) Try to limit your area of selection as much as possible by zooming in on any of the maps that you have created so far.
  b. Go to the NGA's ArcticDEM index and data download site at https://nga.maps.arcgis.com/apps/webappviewer/index.html?id=aff5fa8f5d5548c6bff44cc8be385f61 and find your preferred area. Select it on the map, which will make a CSV file available for download on the **Results** tab. Download this file.
  c. In ArcGIS Pro, import the DEM file that you just downloaded.
  d. Import the combined land-cover and vegetation file (**BAWLD_V1**).
  e. On the **Analysis** tab, in the **Geoprocessing** pane, click **Tools** > **Toolboxes**.
  f. Click **Intelligence Tools** > **Suitability** > **Drop Zone Suitability**.
  g. For **Input Percent Slope Raster**, select the DEM that you created and downloaded.
  h. For **Input Combined Vegetation Features**, choose the **BAWLD_V1** layer.
  i. Set **Clip Features** to the analytic map that you used to identify the area of interest.
  j. Name your new layer something that works for your purposes. (*Tip*: If you see a warning that your DEM file contains more than 4,000 rows and columns, your DEM file is too large. Select a smaller area on the NGA's data portal and download the dataset for the smaller area.)
  k. Click **Run**.

- How suitable for development is the area that you selected?
- Is the area more suited for approaching from air or by sea?
- Are there any Indigenous communities that would be affected by development in your area of interest? Which ones?
- What is the intended reason for development? Commercial interests? Military? How would the development you envision affect the communities that live there?

Data visualization

1. In ArcGIS Pro, acquire the coordinates of at least three places where you believe Russian military deployments are most likely. Create a basic dataset of these locations by creating a table of points in Microsoft Excel and save it as a CSV file. (*Tip*: For your table, you may want

to include unit type, recency of deployment, or other factors as one or more distinct variable in your dataset.)

2. Add this dataset to your map as a new layer.
3. Assign appropriate symbology to your dataset to illustrate priorities that you find important to display to your audiences.
4. Export this map to present to your audience.
   - What does this visualization tell you about your recommendations and conclusions? Does seeing it physically depicted suggest other variables that can be included in future analyses?

Workflow communication

- How would you respond to a highly technical audience member who wants to see your methodological process but does not necessarily seek to replicate your analysis? Rather than send step-by-step instructions, how would you graphically depict the steps of your workflow in a concise manner?
- Canva (www.canva.com) and draw.io (www.drawio.com) are free, open-source tools for illustrating workflow diagrams with other applications. You can use either of these tools to create a workflow diagram that readers with an understanding of ArcGIS and spatial analytics would understand.
- How do you conceptualize your workflow as a diagram?
- What are the various inputs and outputs?

Additional resources

General

- The Arctic Institute (www.thearcticinstitute.org) is a research institution that provides general reference material for state and human security, engagement with Indigenous peoples, and commerce in the Arctic region.
- For an overview of state sovereignty over the Arctic as climate change alters the landscape in the region, see the ArcGIS StoryMaps story "Arctic Sovereignty: What Is All This Fighting About?" (January 16, 2023) by Ayan Rizvi at https://storymaps.arcgis.com/stories/ee8df9c46a7e4a7ea772c04469cbd66f.

Russia and NATO in the Arctic

- See the Carnegie Endowment for International Peace report "Russia in the Arctic: A Critical Examination" by Eugene Rumer, Richard Sokolsky, and Paul Stronski at https://carnegieendowment.org/research/2021/03/russia-in-the-arctica-critical-examination?lang=en. The report details information on the recent history of Russia's military expansion in the Arctic.

- *The Handbook of the Arctic: A Broad and Comprehensive Overview* (2022), edited by Egor V. Pak, Artem I. Krivtsov, and Natalia S. Zagrebelnaya, provides a critical reference for all things related to the Arctic region. It is available online through SpringerNature Link with institutional access at https://link.springer.com/referencework/10.1007/978-981-16-9250-5.
- See the Reuters story "Dark Arctic: NATO Allies Wake Up to Russian Supremacy in the Region" (November 16, 2022) by Jacob Gronholt-Pedersen and Gwladys Fouche at https://www.reuters.com/graphics/ARCTIC-SECURITY/zgvobmblrpd/. The story is particularly useful for maps, charts, graphs, and other visual comparisons between the two actors.
- The World Economic Forum published the article "A Container Ship Is About to Sail an Arctic Sea Route for the First Time—Thanks to Climate Change" (August 27, 2018) by Stine Jacobsen at www.weforum.org/stories/2018/08/maersk-sends-first-container-ship-through-arctic-route.

Academic summaries

- Odgaard, Liselotte. 2022. "Russia's Arctic Designs and NATO." *Survival* 64 (4): 89–104. doi.org/10.1080/00396338.2022.2103259.
- Roberts, Kari. 2020. "Understanding Russia's Security Priorities in the Arctic: Why Canada–Russia Cooperation Is Still Possible." *Canadian Foreign Policy Journal* 27 (2): 211–27. doi.org/10.1080/11926422.2020.1842220.

Interpretations of Russia's intentions since the war in Ukraine

- Visit the Arctic Institute, Center for Circumpolar Security Studies at www.thearcticinstitute.org/arctic-2023-russian-foreign-policy-concept.
- Visit the Center for Strategic and International Studies (CSIS) at www.csis.org/analysis/russian-arctic-threat-consequences-ukraine-war.

Human security concerns and Indigenous peoples of the Arctic

- For a description of some of the effects of militarization on human security in the Arctic and effects on Indigenous peoples, see "Nature Is Being Destroyed: Russia's Arms Buildup in Barents Sea Creating Toxic Legacy" (July 24, 2023) by John Last of *The Guardian* at www.theguardian.com/environment/2023/jul/24/russia-barents-sea-toxic-legacy-military-buildup-mining-shipping-indigenous.
- *The Handbook of the Arctic: A Broad and Comprehensive Overview* (2022), edited by Egor V. Pak, Artem I. Krivtsov, and Natalia S. Zagrebelnaya, contains a section on human security in the Arctic. It is available online through SpringerNature Link with institutional access at https://link.springer.com/referencework/10.1007/978-981-16-9250-5.
- See "Human Security in the Arctic: A Review of the Russian Literature" (January 14, 2022) from the North American and Arctic Defence and Security Network (NAADSN) at www.naadsn.ca/wp-content/uploads/2022/01/21-nov-Plakhov-Policy-Primer.pdf.

- Visit the website for the Russian Association of Indigenous Peoples of the North at https://arctic-council.org/about/permanent-participants/raipon.

Data

- Arctic sea ice data from the National Snow and Ice Data Center (NSIDC). The extent of Arctic sea ice across time, measured every January, April, July, and November. See the National Snow and Ice Data Center at https://nsidc.org/home.
  - Provided as 20 shapefiles and a CSV file titled Sea_Ice_Extent_Monthly_Data_by_Year_G02135_v3.0.csv.
- ArcGIS Living Atlas of the World (Esri)
  - Arctic Ocean Base
  - Arctic Sea Ice Extent
- ArcGIS Online (Esri)
  - Arctic Peoples Subdivided According to Language Families
  - Kola Peninsula Russian Military Bases
  - RussianMilLanduse
- BAWLD V1 (land cover and vegetation). David Olefeldt, Mikael Hovemyr, McKenzie Kuhn, David Bastviken, Theodore Bohn et al. Fractional land-cover estimates from the Boreal–Arctic Wetland and Lake Dataset (BAWLD), 2021. Arctic Data Center: doi:10.18739/A2C824F9X.
  - Provided as a shapefile titled BAWLD_V1.
- Ecological and biological significance areas (EBSAs) in the Arctic Marine Environment contains shipping and natural resource data created by the Ecologic Institute and the Center for Marine Biodiversity and Conservation (CMBC) at the Scripps Institution of Oceanography, University of California, San Diego. Available from the Conservation of Arctic Flora and Fauna (CAFF) working group of the Arctic Council: https://geo.abds.is/geonetwork/srv/eng/catalog.search#/metadata/06bcf22f-fcf5-4f56-b104-23888968f7e0.
  - Provided as a shapefile titled EBSA_0511_JC_area_sort.shp.
- Energy Resources Potential (USGS): https://arctic-nga.opendata.arcgis.com/maps/energy-resources-potential-usgs/about. Polygon shapefile displaying the probability of expected natural resources per geographic area from the NGA. Based on USGS data.
  - Provided as a shapefile titled Probability_of_Undiscovered_Oil_or_Gas_Fields.shp.
- NGA ArcticDEM: https://nga.maps.arcgis.com/apps/webappviewer/index.html?id=aff5fa8f5d5548c6bff44cc8be385f61. Optional dataset that may be helpful in exploring elevation-related applications in this chapter.
- Sea routes, shipping, and hydrography: https://arctic-nga.opendata.arcgis.com/datasets/nga::arctic-sea-routes/explore. Shapefile of polylines depicting the current Russian route, the Northwest Passage, and the hypothetical Transpolar Route from the NGA Arctic Open Data Application.
  - Provided as a shapefile titled Arctic_Sea_Routes.shp.

Acknowledgments

The authors would like to thank Steven L. Lamy, PhD, and Steven D. Fleming, PhD, for their knowledge on Arctic politics, militarization, and human security.

About the authors

Michael R. Pfonner is a human security practitioner in the migration space and a visiting scholar at the University of Southern California, Spatial Sciences Institute. He has a PhD in political science and international relations from the University of Southern California. His professional and academic research interests include human security, armed conflict, nationalism, and forced migration.

Darren Martin Ruddell is a professor and director of graduate studies at the University of Southern California, Spatial Sciences Institute, where he directs academic programs in HSGI. Ruddell earned his PhD from the School of Geographical Sciences and Urban Planning at Arizona State University. His teaching and research efforts use geospatial technologies to investigate and advance issues of human safety and security, climate change, and human-environment interactions.

Chapter 17

Activity-based intelligence analysis: Geospatial-temporal analysis with big, diverse datasets

| ***Patrick Kenney,*** *Whitespace*

Overview

This chapter requires you to learn and apply the basic concepts of a geospatial analytic methodology called activity-based intelligence (ABI). Since around 2005, analysts at the National Geospatial-Intelligence Agency (NGA) have been developing and applying analytic methods under the ABI label. The ABI analysis approach emerged from the efforts of geospatial analysts to answer intelligence questions focusing on individuals and networks in mission spaces where traditional imagery-focused GEOINT analysis was insufficient to solve the problems.

This chapter begins with the "Four Pillars" of ABI, which presents a nontraditional approach to thinking about GEOINT data, GEOINT analysis, and assessments of GEOINT problems. The chapter then introduces a notional but representative analytic challenge, in which you apply these concepts to solve a problem. Your task is to play the part of an investigative crime analyst with the Baltimore Police Department given an anonymous hotline tip. The tipster claims to have seen a man, Jared Combs, who was convicted of dealing drugs but is reportedly out dealing drugs again after his release on parole. You will prepare data for geospatial-temporal analysis, analyze the data to discover the behaviors and relationships related to the person and activity of interest, and then use these discoveries to assess whether Mr. Combs has indeed returned to crime.

Learning objectives

- Explore nontraditional GEOINT data and its key attributes.
- Prepare diverse GEOINT data for geospatial-temporal analysis.
- Analyze multiple datasets to discover activities of interest.
- Correlate data related to multiple persons, events, and objects to discover and map relationships.

Technical components

- Software: ArcGIS Pro 3.4
- Datasets: Worldline synthetic data, sample set, v. 2.5 (available at https://inthewhitespace.com/worldline)

  Note: This synthetic data contains real places, but the people and activities in it are simulated—they are not real. Also, the sensor collection types in this dataset are realistic versions of analogous real-world collection but simplified for ease of use.

Prerequisites

- Prior knowledge or experience with GIS software or spatial data is helpful but not required.
- Technical skill level: Intermediate GIS.
- Time to complete chapter: Two to three hours without intelligence brief, three to four hours with intelligence brief.

Background and situational context

ABI analysis grew out of the experiences of NGA analysts attempting to provide GEOINT support in the early years of the wars in Iraq and Afghanistan. In these conflicts, imagery analysis was of limited use in answering intelligence questions focused on nonstate adversaries with little to no imagery signature and about whom key locations, persons, and behaviors were mostly unknown. To provide the needed GEOINT support to leaders of US and coalition forces, NGA analysts began to rethink what geospatial means in the context of intelligence data and analysis. This new approach was eventually labeled *activity-based intelligence* and was summed up by the Four Pillars of ABI:

- Georeference to discover
- Integrate before exploit
- Data neutrality
- Sequence neutrality

Georeference to discover and integrate before exploit

All the data gathered by an analyst with a data- and sequence-neutral mindset does no good without a means to view it altogether in a single, cohesive analytic framework within which the analyst can find connections among the people, places, events, and objects observable in the data. For geospatial analysts, a map or—better yet—a GIS is an excellent way to view data from a shared perspective, spatially. Everything happens somewhere. Identifying the spatial attributes of the observations within multiple diverse datasets means that these attributes can be plotted together, or georeferenced, to places on the earth and analyzed to see what observations are near others. For instance, a classified collection platform detects a radio call sign being used along a busy road, a local news outlet reports on a car bombing along that road later that day, and a user of social media posts a photo complaining of a stalled car blocking a lane of traffic an hour earlier. These three pieces of information would be merely a few of thousands of signal intercepts, news reports, and social media posts in the data stores available to an analyst. To examine all that data—to "exploit" it, in intelligence jargon—would take forever. However, to plot it all on a map and allow spatial (and temporal) relationships to point toward the needles in the haystack (that is, to "integrate before exploit") allows an analyst to filter large volumes of data to those records whose proximity to some location or event is most likely of analytic value.

Data neutrality and sequence neutrality

In this new GEOINT perspective, any type or source of information that can plot a person, thing, or event to a place (or a time) on the earth is GEOINT data. A human intelligence (HUMINT) report about a target visiting a shop that has the address or coordinates of that visit is GEOINT. A signals intelligence (SIGINT) radio intercept with the ellipse or circle where the device was detected is GEOINT. A news report about an attack on a village with the general location of that village is GEOINT. This data-neutral approach discarded the dependence on observing targets of analysis in imagery and opened a large volume of information of varied sources to GEOINT analysis. Along with being open-minded about what data to use, ABI analysis also meant becoming open-minded, and mindful, about when the data occurred. Discovering and understanding the behaviors and relationships to answer intelligence questions often requires looking at the past, not just the newest information. This sequence-neutral approach to data encouraged analysts to seek and use information about targets and areas of operation that were years or decades old. Being thoughtful about the timing of events captured in the data led analysts to discover and chart time lines of activities of interest to find patterns and dependencies. For example, an insurgent group repeatedly smuggles weapons from a neighboring country, which requires communications, stopovers at safe houses, and so on. Each event is a chance for an analyst to better understand this pattern of interest and discover the next instance of smuggling.

Crime scenario and analysis

The exercise in this chapter covers a simple example of this nontraditional approach to GEOINT data and analysis. You will play the role of a crime analyst with the Baltimore Police Department, tasked with following up on an anonymous tip.

Crime analyst task

Review the tip (SW11NOV2016029.docx) that was received on November 11, 2016, about a parolee reportedly dealing drugs. Do your best to ascertain whether the parolee has indeed returned to crime.

DATE OF REPORT:

11NOV2016

DATE OF INFORMATION:

07NOV2016 – 11NOV2016

INFORMATION:

An individual provided an anonymous tip via phone call claiming that on 11NOV2016 at approx. 11:00 a.m. the individual saw someone who the tipster believed to be Jared Combs loitering on the southeast corner at Helen Mackall Park (39.29533, –76.66002) with a dark-red or brown backpack until an unknown man drove up in a "white, regular-type car." The tipster saw the possible Mr. Combs interact with the driver but could not see clearly what they were doing. Then the driver drove off and the possible Mr. Combs departed the area south on foot. The tipster thinks this may have been a drug deal because Mr. Combs "has been in that dope mess forever." The tipster recalled briefly seeing the possible Mr. Combs in the park earlier in the week, as well, though was not sure what day.

Exercise workflow 17

1 Examine the data

1. Open **ArcGIS Pro.** Under **New Project**, click **Map** to create a new map.
2. In the **New Project** window, apply the following settings:
    - **Name:** Chapter_17
    - **Location:** C:\SecurityFirst
3. Click **OK.**
4. In the **Contents** pane, right-click **World Topographic Map** and click **Remove.** Remove the **World Hillshade** basemap, too.
5. Navigate to **SecurityFirst_ExerciseData/Chapter17.**

6. Open the **Worldline_Sample_Data** folder and explore the data in the folder.

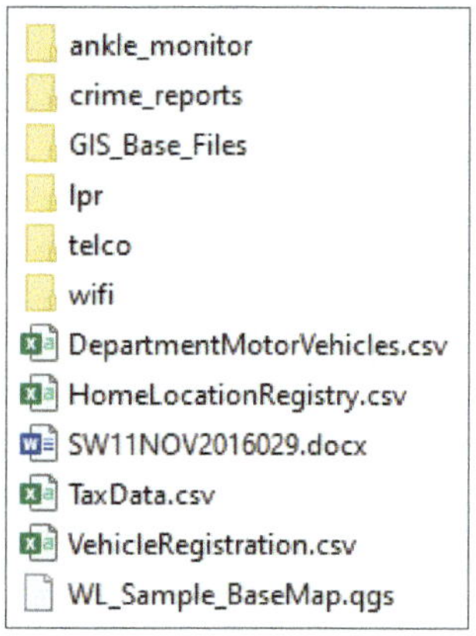

7. In the **Worldline User Docs** folder, glance over **WorldLine Sample Data Catalog.pdf.** As you do, think about the following questions:
    - What data is available?
    - Which files have spatial or temporal attributes?
    - What is the spatial and temporal resolution of each dataset?
    - What identifiers are there for people, places, events, and objects?
    - What challenges might each dataset present during analysis?
    - How are different datasets related to each other?
    - How might each dataset help solve the intelligence problem at hand?

2 Add data and prepare data layers for analysis

In the Worldline sample dataset, all datasets with geocoordinates already have shapefiles created from CSV files. But to be able to use most tools in ArcGIS Pro, you should migrate the geolocated data as feature classes into the project geodatabase and the nongeolocated data as tables into the geodatabase.

Note: Although there are shapefiles of geolocated datasets in the sample data, migrating data as a feature class versus a shapefile will be needed for temporal analysis tools later. Although it is more time consuming to generate feature classes instead of adding existing shapefiles, you must do the former.

1. In the **Catalog** pane, right-click the **Folders** folder and click **Add Folder Connection**. Navigate to **SecurityFirst_ExerciseData/Chapter17** and select the **Worldline_Sample_Data** folder. Click **OK.**

    The next step will be to georeference the CSV data with longitude and latitude coordinates.

2. On the ribbon, click the **Analysis** tab. In the **Geoprocessing** group, click the **Tools** button.
3. In the **Geoprocessing** pane, search for XY Table to Point. In the results, open the tool.

4. In the **XY Table to Point** tool, apply the following parameters:

    - **Input Table:** JaredCombs_Ankle_Monitor.csv (Worldline_Sample_Data > ankle_monitor)
    - **Output Feature Class:** JaredCombs_Ankle_Monitor

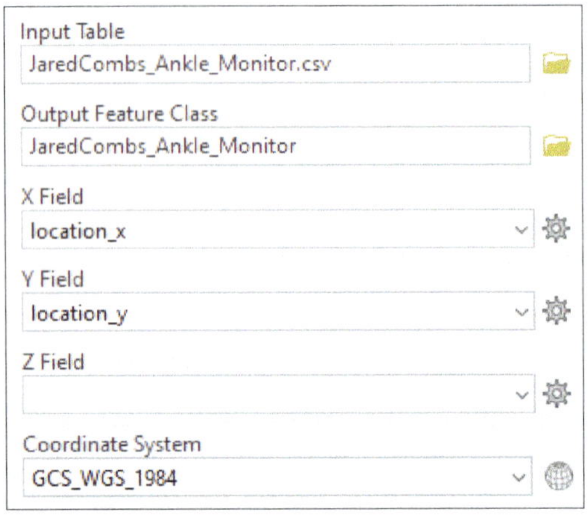

5. Click **Run.**

    When you add a layer to your project, spend some time adjusting the symbology in ways that make it easy for you to distinguish that layer from others. Doing so helps you identify spatial correlations and patterns in the data. The default symbology is often unsuitable.

6. In the **Contents** pane, right-click **JaredCombs_Ankle_Monitor** and click **Symbology**.
7. In the **Symbology** pane, click the point representing the **Symbol.** Click the **Properties** tab.
8. Format the symbology appropriately for the situation, as the following settings do:

    - **Shape fill symbol:** Solid fill with outline
    - **Color:** Arctic White
    - **Outline color:** Ultramarine
    - **Outline width:** 1 pt
    - **Size:** 3 pt

9. Click **Apply**.

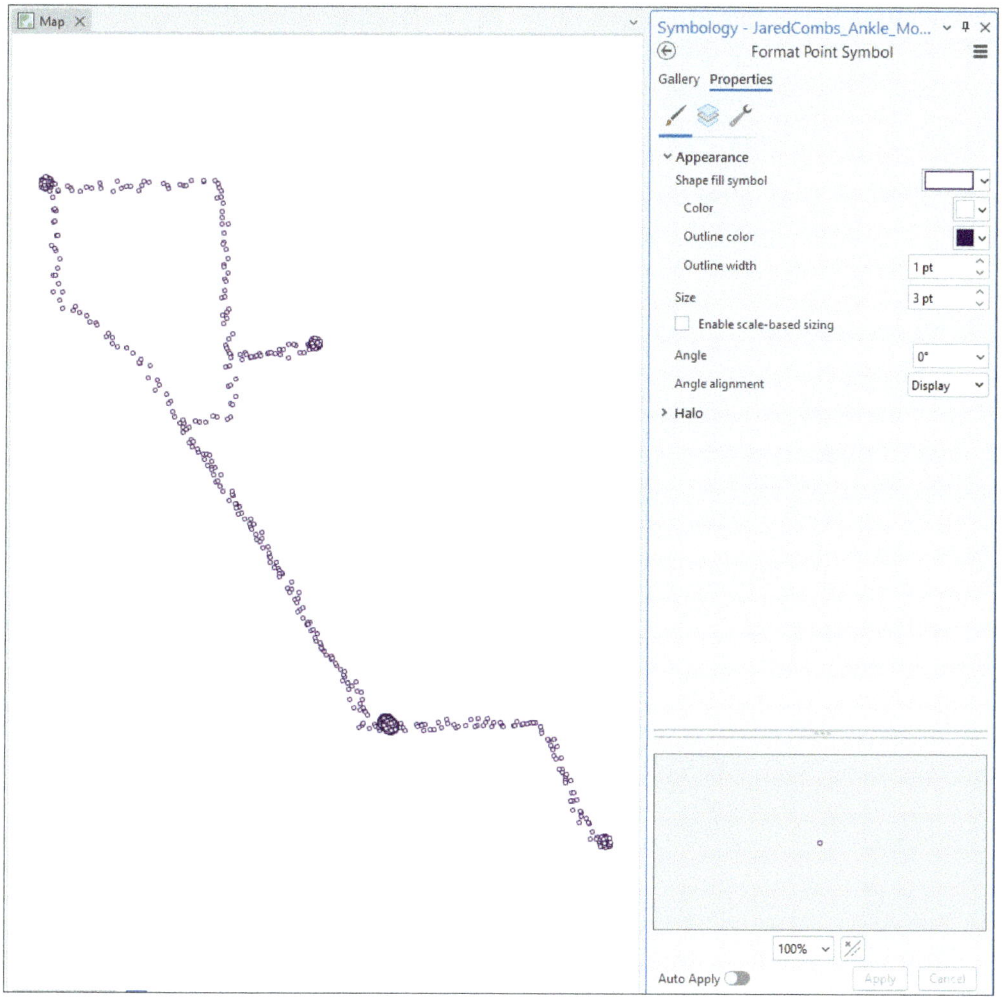

10. In the **Contents** pane, right-click **JaredCombs_Ankle_Monitor** and click **Attribute Table.**
11. Locate the temporal attribute column called **datetime.**

    The ankle monitor data does not merely have geocoordinate fields; it also has a temporal attribute. This information is critical for an intelligence analyst. This field can be used to time-enable a data layer, which will let you use spatiotemporal tools, improving your analyses and insights.

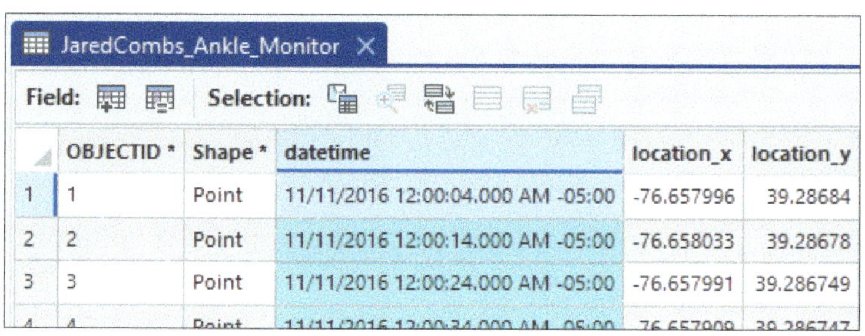

	OBJECTID *	Shape *	datetime	location_x	location_y
1	1	Point	11/11/2016 12:00:04.000 AM -05:00	-76.657996	39.28684
2	2	Point	11/11/2016 12:00:14.000 AM -05:00	-76.658033	39.28678
3	3	Point	11/11/2016 12:00:24.000 AM -05:00	-76.657991	39.286749

The first step for time-enabling a layer with dates and times (date-times) is to confirm that it is a feature class in a geodatabase, not a file-based shapefile. In a shapefile, a date field can hold only date information, whereas a feature class field can hold date and time information. Because the **JaredCombs_Ankle_Monitor** layer is already a feature class, you can proceed.

The **datetime** field uses the **Text** data type, which most ArcGIS Pro tools cannot use for temporal calculations. To use temporal tools, you will use a tool to create a field in the **Date** data type to perform temporal calculations.

12. In the **Geoprocessing** pane, search for Convert Temporal Field and open the tool.

    Note: If you have ArcGIS Pro 3.2 or earlier, use the Convert Time Field tool.

13. In the **Convert Temporal Field** tool pane, apply the following parameters:
    - **Input Table**: JaredCombs_Ankle_Monitor
    - **Input Field**: datetime
    - **Output Field Name**: datetime_Converted
    - **Output Field Type**: Date
14. Click **Run.**
15. Review the results in the attribute table. Scroll to the far right of the table and locate the **datetime_Converted** field. When you have finished, close the table.

datetime_Converted
11/11/2016 12:00:04 AM
11/11/2016 12:00:14 AM
11/11/2016 12:00:24 AM
11/11/2016 12:00:34 AM
11/11/2016 12:00:44 AM

16. To time-enable the layer, right-click the **JaredCombs_Ankle_Monitor** layer and click **Properties.**
17. Click the **Time** tab. Set **Filter using time** to **Filter layer content based on attribute values.**
18. For **Layer Time**, confirm that **Each feature has a single time field** is selected. For **Time Field,** change the field to **datetime_Converted.**
19. Click **OK.**

    When at least one layer is time-enabled, the time slider appears at the top of the map. You can collapse the time slider when not using it by clicking the caret in the slider's top right.

20. Point to the time slider and click **Play.**

    A time series appears of the suspect's movements that were captured on the ankle monitor.

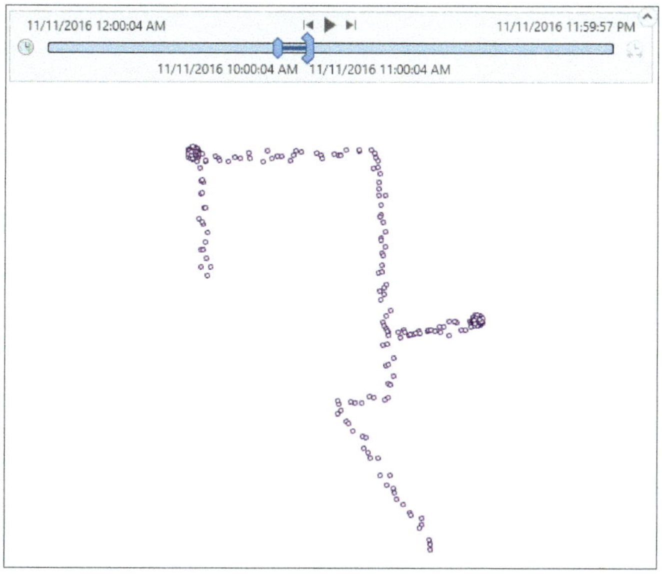

In keeping with the data neutrality pillar of ABI analysis, you will add the remaining data layers to the project.

21. Before adding the layers, click the pause button at the bottom right of the map to pause rendering in the map as each layer is added.

    While rendering is paused, you can uncheck the boxes in the **Contents** pane for large layers (for example, **Telco_2016_11_11.shp**) as you add them. You can leave them unchecked until you need to visualize the larger data layers in the map. This is helpful when working with large data layers.

22. Open the **Feature Class to Geodatabase** tool.

23. In the tool pane, for **Input Features**, click the **Browse** button and navigate to the **Worldline_ Sample_Data** folder connection. Add each of the following shapefiles to convert them into file geodatabase feature classes:

    - **Buildings_Baltimore.shp**
    - **LPR_FieldOfView.shp**
    - **LPR_Sites.shp**
    - **Streets_Balt.shp**
    - **WiFiSensorsLaydown.shp**
    - **Telco_2016-11-11.shp**
    - **DepartmentMotorVehicles.csv**
    - **HomeLocationRegistry.csv**
    - **TaxData.csv**
    - **VehicleRegistration.csv**

24. For **Output Geodatabase**, select **Chapter_17.gdb**. Click **Run**.

25. In the **Catalog** pane, expand the **Databases** folder and open **Chapter_17.gdb**. Right-click each converted geodatabase feature class and click **Add To Current Map**.

26. For the following CSV files, use the **XY Table to Point** tool to convert the latitude and longitude fields into points on the map.

    Note: From the folder connection, view the tables for each of the CSV files to see the coordinate point fields. In the XY Table to Point tool, the x field represents longitude (~ –79) and the y field represents latitude (~39).

    - **crime reports.csv**
    - **Baltimore_Businesses.csv**
    - **CellSiteLocations.csv**
    - **LPR_2016-11-11.csv**
    - **WIFI_2016-11-11.csv**

    Tip: Several datasets are large and may take several minutes to load (the Telco cell phone data layer has hundreds of thousands of records). Being patient with large datasets is a normal part of modern datacentric GEOINT analysis.

27. Click the **Pause drawing** button again to load the data in the map.

    You will adjust the symbology of each layer so that you can visually distinguish the layers on the map.

28. In the **Contents** pane, for each layer, right-click it and click **Symbology**. In the **Symbology** pane, use the **Gallery** tab to select an appropriate **Symbol** and the **Properties** tab to configure it.

    Tip: Use the following image as your own model, if you want.

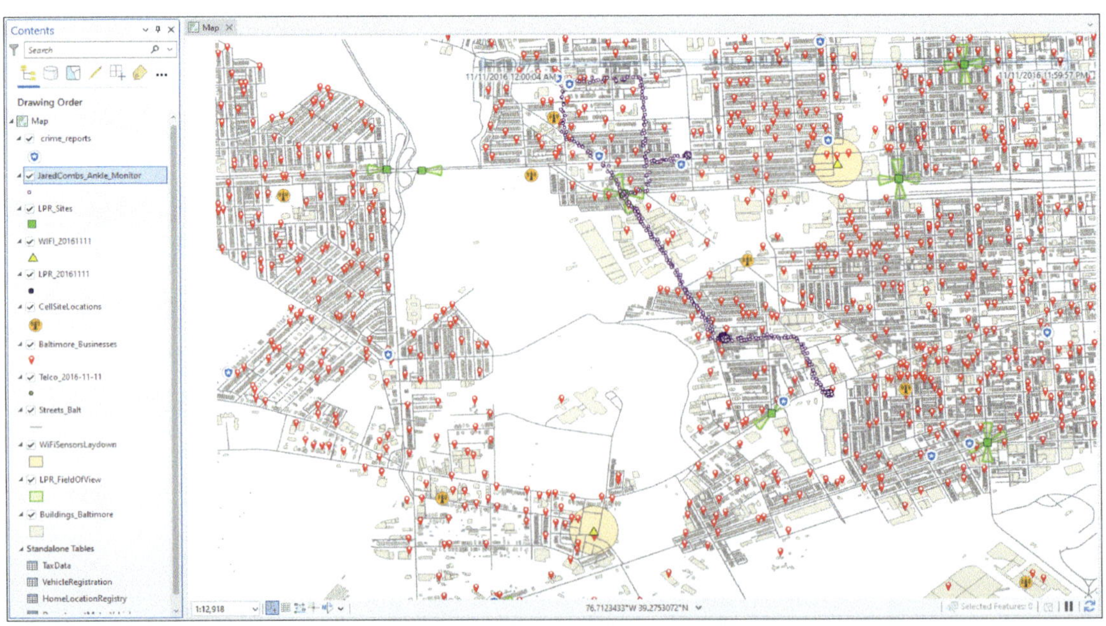

29. On the **Quick Access** toolbar, click **Save Project**.

3 Perform an initial analysis of the suspect's activities

The hotline tip stated that Jared Combs was seen at the southwest corner of Helen Mackall Park (IVO 39.29533, –76.66002) at about 11:00 a.m. yesterday (November 11, 2016), allegedly dealing drugs. You need to verify that Combs was there by using the ankle monitor he wears as a condition of his parole. The tip gives you a spatiotemporal reference for this activity.

1. On the **Map** tab, in the **Navigate** group, click the **Go To XY** tool.

   The tool pane appears at the bottom of the map.

2. Type the coordinates from the tip:

   - **Long:** –76.66002
   - **Lat:** 39.29533

3. Click the **Pan to Location** button to jump to the location of Combs's reported drug dealing.

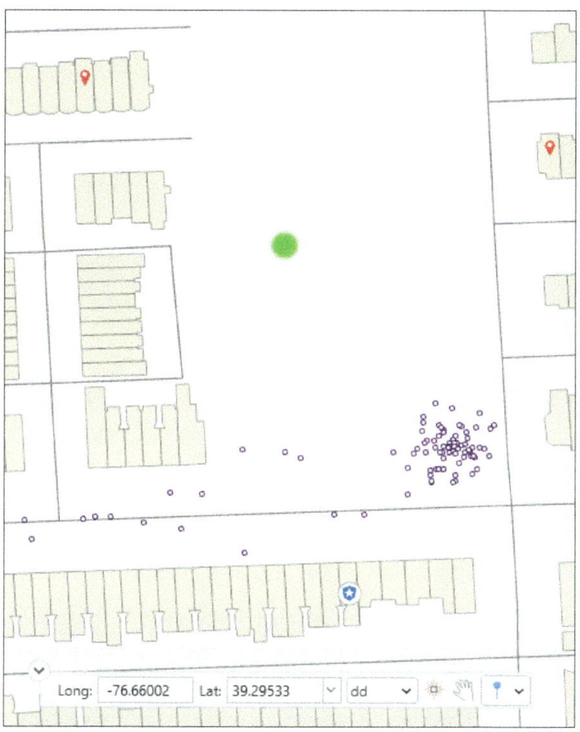

   The GPS ankle monitor points in the bottom right of the map show a spatial correlation between Combs's movements and the reported drug dealing. You want to know whether there is a temporal correlation, too.

4. On the ribbon, click the **Time** tab. In the **Current Time** group, enter the following settings:

   - **Start:** 11/11/2016 10:30:00AM
   - **Span:** 5 **Minutes**

5. Play the time slider to see whether Combs entered the area at around the reported 11:00 a.m. time, and if so, how long he stayed.

> **Tip:** When working with time-enabled data, pay attention to whether the time slider is active because the time slider will filter out data on the map and in the attribute table. To enable or disable the time slider, click the clock icon on the left side of the slider.

The suspect's ankle monitor does, indeed, show a temporal correlation between his activity and the area of the park where the tipster reported a possible drug deal—between 10:35 and 10:50 a.m.

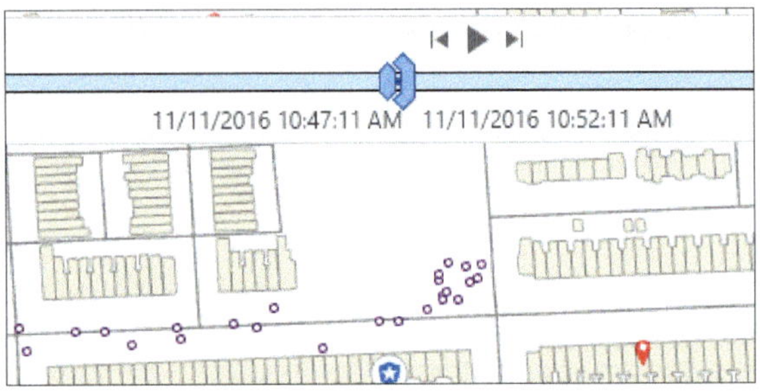

Now that you have discovered that Combs was at the place and time of the reported dealing, you will apply the sequence neutrality ABI pillar and explore his activities before and after the event of interest to learn more about what happened.

6. In the **Current Time** group, change **Span** to 1 **Minutes**.
7. Move the time slider forward and backward in time to examine the suspect's movements around the time of his reported drug dealing.

   According to his ankle monitor data, Combs came from the northwest, loitered in the park for about 10 minutes, and then went southwest, which would not leave much time for selling drugs to passersby.

   Because the tipster was correct about the suspect's presence, the transaction with the white car may be true, too. This meeting during the suspect's brief stay in the area might have been a chance encounter or have been planned ahead. If it was planned, you could query the suspect's cell phone data for call or text recipients before he got to the park. The phone data in the Telco files is quite large—rendering it on the map can greatly slow the software. However, you are not looking for a spatial correlation—just a temporal correlation between his time in the park and his possible phone activity, so you can leave the layer unchecked and unrendered. Instead, you need to query phone records only temporally in the attribute table without rendering the data on the map.

8. Run the **Convert Temporal Field** tool on the **Telco_2016-11-11** layer using the following settings:
   - **Input Table:** Telco_2016-11-11
   - **Input Field:** DTG
   - **Input Format:** yyyy-MM-ddTHH:mm:ss.s
   - **Output Field Name:** DTG_New
   - **Output Field Type:** Date
9. In the **Convert Temporal Field** tool, change the **Input Field** to **DTG_End** and rename its output DTG_End_New. Run the tool again.

> **Tip:** Running this tool on the Telco layer will take several minutes.

Because of the format of the time fields, the result is currently in the UTC time zone. You want the temporal fields to be in the US Eastern time zone to match the location of the report.

10. Search for and open the **Convert Time Zone** tool. Run the tool with the following settings:
    - **Input Table:** Telco_2016_11_11
    - **Input Time Field:** DTG_N
    - **Input Time Zone:** UTC
    - **Output Time Field:** DTG_Converted
    - **Output Time Zone:** US Eastern Standard Time
11. Rerun the tool with the **DTG_End_New** time field. Name the output DTG_End_Converted.

DTG	DTG_END	DTG_Converted	DTG_END_Converted
2016-11-11T00:00:00.000-05:00	2016-11-11T01:01:21.000-05:00	11/11/2016	11/11/2016 1:01:21 AM
2016-11-11T00:00:00.000-05:00	2016-11-11T01:21:41.000-05:00	11/11/2016	11/11/2016 1:21:41 AM
2016-11-11T00:00:00.000-05:00	2016-11-11T01:22:02.000-05:00	11/11/2016	11/11/2016 1:22:02 AM
2016-11-11T00:00:00.000-05:00	2016-11-11T02:16:57.000-05:00	11/11/2016	11/11/2016 2:16:57 AM
2016-11-11T00:00:00.000-05:00	2016-11-11T02:59:20.000-05:00	11/11/2016	11/11/2016 2:59:20 AM
2016-11-11T00:00:00.000-05:00	2016-11-11T03:36:30.000-05:00	11/11/2016	11/11/2016 3:36:30 AM
2016-11-11T00:00:01.000-05:00	2016-11-11T00:00:01.000-05:00	11/11/2016 12:00:01 AM	11/11/2016 12:00:01 AM
2016-11-11T00:00:01.000-05:00	2016-11-11T00:00:21.000-05:00	11/11/2016 12:00:01 AM	11/11/2016 12:00:21 AM
2016-11-11T00:00:01.000-05:00	2016-11-11T00:00:23.000-05:00	11/11/2016 12:00:01 AM	11/11/2016 12:00:23 AM
2016-11-11T00:00:01.000-05:00	2016-11-11T00:00:27.000-05:00	11/11/2016 12:00:01 AM	11/11/2016 12:00:27 AM

Now you will time-enable the layer.

12. In the **Contents** pane, right-click **Telco_2016-11-11** and click **Properties**.
13. In the **Properties** window, click the **Time** tab. Under **Filter using time**, enable timing.
14. For **Layer Time**, select **Each feature has a start and end time fields**.

15. For **Time Extent**, the start time will be 11/11/2016 12:00:00 AM. The end time will be 11/11/2016 11:59:59 PM.

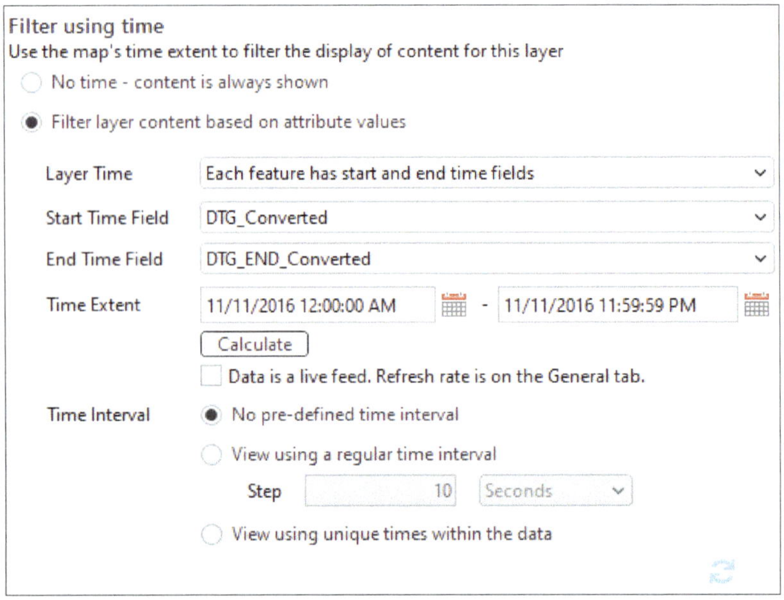

To filter the dataset to only the suspect's calls and text messages, you will set up a filter using a definition query.

16. Click the **Definition Query** tab. Click **New definition query**.

17. Turn on **SQL** mode. Create a query for records with the suspect's international mobile subscriber number (IMSI)—0310070897556017—and that are either a call or SMS using the following SQL expression:

```
IMSI = '0310070897556017' And (Type = 'CALL' Or Type = 'SMS')
```

18. Click **Apply** and then **OK.**

    You can now analyze Combs's movements and phone activity through space and time. The Telco records are filtered by the definition query and the time slider, just as the features on the map would be.

19. In the **Contents** pane, right-click **Telco_2016-11-11** and click **Attribute Table.**

20. On the ribbon, click the **Time** tab. In the **Current Time** group, change the **Start** to 11/11/2016 10:30:00 AM and set **Span** to 15 minutes.

    Examining the filtered datasets together, you can see a call from Combs to another phone (IMSI: 0310012708536005) at 10:36 a.m. and a text from the same number back to Combs at 10:41 a.m. as he was moving southeast toward the site of his reported interaction with the white car.

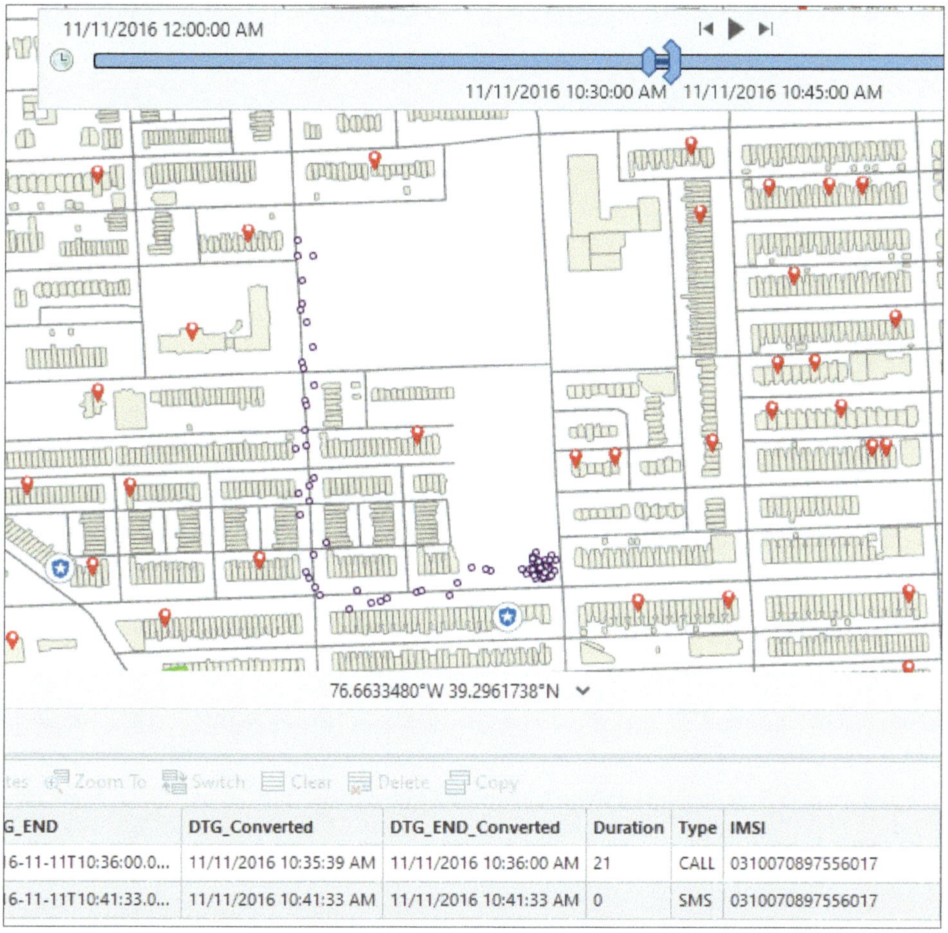

4 Further analyze the activities of Combs and his associate

You have discovered a correlation between Jared Combs's phone communications (IMSI: 0310012708536005) and his loitering at the scene of the reported drug dealing. However, a correlation is not proof. To see whether the owner of that phone (a possible associate) was involved with a reported crime, you will further examine the data to place this other individual at the scene of the crime.

1. Reopen the properties for **Telco_2016-11-11**. On the **Definition Query** tab, turn off **SQL** mode. Update the query to filter the layer to all the records for the new phone using the following expression:

   ```
   Where IMSI is equal to 0310012708536005
   ```

2. Click **Apply** and click **OK.**

   This query includes not only the calls and texts but also the GPRS records, which are the constant background data communications between the device and cell towers. Because these GPRS communications are always occurring, you can view a persistent though imprecise geolocation of cell phones.

3. On the ribbon, click the **Time** tab. Change the **Span** to 1 **Minutes**. Set the **Start** to 11/11/2016 10:30:00 AM.
4. Using the time slider, move forward through the time frame of interest to observe the location of the phone during the events. Watch the location of the phone before and after Combs was loitering in the park.

   Your analysis reveals that the phone was active on one cell tower before Combs's activity, transferred to another on the other side of Combs's location after their phone communication, and then returned to its original cell tower. This may indicate the phone traveling to and then from a meeting with Combs, but the geolocation of this data is not precise enough to assess a physical meeting with any certainty. However, you can look for further data to corroborate a possible physical interaction.

   Within the synthetic Worldline data, there is a **HomeLocationRegistry** file and a **VehicleRegistration** file. The former is essentially data on cell phone accounts, and the latter holds records on vehicle ownership. Before reading the next steps, think about what might be done with these datasets to learn more about the activities of Combs's associate, considering the information you already have.

   You have the associate's phone IMSI, so you can try to query his account info in the **HomeLocationRegistry** and discover identifying information about the associate. If you find such information, you can then use it to query the **VehicleRegistration** data to see if the associate owns a vehicle, such as a white car.

5. Open the **HomeLocationRegistry** stand-alone table.
6. At the bottom left of the table, click the **Show Selected Records** button. Next to **Selection**, click the **Select By Attributes** button.

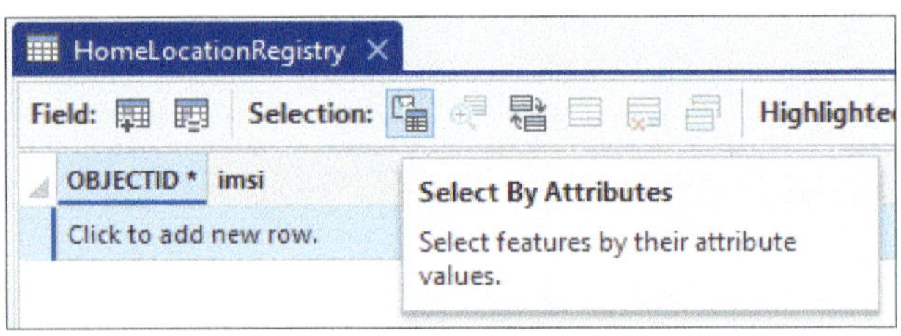

7. To select the record with an IMSI of 0310012708536005, create the following expression:

   `Where imsi is equal to 0310012708536005.`

8. Click **OK**.

	OBJECTID *	imsi	msisdn	make	model	mcc	mnc	mac	firstname	lastname	middlename	ssn
1	20613	310012708536005	443-209-3161	Motorola	A780	310	12	null	ARTHUR	COMBS	ALBERT	495-CP-4562

Jared Combs's phone contact is Arthur Albert Combs. The shared last name indicates a possible family relationship. Now that you know his name, you can search for more information on Arthur Combs in other datasets.

9. Open the **VehicleRegistry** stand-alone table. Click the **Show Selected Records** button and click the **Select By Attributes** button. To select the record with Arthur Albert Combs, create the following expressions:

    `Where Owner First Name is equal to ARTHUR`

    `Where Owner Middle Name is equal to ALBERT`

    `Where Owner Last Name is equal to COMBS`

10. Click **OK**.

	OBJECTID *	Vehicle Make	Vehicle Model	Vehicle Year	Vehicle Color	State Registered	License Plate	VIN	Owner Last Name	Owner First Name	Owner Middle Name
1	80797	CHEVY	IMPALA	2005	WHITE	MARYLAND	249366B/C	11627486858674438	COMBS	ARTHUR	ALBERT

According to the vehicle registry data, A. Combs owns a white Chevrolet Impala, which matches the description of the vehicle from the hotline tip that J. Combs interacted with in the park. If you look at the area of interest, you can see features for the license plate reader (LPR) sensors located between the cell towers used by A. Combs during the time frame of interest.

You can investigate whether A. Combs's white car was observed passing these sensors during that time, which would give you more spatially precise observations of A. Combs approaching and leaving from the meeting location in the park.

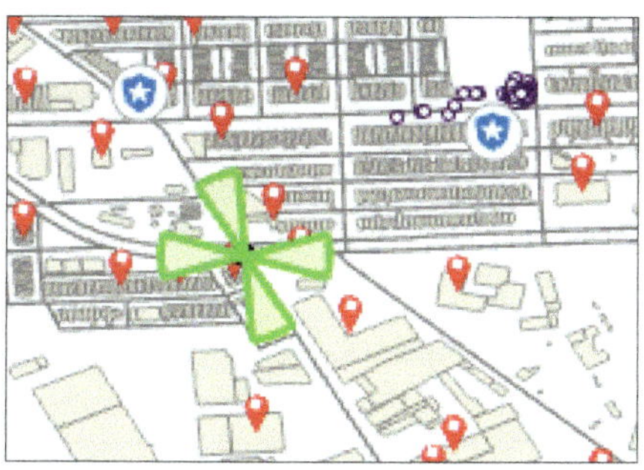

11. Run the **Convert Temporal Field** tool on the **LPR_2016-11-11** layer using the following settings:
    - **Input Table:** LPR_20161111
    - **Input Field:** datetime
    - **Output Field Name:** datetime_Converted
    - **Output Field Type:** Date
    - **Time Zone Offset:** String, US Eastern Standard Time

12. In **Properties**, time-enable the layer.

13. On the **Definition Query** tab, create a query to filter the LPR data to records for license plate 249366B/C using the following expression:

    `Where licenseplate is equal to 249366B/C`

14. Click **Apply**.

15. Set the time slider **Span** to 10 **Minutes** and **Start** to 11/11/2016 10:40:00 AM to filter the data to the time frame of interest.

16. If needed, in the **Contents** pane, move the **LPR_2016-11-11** layer higher in the drawing order so that the features in that layer are visible and not covered by other layers.

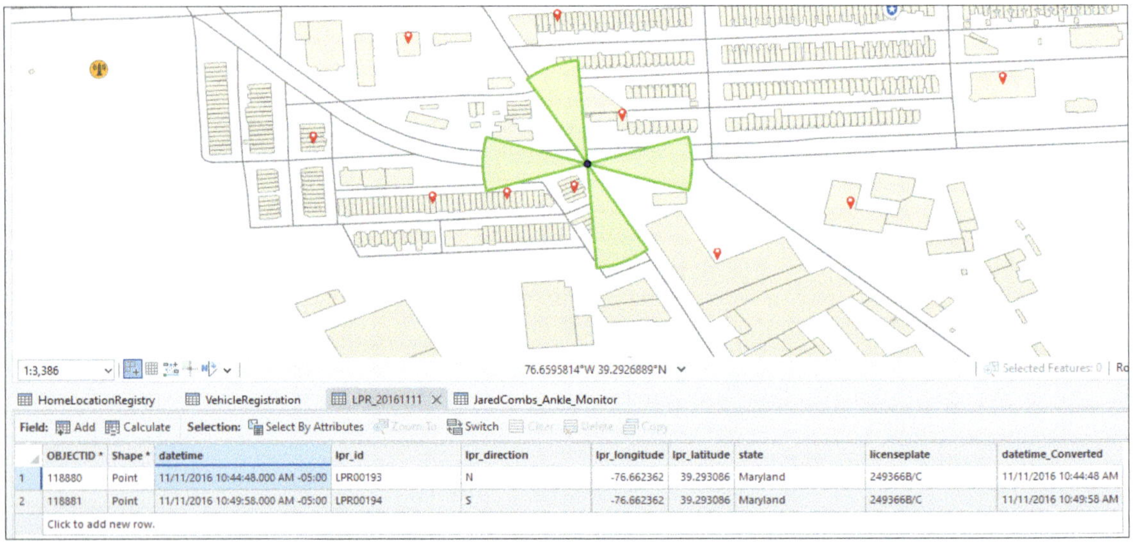

Now you can see that the vehicle indeed traveled northbound at 10:44 a.m. on 11/11/2016 in the general direction of J. Combs's ankle monitor points and then quickly returned southbound five minutes later. These records show that A. Combs's phone and vehicle moved from a point of origin to the south after a phone communication at 10:41 a.m. and then returned south soon after. Finally, J. Combs's GPS tracks likewise departed the meeting location.

These observations together indicate that A. Combs and J. Combs possibly had a coordinated meeting in the park and then departed separately.

5 Conduct a final investigation

You may have noticed that J. Combs's ankle monitor tracks, before he headed to the park, spatially correlated with a feature from the crime reports layer.

1. Open the attribute table for the **crime_reports** layer.
2. On the **Map** tab, in the **Selection** group, click the **Select** tool.
3. In the **crime_reports** attribute table, click the **Show Selected Records** button. Select the crime report in the map spatially correlating to the ankle monitor tracks.

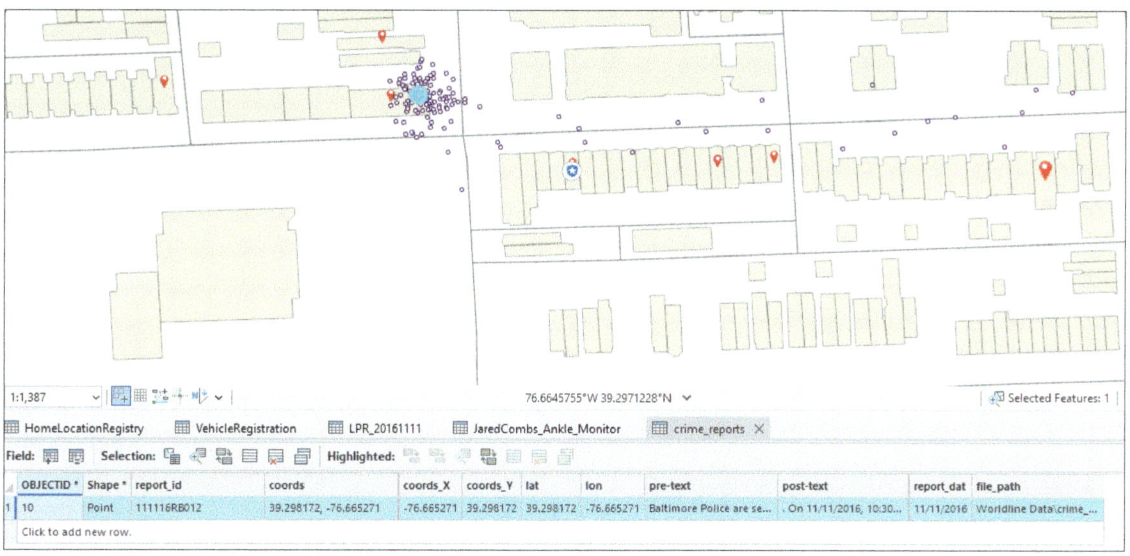

The crime reports layer was created by a tool that georeferences unstructured textual data by scanning for geocoordinates and generating records for each set of coordinates found, with some associated data pulled from the source file. You can reference the **report_id** field to look up the full report in its DOCX file.

> Baltimore Police are seeking the public's help in identifying the suspect involved in the robbery of the On Stack Jewelry located at 904 Poplar Grove St., Baltimore, MD 21216, United States (39.298172, –76.665271). On 11/11/2016, 10:30 AM, the suspect entered the business, displayed a box cutter knife, and demanded the employees lie on the floor. The employees complied and the suspect took an undisclosed amount of merchandise and fled the store. He was last seen on foot, heading southbound. There were no injuries.
>
> The suspect is described as a White male, late teens, and about 5′ 8″ tall. He was wearing dark pants, a white shirt, black cap, and a gray hooded sweatshirt.

You can see that the crime reportedly occurred at 10:30 a.m. November 11, which correlates closely with J. Combs's ankle monitor data. This additional information shines new light on Jared Combs's activities on the morning of November 11 and his possible interaction in the park with Arthur Combs.

Interpretation of analysis

After completing this exercise, answer the following questions in detail. The answers to these questions, as well as the analysis results from the exercise workflow, will inform the intelligence brief.

1. Based on this analysis, can you report with confidence that Jared Combs participated in any illegal activities or violated his parole?
2. Based on this analysis, what patterns or observations can you report with confidence regarding the activities of Jared Combs on November 11, 2016?
3. Did the integration of spatial and temporal analyses offer additional insight into Jared Combs's movements? Are the results enough to exonerate or convict him as a suspect in this case?
4. What actionable information can be derived from this analysis? How might the Baltimore Police Department use this information to investigate this case?
5. What other factors or data might be explored to provide additional insight into this case?

Intelligence brief

Based on the analysis from the exercise workflow, produce a three- to five-page intelligence brief by responding to the following question prompts. Incorporate your geospatial outputs from the exercise to enrich and illustrate your conclusions.

- **BLUF:** List three to five primary observations from your analysis in bulleted-list format that summarize findings of this work.
- **Top-level analysis:** Report the primary data points supporting your BLUF.
- **Secondary analysis:** Provide the data sources and deeper analytic products from this analysis.
- **Impact:** Describe the impacts of your findings on the department or organization.
- **Confidence:** Report the level of confidence of your assessment based on available data, accuracy, and so on.

- **Bias:** Explain any background or information you have that could have affected your analysis.
- **Conclusions:** Determine what type of analytic conclusions there may be and how the data requirements, analysis, and brief writing may change based on what analytic framework you are following (descriptive, predictive, prescriptive).

Applying and extending your skills

This exercise involved integrating a variety of datasets within a GIS analytic environment to discover and understand activities of interest that would not be visible through a single dataset and spatio-temporal methods. This process—using spatial and temporal tools to discover correlations between people, events, activities, and locations—is iterative and open-ended. You could have continued the workflow in various ways with the data from the Worldline sample dataset. Consider the following ways to pursue the case further and try to implement them using the remaining four days' worth of data.

- In a general sense, what do you not know about Jared or Arthur Combs, their activities, and their relationships (to people, places, or things)?
- Given the unknowns about their relationships, do you have datasets that may reveal that information?
- Who else is in the social network of the two men, and can you identify connections between those persons and the activities of interest?

Data sources and analyses

Return to the Worldline sample data folder and review the types of data that you have not used. Think about what you might discover in those files. Likewise, repeat the methods of the chapter workflow on the earlier four days of data to search for further evidence of possible criminal activity.

Data visualization

- In ArcGIS Pro, consider how you symbolized the layers and arranged them in rendering order to see whether there were other ways that better maximized the data on the map.
- In ArcGIS Pro, experiment with creating a link chart to capture and visualize the links that you discovered in the datasets.
- If you have access to ArcGIS Pro Intelligence or ArcGIS AllSource, integrate data that is relevant to your analysis within a time line.

Workflow communication

Consider how you would visualize the chapter workflow (and any further steps you took to analyze the preceding four days of data). Modeling the process by illustrating what datasets, tools, and methods yielded certain findings would enable repetition of the workflow in whole or in part to solve other intelligence problems.

Additional resources

- Biltgen, Patrick, Todd S. Bacastow, Thom Kaye, and Jeffrey M. Young. 2015. "Activity-Based Intelligence: Understanding Patterns-of-Life." Artech House. https://medium.com/the-state-and-future-of-geoint-2017-report/activity-based-intelligence-understanding-patterns-of-life-481c78b7d5ae.
- Esri National Government Video. 2017. "Activity-Based Intelligence: A Perilous Journey to Intelligence Integration." https://mediaspace.esri.com/media/t/1_vakm42uc.
- UK Ministry of Defence Joint Doctrine Note 1/23. 2023. "Intelligence, Surveillance, and Reconnaissance." www.gov.uk/government/publications/intelligence-surveillance-and-reconnaissance-jdn-123.
- Worldline synthetic activity data. Whitespace. https://inthewhitespace.com/worldline.

Acknowledgments

The author would like to thank Whitespace for letting him use the Worldline synthetic sample data in this chapter.

About the author

Patrick Kenney enlisted in the US Marine Corps in 2000 serving as a combat engineer and civil affairs specialist. Later, Patrick earned his BA in international affairs (and a certificate in GIS) from the University of Mary Washington (Fredericksburg, Virginia) in 2009 and his MA in security studies from Georgetown University in 2012. Patrick works at Whitespace in Alexandria, Virginia. He spends part of his time working on various projects analyzing human behaviors and activity and part of his time creating and instructing courses in intelligence analysis skills, ABI, and data analytics. He has taught at the NYPD, US National Geospatial-Intelligence Agency, Danish Defense College, UK National Centre for Geospatial Intelligence (NCGI), and elsewhere.

Chapter 18

Quantifying and mapping land use and land cover change for HSGI

Amélie Y. Davis, *US Air Force Academy*
Madeline A. Williams, *US Air Force Academy*

Overview

In this chapter, you will use various raster overlays to obtain spatial information pertinent to land use land cover (LULC) changes in a small region of Uganda in East Africa. The data covers the same geographic region but for three time periods. You will learn how to process the data to identify LULC transitions in the time series and, perhaps most important, where these changes have taken place.

Most datasets for this type of analysis are in raster format because they are based on remotely sensed data (aerial photos and satellite images). After completing this practical exercise, you should be able to perform basic raster analyses, especially reclassifications and overlays. One of the end products of the exercise will be a map that constitutes actionable intelligence for where changes have occurred in the study area. You will then practice analyzing these changes to create a geospatial intelligence brief. This approach applies to food-water-energy nexus issues that are critical for attaining many UN Sustainable Development Goals (United Nations n.d.).

Learning objectives

- Describe how landscapes are represented in a raster format and what pixels represent.
- Conduct spatial operations on rasters using raster tools such as Reclassify and Raster Calculator.
- Add a new field and calculate a percentage difference using the Field Calculator.

- Analyze spatial patterns and interpret results of spatial operations to provide actionable intelligence that determines potential ramifications that LULC changes might have on human security (in other words, food and water scarcity, depletion of resources, and increased potential for conflict and disease).

This knowledge could help guide efforts to implement preventive measures and programs before resources are depleted or conflict or disease occurs.

Technical components

- Software: ArcGIS Pro 3.4
  - Spatial Analyst extension or 3D Analyst extension required
- Dataset: LULC data from a farming region in Uganda at 100-meter spatial resolution from Mugisha (2002) and Olson et al. (2008)

Prerequisites

- Prior knowledge of or experience with GIS software or spatial data is helpful but not required.
- Technical skill level: Introductory GIS.
- Time to complete chapter: One to two hours without intelligence brief, three to five hours with intelligence brief.

Background and situational context

As people convert land to support a growing population and increased resource needs, the earth's surface is modified. Understanding how and why those changes occur, what their ramifications are, and where they occur is important from an HSGI perspective. In this exercise, you will focus on examining where these changes occur and interpreting the results.

The biophysical characteristics of the earth's surface can be grouped into land types that are collectively referred to as land cover. The applications or uses of these land types can be categorized into land uses. Examples of land covers include forest, built (or developed), and water. Examples of land uses stemming from forested land covers include palm oil plantations, Christmas tree farms, and forest preserves. Land uses and covers are typically spatially represented—that is, on a map—and can be obtained (or created) for different periods and at various intervals to monitor and provide insight into how these elements are changing over time. These can also be fed to spatially explicit models that use artificial neural networks or cellular automata (Pijanowski et al. 2002; Omrani et al. 2017) to predict where LULC changes might occur in the future (or have occurred in the distant past). The science of LULC change—that is, land change science—examines the type of change, magnitude, and locations (Rindfuss et al. 2004).

Land change science is critical to HSGI issues. Human security and safety depend on food and water availability, both of which can be detected at broad and fine scales using LULC change detection approaches. Access to reliable energy also affects human security and safety, and energy infrastructure has a spatial footprint, especially renewable energies—solar and wind farms. As nations move to provide more of their energy from renewables, detecting where and what kind of renewables are or might be installed (on land and water) is important from a geospatial intelligence perspective. Other examples of change detection affecting human security and safety would include mapping and monitoring of any natural resources (for example, mangroves and forests) as they provide protection and resources. Depletion of these resources is often a factor that can lead to instability and thus increases the likelihood of conflict in a region.

The abundance of raster data available to the intelligence community makes this approach useful and adaptable to a wide variety of situations. This type of analysis can be used to solely identify where LULC changes have occurred, but it can also be combined with other information to understand why these changes have occurred; perhaps local population growth is driving the increase in urbanization or agricultural demand, or perhaps far-removed drivers exist such as deforestation due to urban growth and international demand (DeFries et al. 2010).

Preventing far-reaching and unintended consequences of these LULC changes starts with being aware of where they are occurring. This can be accomplished by instituting monitoring programs using remote sensing technology and geospatial analyses similar to the processes you will be completing in the exercise. In addition to surveillance, it is important to analyze what is driving these LULC changes and, if necessary, to determine how to slow or halt the drivers of undesirable LULC changes.

This workflow enables you to identify specific locations of LULC changes within a landscape around Sango Bay in Uganda, a region where land cover is rapidly being modified.

Exercise workflow 18

These aerial pictures of the landscape (figures 18-1 to 18-3) will help familiarize you with the area (these may not be of Sango Bay exactly but were taken in East Africa by Dr. Bryan C. Pijanowski).

Figure 18-1. Scrub/woodland landscape.

Figure 18-2. Small-scale cultivation landscape.

Figure 18-3. Short grass/bare ground landscape.

1 Calculate a raster change analysis manually

In this section, you will conduct a practical paper and pencil exercise to help you understand the raster manipulations that you will complete later in the exercise using ArcGIS Pro. These landscapes are a simplified version of the land cover rasters you will manipulate later.

1. Navigate to **SecurityFirst_ExerciseData/Chapter18** (the chapter data folder) and open **Chapter_18_Raster_LULCC_Worksheet**.
2. Follow the directions in the worksheet to complete manual calculations of an LULC change analysis.

An answer key is provided at the end of the worksheet if you need assistance or want to verify your work. Once you have finished the worksheet, return to this tutorial to complete the remaining steps. The following instructions assume that you have completed the manual calculations.

2 Calculate a raster change analysis using ArcGIS Pro

In this section, you will become more familiar with the raster data format and learn to calculate a percentage difference. With this raster data, you have more land cover classes than you had in the manual calculations example, but you will use spatial analysis to perform the same type of analyses using ArcGIS Pro.

Enabling the Spatial Analyst extension

If you need to enable the Spatial Analyst extension (that is, if the Spatial Analyst toolset is unavailable), follow these steps.

1. In ArcGIS Pro, click the **Project** tab on the main ribbon.
2. Scroll down and click **Licensing**.
3. In the pop-up menu, check the box for **Spatial Analyst** and click **OK**.
4. Return to the normal view in ArcGIS Pro by clicking the back arrow.

1. Open **ArcGIS Pro**. Under **New Project**, click **Map** to create a new map.
2. In the **New Project** window, apply the following settings:
   - **Name:** Chapter_18
   - **Location:** C:\SecurityFirst
3. In the **Catalog** pane, click the **Computer** tab.
4. Navigate to the **SecurityFirst_ExerciseData/Chapter18** folder and open it. Right-click the **SangoBay** folder and click **Add To Project.**
5. Click the **Project** tab and expand **Folders.** Expand **SangoBay.**
6. For the three raster layers in the folder, right-click each layer and click **Add to Current Map.** If you are prompted to **Calculate statistics,** click **Yes.**

   To set up your map, the first step will be to set LULC class symbology. An appropriate color scheme has already been created; you will apply it to all three raster layers.
7. In the **Contents** pane, right-click **lulc_2000** and click **Symbology.**
8. In the top right of the **Symbology** pane, click the **Options** button and click **Import from layer file.** Navigate to the **SangoBay** folder and click **lulc_2000.lyr.** Click **OK.**
9. Repeat these steps to update the symbology for **lulc_1985** and **lulc_1955.**

Next, you will get oriented with the study area.

10. On the ribbon, click the **Map** tab. In the **Inquiry** group, click **Locate**.
11. In the **Locate** pane, click the search bar and type Uganda. Click the first entry in the search results.
12. Observe the scale of the data, focusing on the **lulc_2000** feature class, within the context of Uganda and within the world, continent, and region.

Question

Where does it sit geographically (latitude, longitude of approximate center point), and what specific features or boundaries delimitate the data: a country boundary, an ocean, local roads, and so forth? Consider where the equator is in relation to the **lulc_2000** raster.

You may want to zoom in to or out from the study area by using the scroll wheel on your mouse, two fingers on your touch pad, or the Explore and fixed Zoom In/Zoom Out buttons in the Navigate group on the Map tab.

Answer

- **Scale:** local—Sango Bay, Uganda, and surrounding areas.
- **Geography:** 31.59 longitude, –0.741 latitude.
- **Boundaries:** Lake Victoria on the east side, the Tanzania border on the south side, and a major road on the west side.
- **Equator:** Just north of the area depicted by the **lulc_2000** raster.

Question

In what coordinate system is **lulc_2000**? Briefly justify the use of this coordinate system for this scale and region of analysis. You can check the coordinate system by right-clicking the layer > Properties > Source and scrolling down to Spatial Reference to view the Projection row.

Answer

- **Coordinate system:** Universal Transverse Mercator (UTM) Projection (Zone 36S).
- This projection is appropriate because the entire **lulc_2000** raster is in the Southern Hemisphere, and that part of Uganda is in Zone 36. You will be performing local scale analyses (the raster covers a small area), so UTM is suitable.

Next, you will examine the data used in this analysis.

13. In the **Contents** pane, right-click **lulc_2000** and click **Zoom To Layer.**
14. Right-click **lulc_2000** again and click **Attribution Table.**

    lulc_2000 is a raster representation of the landscape around Sango Bay, Uganda. The colors represent land uses and land cover classes.

    - **VALUE** provides the code for a given land use category.
    - **COUNT** represents the number of cells (or pixels) within a given land use category.
    - **CLASS** describes the predominant land use for a given category.

    Consider this example: **VALUE** = 2 represents **Tall Grass** within the study area.

    Because this is a raster representation of the landscape, the image is composed of continuous cells (pixels); each cell has a code associated with it or is classified as **NODATA.**

15. In the attribute table, click the numbers in the column left of **Rowid** to select one of the land use categories, such as **Tall Grass.**
16. On the map, zoom in closely to examine a single pixel of **Tall Grass**, which is highlighted.
17. On the ribbon, click the **Map** tab. In the **Inquiry** group, click **Measure** > **Measure Distance.** Measure the distance in meters of one side of a pixel.

    The cell size is 100 by 100 meters.

18. Open the attribute table for **lulc_2000** and **lulc_1985**. Compare them.

Questions

1. Which LULC class is most prevalent in the landscape in both layers?
2. What percentage of the total landscape in 2000 does the **Cultivation (Small Scale)** class occupy? In 1985? Show your calculations by hand or by calculator—this is not a GIS operation.
3. Compare the data in the two attribute tables. Which LULC classes had no change between the two time periods?

Answers

- Most prevalent: **Cultivation (Small Scale).**
- Divide cultivation count by total count.
  - 1985: (114,362 divided by 245,153) times 100 = 46.6 percent.
  - 2000: (115,883 divided by 245,103) times 100 = 47.3 percent.
- No change: **Scrub/Grassland**, **Woodland/Seasonal Swamp**, **Thickets**, and **Cultivation (Large Scale)**.

3 Calculate an automated raster change analysis

In the following steps, you will learn how to export a raster attribute table, combine it with another attribute table that contains data you want to analyze, add a field (column), and use attributes from both tables in calculations.

1. Open the attribute table of the **lulc_2000** raster and click the **Options** button. Click **Export** and use the default name for the **Output Table**. Click **OK**.
2. Repeat these steps for the **lulc_1985** raster.
3. Close the attribute table of the two rasters (**lulc_2000** and **lulc_1985**) and open the two stand-alone tables you just created, located at the bottom of the **Contents** pane.

   You will now combine the two tables you exported into one table that contains the information from both tables. This will allow you to compare the number of cells (pixels) in each LULC class. Because the files have a unique identifier that, by definition, is common to both (with a one-to-one correspondence), you can perform a join operation.
4. In the **lulc_2000_ExportTable**, click the **Options** button.

   Note: You can use either table to execute the join in this scenario.
5. Click **Joins and Relates** > **Add Join.**
6. In the **Add Join** window, apply the following parameters:
   - **Input Table:** lulc_2000_ExportTable
   - **Input Join Field:** VALUE
   - **Join Table:** lulc_1985_ExportTable
   - **Join Table Field:** VALUE

7. Click **OK.**
8. Check the **lulc_2000_ExportTable** to confirm that the join was successfully performed.

   You will now add a new field and calculate the percentage difference in the number of cells per class between 1985 and 2000 in Sango Bay.
9. In the attribute table for **lulc_2000_ExportTable**, click the **Add Field** button.

   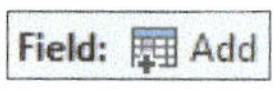

10. Enter the following parameters:
    - **Field Name:** Per_Diff
    - **Alias:** LULC Difference (%)
    - **Data Type:** Float

    Learn more about the various data types for fields at links.esri.com/SF-FieldDataTypes.

Question

Why is the **Double** field type not appropriate in this case?

Answer

The **Per_Diff** field will contain a percentage change. The data storage needs for that number with decimals will be between 3.4E38 and 1.2E38, which takes half the storage that the **Double** field type would need. In this small table, that does not matter. But in a table with a large number of rows and columns, those decisions can add up in file size, so you should be judicial about field type choice.

Consider the other field names. Two of the aliases have the name **VALUE**, which is a bit confusing because one of those is the number of cells (or pixels) from 1985 and the other from 2000. You will change the **Alias** for these two fields so it is clear which is which.

11. In the **Alias** column, for the first (top) occurrence of **COUNT**, double-click and rename it COUNT2000. For the second occurrence, rename it COUNT1985.
12. On the ribbon, click the **Fields** tab. In the **Manage Edits** group, click **Save.**
13. In the table for **lulc_2000_ExportTable**, scroll to the right until you see your new field filled with **<Null>**, which is normal at this stage.
14. Right-click the header for the **LULC Difference (%)** field and click **Calculate Field.**
15. Examine the **Calculate Field** tool that appears.

    You will build an expression—that is, create the formula needed to calculate the percent difference in LULC for each class between the two time steps (1985 and 2000). The calculations will go into the **lulc_2000_ExportTable.Per_Diff** field. You are telling the tool what you want it to calculate using the available fields.

Review how percentage difference is calculated. Consider the **COUNT** field names under the **Fields** menu. The first (top) occurrence of **COUNT** represents the **lulc_2000**, whereas the second occurrence represents **lulc_1985**.

16. In the **Calculate Field** tool, apply the following parameters:
    - **Field Name (Existing or New)**: LULC Difference (%)
    - **lulc_2000_ExportTable.Per_Diff**: `((!lulc_2000_ExportTable.COUNT! - !lulc_1985_ExportTable.COUNT!) / !lulc_1985_ExportTable.COUNT!)*100`

Tip: In the Fields list, double-click the field that you want to add to your expression.

Your field name is preceded by a dot (.) and the name of the table from which it came (**lulc_2000_ExportTable** or **lulc_1985_ExportTable**). This helps you keep track of which fields (or columns of data) come from which table. You changed only the **Alias** (not the **Field Name**), so the formula does not use the **Alias** name (**COUNT2000** or **COUNT1985**) because aliases are merely for display purposes and do not change the actual field name.

17. Click **OK**.

Your table should now include values in the **LULC Difference (%)** field.

	OBJECTID *	VALUE *	COUNT2000	CLASS	OBJECTID	VALUE	COUNT1985	CLASS	LULC Difference (%)
1	1	1	22970	Papyrus	1	1	22977	Papyrus	-0.030465
2	2	2	3591	Tall Grass	2	2	4409	Tall Grass	-18.55296
3	3	3	26315	Tropical High Forest	3	3	26463	Tropical High Forest	-0.559272
4	4	4	115883	Cultivation (Small Scale)	4	4	114362	Cultivation (Small Scale)	1.329987
5	5	5	7062	Seasonal Swamp	5	5	7165	Seasonal Swamp	-1.437544
6	6	6	40665	Scrub/Seasonal Swamp	6	6	40781	Scrub/Seasonal Swamp	-0.284446
7	7	7	6360	Short Grass/Bare Grou...	7	7	6551	Short Grass/Bare Grou...	-2.915586
8	8	8	16643	Scrub/Woodland	8	8	16814	Scrub/Woodland	-1.01701
9	9	9	2680	Scrub/Grassland	9	9	2680	Scrub/Grassland	0
10	10	10	955	Open Water	10	10	959	Open Water	-0.417101
11	11	11	1593	Woodland/Seasonal S...	11	11	1593	Woodland/Seasonal S...	0
12	12	12	137	Thickets	12	12	137	Thickets	0

If you want to change the number of decimals displayed, read the section on numeric field formatting at links.esri.com/SF-FieldDataTypes.

Join operations are temporary changes to a table and appear only in your project file unless you export the table with the join as a new table. Any fields you added using the **Calculate Field** tool, however, are permanent.

Questions

1. What do the results of the calculation indicate? Do the negative values in the new **Per_Diff** column indicate a gain or a loss for a particular LULC class between 1985 and 2000? Provide an example from the table to support your response.
2. Based on the new **Per_Diff** column, which class gained the most area? How many cells did that class gain?
3. Compare the **COUNT1985** and **COUNT2000** columns. Which class lost the most area? How many cells did that class lose? How much area of this class was lost in square kilometers?

Answers

- A negative value represents a loss of area for a particular LULC class. One example of LULC loss occurs with the **Papyrus** land cover as it decreases from 22,977 cells in 1985 to 22,970 cells in 2000. For example, between 1985 and 2000, **Tall Grass** lost 818 cells, or a –18 percent change.
- Class: **Small Scale Cultivation**. Cells gained: 1,521. Square kilometers: 1,521 times 100 times 100 = 15,210,000 square meters = 15,210,000 times 0.01 km^2 = 0.01 km^2.
  - Note: Cell size x = 100 meters and cell size y = 100 meters, so cell area = 100 times 100 = 10,000 m^2 = 0.01 km^2.
- Class: **Tall Grass**. Cells lost: 818. Kilometers squared: 818 times 0.01 km^2 = 8.18 km^2.

4 Reclassify data and visualize results

You have just studied the two time series on a landscape scale. Now you will identify exactly where the gains and losses in certain LULC classes occurred in the landscape.

To do this, you will perform a raster overlay analysis.

1. Close all tables and focus on the **Map** view.
2. On the ribbon, click the **Analysis** tab. In the **Geoprocessing** group, click **Tools**.
3. In the **Geoprocessing** pane, click the **Toolboxes** tab and find the **Spatial Analyst Tools** toolbox. Expand the toolbox and click **Reclass**. Double-click the **Reclassify** tool.

   Tip: To understand what any tool does, click the Help question mark on the top right of the tool.

4. In the **Reclassify** tool pane, set your **Input raster** to **lulc_2000** and make sure the **Reclass field** is set to **VALUE**.

   This table lists the numeric values and corresponding class.

VALUE	CLASS
1	Papyrus
2	Tall Grass
3	Tropical High Forest
4	Cultivation (Small Scale)
5	Seasonal Swamp
6	Scrub/Seasonal Swamp
7	Short Grass/Bare Ground
8	Scrub/Woodland
9	Scrub/Grassland
10	Open Water
11	Woodland/Seasonal Swamp
12	Thickets
13	Woodland
14	Cultivation (Large Scale)

You will change (reclassify) the values to create a new grid (raster) so that you can identify the location of the LULC changes that have occurred in the study area.

5. For the **Reclassification** table, change all the values under the **New** heading to 0 except for the **Cultivation (Small Scale)** class (code = **4**), which will be changed to 1. Leave the **NODATA** class as **NODATA**.
6. Name the **Output raster** lulc_2000rc and make sure that it is saved in the folder for this chapter's exercises.
7. Click **Run**.

Question

In the output raster, what do the 1 and 0 stand for?

Answer

1 stands for land areas with small-scale cultivation land cover in 2000. 0 stands for any land area without small-scale cultivation land cover in 2000.

8. Change the **Input raster** to **lulc_1985**. Change all values under the **New** heading to 0, except for the **Cultivation (Small Scale)** class, which will be changed to 100. Name the **Output raster** lulc_1985rc.
9. Click **Run**.

Question

In the output raster, what do the 100 and 0 stand for?

Answer

The 100 stands for land areas with small-scale cultivation land cover in 1985. The 0 stands for any land area without small-scale cultivation land cover in 1985.

10. In the **Spatial Analyst Tools** toolbox, click **Map Algebra**. Double-click the **Raster Calculator** tool and read its description.
11. Using the two newly created reclassed LULC raster files for 1985 and 2000, type the following **Map Algebra** expression:

```
"lulc_1985rc" + "lulc_2000rc"
```

12. Name the **Output raster** cultiv_sm_85_00.
13. Click **Run**.

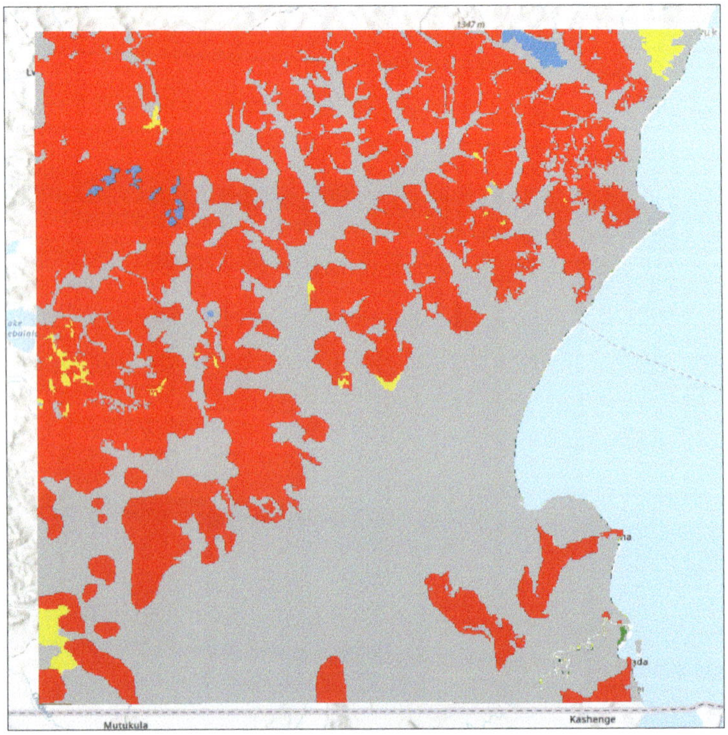

14. Open the attribute table of the newly created file.

Questions

What do 0, 1, 100, and 101 represent? How many cells are in each class? You can check the cell numbers by looking at the attribute table of cultiv_sm_85_00 raster.

Answers

0 represents cells that were not cultivated in 1985 and 2000 (that is, cells that were any LULC class other than **Cultivation** in both time periods). The number of cells for 0 are 128,041. 1 represents cells in which small-scale cultivation was present in 2000 but not 1985 (that is, new small-scale cultivation). The number of cells for 1 is 2,700. 100 represents cells in which small-scale cultivation was present in 1985 but not 2000 (that is, any area that changed from small-scale cultivation in 1985 to another class in 2000). The number of cells for 100 is 1,179. Finally, 101 represents cells in which small-scale cultivation was present in both 1985 and 2000. The number of cells for 101 is 113,183.

5 Reclassify results and identify change

The next step in the analysis is to reclassify results and identify where small-scale cultivation expanded between 1985 and 2000.

1. In the **Contents** pane, right-click the **cultiv_sm_85_00** raster and click **Symbology**.

2. In the **Symbology** pane, for the **Values** table, change the values in the **Label** using the following names:

    - **0**: No cultivation
    - **1**: New small-scale cultivation
    - **100**: Change in small-scale cultivation from 1985 to 2000
    - **101**: Cultivation

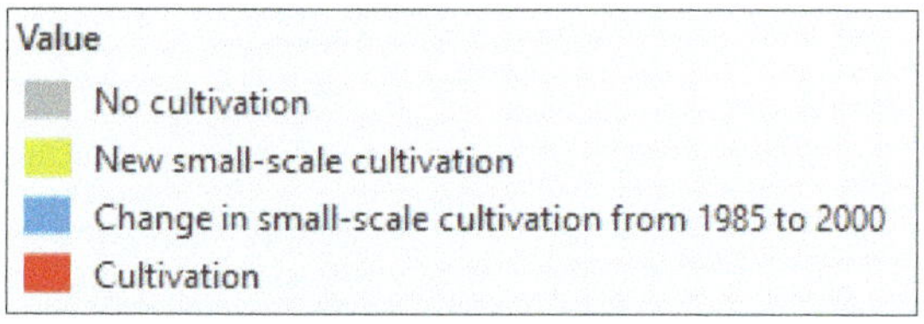

3. Use the **Reclassify** tool with **cultiv_sm_85_00** so that values of **0**, **100**, and **101** have a new value of 0. **1** should have a value of 1. Name the reclassed map rc_cultiv_sm00.

4. Use the **Raster Calculator** tool to create a new raster called old_cultiv_sm00 using the following expression:

    ```
    "rc_cultiv_sm00" * "lulc_1985"
    ```

    The result is a map of where small-scale cultivation expanded from 1985 to 2000. The pixels from the various LULC classes represent the LULC classes that were converted to small-scale cultivation from 1985 to 2000 and the spatial location of the changes.

    Next, you will update the symbology to match the colors in the original LULC rasters.

5. In the **Contents** pane, right-click the **old_cultiv_sm00** layer and click **Symbology**. Click the **Options** button and select **Import from layer file.**

6. Navigate to the **SangoBay** data folder and select **lulc_2000.lyr**. Click **OK.**

    You can now visualize how the landscape changed for small-scale cultivation expansion from 1985 to 2000.

Questions

1. Which class was converted to small-scale cultivation the most from 1985 to 2000? How many square kilometers of it were lost to cultivation?
2. What were the four main classes that were lost to small-scale cultivation between 1985 and 2000 (including the one from the previous answer)? Use a complete sentence for your answer.

Answers

- Class: **Scrub/Woodland** (LULC class code 8). Square kilometers: 854 times 0.1 km^2 = 8.54 km^2.
- The four main classes lost to small-scale cultivation from 1985 to 2000 in order were **Scrub/ Woodland**, **Tall Grass**, **Short Grass/Bare Ground**, and **Tropical High Forest**.

3. Use the **Reclassify** tool with **cult_sm_85_00** as the **Input raster** and reclassify the values so that **0**, **1**, and **101** have a new value of `0`. **100** should be reclassified to `1`. Name the reclassed map `rc_cultiv_sm85`.
4. Use the **Raster Calculator** tool to create a new raster called `old_cultiv_sm85` using the following expression:

```
"rc_cultiv_sm85" * "lulc_2000"
```

Interpretation of analysis

After completing this exercise, answer the following questions in detail. The answers to these questions, as well as the analysis results from the exercise workflow, will inform the intelligence brief.

1. Based on this analysis, what patterns do you observe in LULC change between 1985 and 2000 in the study area? In short, to what extent has LULC changed in the study area?
2. Which LULC categories experienced the most and least change between 1985 and 2000?
3. What might be some of the drivers or factors behind the LULC changes that you identified in Sango Bay between 1985 and 2000?
4. What is the significance of understanding how people use the land? What implications can this information have on the security environment?
5. What impacts, positive and negative, might the specific LULC changes identified in this workflow have on the surrounding population? Brainstorm ways that the negative effects can be mitigated and the positive effects amplified.
6. What actionable information can be derived from this analysis? How might Sango Bay leadership use this information to develop a plan that serves economic and conservation interests?
7. What additional geospatial data would be useful to incorporate with the LULC data to conduct a comprehensive assessment? What datasets would help assess future LULC changes in this area and the associated ramifications of these changes?

Intelligence brief

Based on the analysis from the exercise workflow, produce a three- to five-page intelligence brief by responding to the following question prompts. Incorporate your geospatial outputs from the exercise to enrich and illustrate your conclusions.

- **BLUF:** List three to five primary observations from your analysis in bulleted-list format that summarize findings of this work.
- **Top-level analysis:** Report the primary data points supporting your BLUF.
- **Secondary analysis:** Provide the data sources and deeper analytic products from this analysis.
- **Impact:** Describe the impacts of your findings on the department or organization.
- **Confidence:** Report the level of confidence of your assessment based on available data, accuracy, and so on.

- **Bias:** Explain any background or information you have that could have affected your analysis.
- **Conclusions:** Determine what type of analytic conclusions there may be and how the data requirements, analysis, and brief writing may change based on what analytic framework you are following (descriptive, predictive, prescriptive).

Applying and extending your skills

Here are some suggestions for diving deeper into this topic or modifying the geospatial workflow.

- The **Tall Grass** land use lost area from 1985 to 2000. You can find which LULC classes replaced the **Tall Grass** areas by performing these steps:
    a. Reclassify the **1985_lulc** raster so that 1 represents tall grass and 0 represents everything else, and reclassify the 2000 raster so that 100 represents tall grass and 0 represents everything else. Use the **Reclassify** tool.
    b. Add the two reclassified rasters together (that is, perform a raster overlay of the two rasters). In the resulting raster, 1 will represent areas that were tall grass in 1985 but not in 2000 (that is, a gain of tall grass). Use the **Raster Calculator** and **Add** tools.
    c. Reclassify this union raster so that **0**, **100**, and **101** are assigned zero (0) but **1** stays one (1). This new raster will show all areas in which tall grass was lost as 1. Use the **Reclassify** tool.
    d. Multiply this raster times the **2000_lulc** raster. The resulting raster will show which LULC classes replaced the tall grass areas (and where). Use the **Raster Calculator** and **Times** tools.
    e. Make a map and interpret the results.

Now that you understand how landscapes are represented using rasters and how to use raster overlays to conduct LULC change analysis (or other raster manipulations that allow you to extract location-based knowledge from data represented in rasters), you can explore the **Change Detection** tool. For more information, see links.esri.com/SF-ChangeDecTool.

Additional resources

- Biryabarema, Ellas. 2020. “Uganda Reverses Forest Destruction by Inviting in Loggers.” *Reuters*. www.reuters.com/article/us-uganda-deforestation-idUSKBN2681IB.
- Hood, Marlowe. 2016. “Africa’s Forests Menaced by Palm Oil Rush: NGO.” *PhysOrg*. https://phys.org/news/2016-02-africa-forests-menaced-palm-oil.html.
- Kiley, Sam. 2022. “Kenya: Drought and Soaring Grain Prices Fuel Insecurity.” *CNN*. www.cnn.com/2022/08/02/africa/kenya-drought-insecurity-intl-cmd/index.html.
- Kutamba, Wilson. 2022. “Govt Starts Forceful Evictions in Sango Bay Estate to Pave Way for Oil Palm Project.” *Monitor News*. www.monitor.co.ug/uganda/news/national/govt-starts-forceful-evictions-in-sango-bay-estate-to-pave-way-for-oil-palm-project-3754120.

- National Ocean Service at NOAA. 2024. "What Is the Difference Between Land Cover and Land Use?" https://oceanservice.noaa.gov/facts/lclu.html.
- OECD. 2024. "Monitoring Land Cover Change to Understand Biodiversity Pressures." www.oecd.org/en/publications/monitoring-land-cover-change-to-understand-biodiversity-pressures_441a7a6c-en.html.
- Potapov, Peter, Matthew C. Hansen, Amy Pickens et al. 2022. "The Global 2000–2020 Land Cover and Land Use Change Dataset Derived from the Landsat Archive: First Results." *Frontiers in Remote Sensing* 3: 856903.
- Sleeter, Benjamin M., Thomas Loveland, Grant Domke, Nate Herold, James Wickham, and Nathan Wood. 2018. "Land Cover and Land-Use Change." From the report *Impacts, Risks, and Adaptation in the United States* of the Fourth National Climate Assessment, volume 2: 202–31. https://nca2018.globalchange.gov/chapter/5.
- USGS. n.d. "Land Use Land Cover Modeling." www.usgs.gov/special-topics/land-use-land-cover-modeling.
- Winkler, Karina, Richard Fuchs, Mark Rounsevell, and Martin Herold. 2021. "Global Land Use Changes Are Four Times Greater than Previously Estimated." *Nature Communications* 12 (1): 2501.

References

- DeFries, R. S., T. Rudel, M. Uriarte, and M. Hansen. 2010. "Deforestation Driven by Urban Population Growth and Agricultural Trade in the Twenty-First Century." *Nature Geoscience* 3 (3): 178–81.
- Mugisha, S. 2002. "Root Causes of Land Cover/Use Change in Uganda: An Account of the Past 100 Years." LUCID Working Paper, no. 14. Nairobi, Kenya: ILRI. www.ilri.org/knowledge/publications/root-causes-land-coveruse-change-uganda-account-past-100-years.
- Olson, J. M., G. Alagarswamy, J. A. Andresen, D. J. Campbell, A. Y. Davis, J. Ge, and J. Wang. 2008. "Integrating Diverse Methods to Understand Climate–Land Interactions in East Africa." *Geoforum* 39 (2): 898–911.
- Omrani, H., A. Tayyebi, and B. Pijanowski. 2017. "Integrating the Multilabel Land-Use Concept and Cellular Automata with the Artificial Neural Network-Based Land Transformation Model: An Integrated ML-CA-LTM Modeling Framework." *GIScience & Remote Sensing* 54 (3): 283–304.
- Pijanowski, B. C., D. G. Brown, B. A. Shellito, and G. A. Manik. 2002. "Using Neural Networks and GIS to Forecast Land Use Changes: A Land Transformation Model." *Computers, Environment and Urban Systems* 26 (6): 553–75.
- Rindfuss, R. R., S. J. Walsh, B. L. Turner, J. Fox, and V. Mishra. 2004. "Developing a Science of Land Change: Challenges and Methodological Issues." *Proceedings of the National Academy of Sciences* 101 (39): 13,976–81.
- United Nations. n.d. "Water, Food and Energy: UN Water." www.unwater.org/water-facts/water-food-and-energy.

Disclaimer

The views expressed in this chapter are those of the author and do not necessarily reflect the official policy or position of the US Air Force Academy, the Air Force, the Department of Defense, or the US Government. Approved for public release: distribution unlimited. PA#: USAFA-DF-2023-527.

About the authors

Amélie Davis is an assistant professor of sustainability, planning, and geospatial sciences at the US Air Force Academy. She has more than 10 years of experience teaching GIS at the undergraduate and graduate levels, as well as using GIS to conduct research in environmental geography, urban planning, sustainability, and urban ecology. She is certified as a GISP and an ENVISION sustainability professional. She received her BA in biology from Earlham College, her MS degree in oceanography with a focus on remote sensing from the University of Delaware at the former Center for Marine Studies, and her PhD from Purdue University in the Department of Forestry and Natural Resources.

1st Lt. Madeline Williams is an intelligence officer in the US Air Force. She has a BS in geospatial science from the US Air Force Academy, where she received the USGIF Certificate in Geospatial Intelligence in 2023.

Chapter 19

Predictive military geography of Poland

Logan Bolan, *Utah State University, Center for Anticipatory Intelligence*
Nathan Kozlowski, *Utah State University, Center for Anticipatory Intelligence*

Overview

In this chapter, you will learn how to use GIST tools and methods regarding HSGI threats to defensible areas within and adjacent to Poland's borders. The workflows presented here will provide insights and actionable information on defensible areas within Poland. The results of this analysis will then be used to produce a final deliverable in the form of an ArcGIS StoryMaps^{SM} story.

Learning objectives

- Perform exploratory data visualization.
- Apply geospatial filters to sort and analyze patterns.
- Use ModelBuilder to integrate and streamline workflows.
- Model the defensibility of a particular geographic region using standard geospatial or cartographic modeling methods.
- Create an ArcGIS StoryMaps story to share and illustrate workflow results.

Technical components

- Software:
  - ArcGIS Pro 3.4
  - ArcGIS Online (at least Publisher level access)
  - ArcGIS StoryMaps

- Datasets:
  - World global multiresolution terrain elevation data (GMTED), World GMTED Slope, ESA Sentinel WorldCover 2020
  - European Space Agency (ESA) 2023; WorldCover 2020 (available from https://esa-worldcover.org/en)
  - ESA Land Cover CCI Product User Guide version 2; Tech. Rep. 2017 (available at maps.elie.ucl.ac.be/CCI/viewer/download/ESACCI-LC-Ph2-PUGv2_2.0.pdf)
  - World Elevation GMTED, World Slope GMTED, Human Geography Label, Europe map

Prerequisites

- Prior knowledge of or experience with GIS software or spatial data is recommended.
- Technical skill level: Intermediate GIS.
- Time to complete chapter: Two to three hours without an intelligence brief, three to five hours with an intelligence brief.

Background and situational context

What would the most defensible areas of Poland be if it were invaded?

The authors have always been fascinated by the strategic significance of Poland (and other Eastern European countries). Poland's geographic location between Russia and Germany has significantly affected its history. During World War II, Russia and Germany used Poland as a launching pad for their respective ground campaigns. Delving into Poland's geography helps answer two essential questions: (1) Is Poland an easy invasion target considering its position in Central Europe? (2) Where should the geographic strengths and vulnerabilities for defense purposes be along Poland's borders?

The John Collins book *Military Geography: For Professionals and the Public* (2012) gave the authors a deep appreciation for how geography plays a defining role in warfare. The book explains how different physical features, terrain, climate, and vegetation can significantly affect military operations and strategies. Inspired by this knowledge, the authors researched and gathered relevant maps and geospatial data on Poland, such as terrain, slope, and land cover, to gain insight into its unique geography, challenges, and opportunities. The result of this effort revealed that Poland's location did make it vulnerable to invasion but also provided opportunities for strategic defense.

Exercise workflow 19

You will use the following geoprocessing tools in **ModelBuilder** to gain insight into Poland's defensible area:

- Merge
- Buffer

- Clip
- Slope
- Mosaic to New Raster
- Extract by Mask
- Reclassify
- Focal Statistics
- Rescale by Function
- Weighted Sum
- Buffer

The exercise steps will help with this analytic workflow.

1 Add data and isolate the study area

1. Open ArcGIS Pro. Under **New Project**, click **Map** to create a new map.
2. In the **New Project** pane, apply the following settings:
    - **Name:** Chapter_19
    - **Location:** C:\SecurityFirst
3. In the **Catalog** pane, click the **Portal** tab. Click **Living Atlas** to access ArcGIS Living Atlas.

    Note: If your Catalog pane is not open, click the View tab on the ribbon. In the Windows group, click Catalog Pane.

4. In the search bar, type Poland Country Boundary. Right-click the first result and click **Add To Current Map**.

    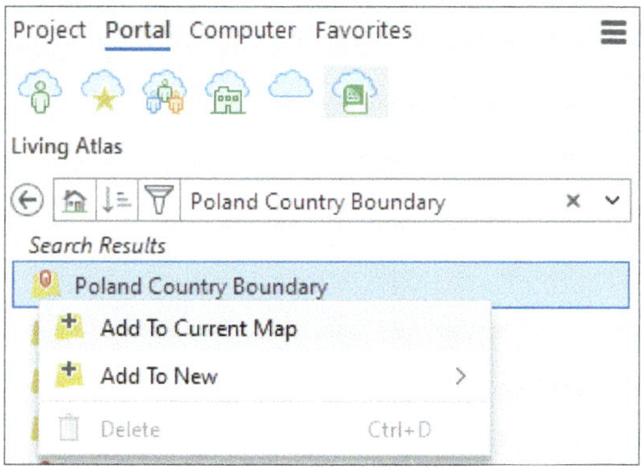

5. Repeat the process and add **Russia Federation Raioni Boundaries** and **Belarus Country Boundary**.
6. Zoom out to show the full study area.

7. Locate and zoom to the Kaliningrad area (**RUS_Raioni**), which is between Lithuania and Poland, on the western end of Russia.
8. On the ribbon, click the **Map** tab. In the **Selection** group, open the **Select** menu and click the **Select by Polygon** tool.
9. On the map, click all polygons from the Kaliningrad area and Belarus to select them.

   There are 22 features selected within Kaliningrad.

> **Tip:** You can confirm this number by looking at the bottom of the map and finding the Select Features section, which will list the number of current selected features.

10. Right-click the **RUS_Raioni** layer and click **Selection** > **Make Layer From Selected Features**.
11. Double-click the new **RUS_Raioni selection** layer and change the **Name** to Kaliningrad. Click **OK**.
12. On the ribbon, click the **Analysis** tab. In the **Geoprocessing** group, click **Tools** to open the **Geoprocessing** pane.
13. In the **Geoprocessing** pane, search for the Merge tool. Open the tool and apply the following settings:
    - **Input Datasets:** Kaliningrad, BLR_County
    - **Output Dataset:** non_NATO_combatants
14. Click **Run**.

    The next step is to buffer the newly created **non_NATO_combatants** shapefile by 100 statute miles, which is your region of interest. This will reduce the size of the study area and computing time while simulating Russian logistics limits.
15. In the **Geoprocessing** pane, search for the Pairwise Buffer tool. Open the tool and apply the following settings:
    - **Input Features:** non_NATO_combatants
    - **Output Feature Class:** non_NAT_combatants_buffer
    - **Distance [value or field]:** 100
    - **Linear Unit:** US Survey Miles
    - **Dissolve Type:** Dissolve all output features into a single feature
16. Click **Run**.

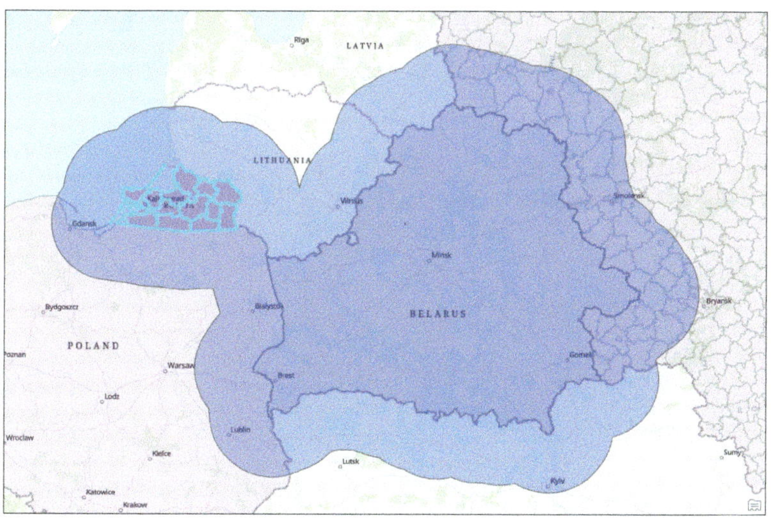

Next, you will use the **Pairwise Clip** tool to isolate the 100-mile corridor between the Polish, Russian Federation, and Belarusian borders.

17. Search for the Pairwise Clip tool and open the tool. Enter the following settings:
    - **Input Features**: non_NAT_combatants_buffer
    - **Clip Features**: POL_Country
    - **Output Feature Class**: Frontier
18. Click **Run**.
19. In the **Contents** pane, uncheck the box for each layer to turn it off, except the newly created **Frontier** layer.

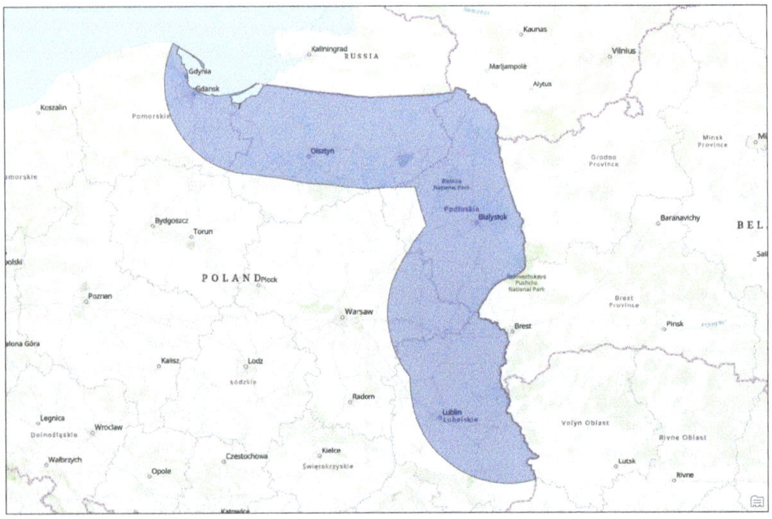

This will be referred to as the frontier for the remainder of the chapter.

20. In the **Catalog** pane, open the **Living Atlas** portal. Search for World Elevation GMTED. Add **World Elevation GMTED** to your map.
21. In the **Contents** pane, drag the **World Elevation GMTED** layer below the **World Topographic Map** layer.

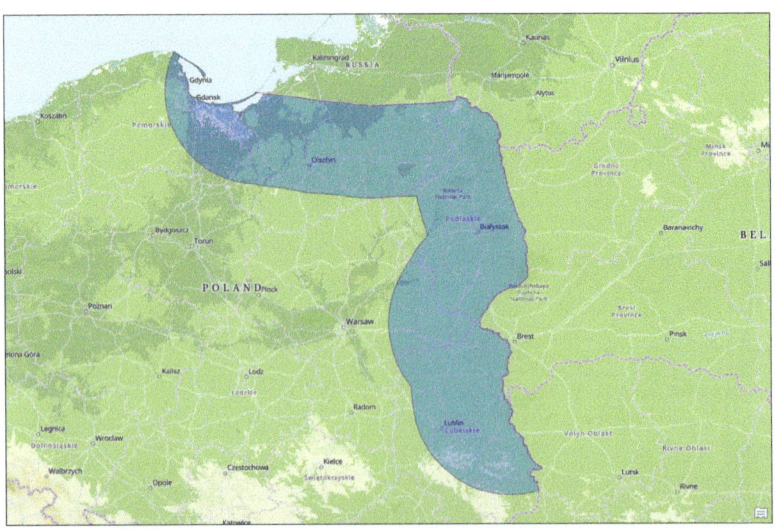

22. Save your project.

2 Add data and convert it to raster format

1. In the **Contents** pane, right-click the **World Elevation GMTED** layer and select **Data** > **Export Raster**.
2. In the **Export Raster** pane, enter the following settings:
    - **Output Raster Dataset:** frontier_elev.tif
    - **Geographic Transformation**: None
    - **Clipping Geometry**: Frontier
3. Click **Export** to create the output TIFF file and add it to the project's **Contents** pane.
4. Turn off the **Frontier** and **World Elevation GMTED** layers.

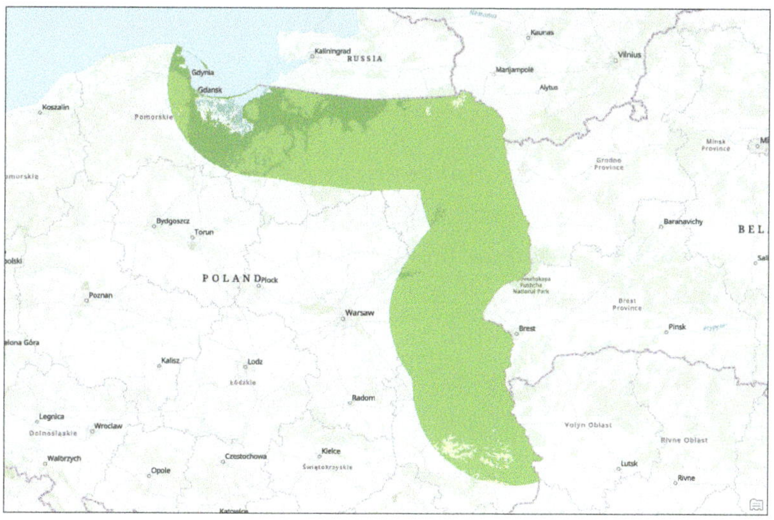

You will repeat the previous steps, but this time, you will add land cover to the map.

5. Return to the **Living Atlas** portal. Search for Global Land Cover 1992-2020. Add **Global Land Cover 1992-2020** to the map.
6. In the **Contents** pane, right-click the **Global Land Cover 1992-2020** layer and click **Data** > **Export Raster**. Enter the following settings and click **Export**:
    - **Output Raster Dataset:** frontier_lc.tif
    - **Geographic Transformation**: None
    - **Clipping Geometry**: Frontier
7. Click **Export** to create the output TIFF file and add it to the project's **Contents** pane.
8. Turn off the **Global Land Cover 1992-2020** layer.

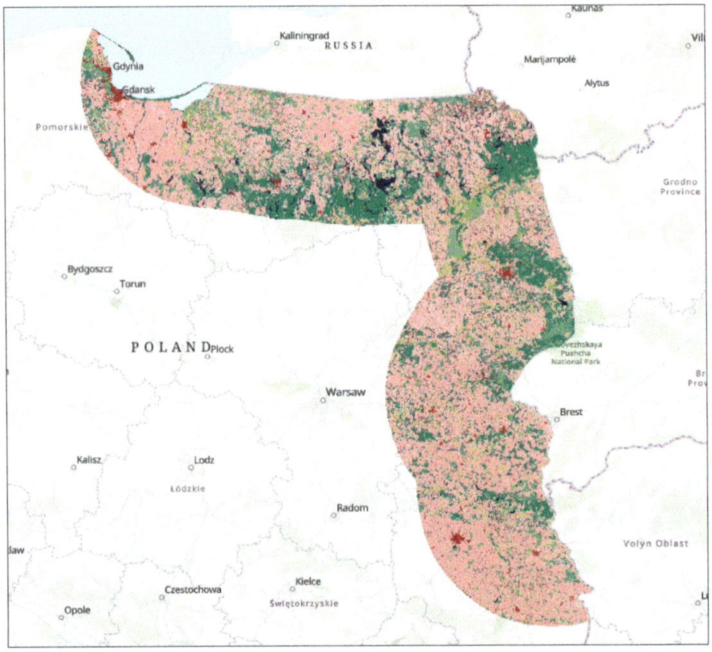

To reduce computing time, the study area is limited to the Frontier region, which is the 100-mile buffer with the Russian border clipped by the boundaries of Poland. Every raster used in the weighted model must be exported and clipped to the **Frontier** layer. Once you master the process, this workflow can be adapted for larger or more focused study areas as needed.

3 Perform a spatial analysis of a study area

You will now perform basic analysis on the clipped Frontier region of interest while also preparing the data for more advanced analyses.

An important aspect in conflict scenarios is mobility; understanding terrain is an essential part of military planning. For example, flat (0-degree plane) is much easier to attack and more challenging to defend compared with attacking an opponent on top of a 60-degree mountain slope, where the defender has the advantage of high ground. Thus, calculating the slope of the landscape is vital for determining ease of movement or preferred battle position.

In this section, you will conduct spatial analyses of multiple variables to gain insight into terrain. You will use the **Range** option in the **Focal Statistic** tool settings. This tool also helps you understand the "trafficability" of the terrain—the ease of movement over such terrain by an infantry soldier or a tank. For example, a flat plain is easier to move over than a rocky field of lava boulders. The lava boulders also provide more cover for an infantry soldier, and there is greater variability between types of terrain—sometimes it is flat, sometimes it is of medium height, and sometimes it ascends very quickly. The range option measures the difference, range, or variability between the elevation cells.

For local terrains, such as peaks, valleys, foot slopes, surface landscapes, and other small rises essential for the tactical (small unit) level of military operations, you will use a tool called **Geomorphon**

Landforms. This tool allows you to identify finer terrain that may not be visible on a large scale but is clear to the infantry units fighting on the front lines. A small peak can provide greater visibility of the entire area and fire control. In contrast, a shallow pit can provide cover, concealment, and protection against small arms fire and artillery.

You will also reclassify land cover for your model because different land cover types provide different offensive and defensive opportunities. Plains and grasslands are great for ranged and ballistic attacks from tanks and snipers but terrible for the infantry defender (unless you dig trenches and tank traps). On the other hand, rain forests and jungles, especially in muddy, mountainous terrain, are better suited for small units fighting in proximity without air support or heavy artillery or tanks. Water also limits opportunities for specific engagement types, whether through river crossings or lakes, that limit logistical mobility. In this model, you are focused on the needs of the infantry defender. Effects of mud and caves were excluded, increasing the terrain's defensibility because it is too complicated to model here (this can be done but requires additional tools and processing). The aim is to understand how defensible the terrain is on a surface landscape level.

The first tool to use is **Slope**. A steep or sloped gradient makes offensive activity difficult.

1. In the **Geoprocessing** pane, search for the Slope (Spatial Analyst Tools) tool. Open the tool and enter the following settings:
   - **Input Raster**: frontier_elev.tif
   - **Output Raster**: Frontier_slope
   - **Output Measurement**: Degree
   - **Method**: Planar
   - **Z Factor**: 1
   - **Target Device for Analysis**: GPU then CPU
2. Click **Run**.

The second tool is **Focal Statistics**, which calculates statistics for input cells within a set of overlapping windows or neighborhoods. (For information about how the tool works, see links.esri.com/SF-FocalStatistics.)

In this case, you will run the **Range** option in the **Statistics type** setting to help determine the ruggedness of the local terrain. The **Range** method calculates the difference between the highest and lowest elevation values within a defined neighborhood around each cell. This provides a measure of topographic relief, with higher values indicating more rugged terrain and lower values indicating flatter terrain.

3. Open the Focal Statistics (Spatial Analyst Tools) tool and enter the following settings:
    - **Input Raster**: frontier_elev.tif
    - **Output Raster Dataset**: Frontier_range
    - **Statistics type**: Range
4. Click **Run**.
5. In the **Contents** pane, right-click the **Frontier_range** layer and click **Symbology**.
6. In the **Symbology** pane, change the **Primary symbology** to **Stretch**.

    This will change the colors to black and white.
7. Open the **Colors scheme** list and change it to **Prediction**.

> **Tip:** At the bottom of the Colors scheme list, check Show names to see the names of the colors.

The **Frontier_range** layer shows the red, yellow, and blue **Prediction** symbology for **Range** in **Focal Statistics**. Red represents the most rugged terrain, yellow represents moderately rugged terrain, and blue represents the least rugged terrain.

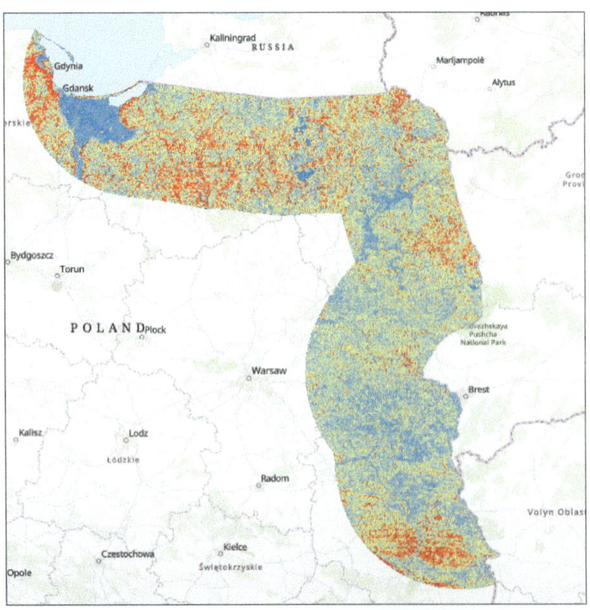

The third process to examine is **Geomorphon Landforms**. The **Geomorphon Landforms** tool can identify landscape features, such as flats, peaks, ridges, shoulders, spurs, slopes, hollows, foot slopes, valleys, and pits. For more information about this tool, see links.esri.com /SF-GeomorphLandforms.

The **Geomorphon Landforms** tool requires an elevation raster dataset such as GMTED. This dataset generates two types of output rasters: a landforms raster and a geomorphon raster. You will use the landforms raster for the model.

8. In the **Contents** pane, turn off **Frontier_range** and **frontier_lc.tif**. Turn on the **frontier_elev.tif** layer.
9. In the **Geoprocessing** pane, open the Geomorphon Landforms tool and enter the following settings:
    - **Input Surface Raster:** frontier_elev.tif
    - **Output Landforms Raster:** frontier_landforms
    - **Output Geomorphons Raster:** <leave blank>
10. Click **Run**.

    A new rainbow-hued raster named **frontier_landforms** should be listed in the **Contents** pane, with a breakdown of landform class types.

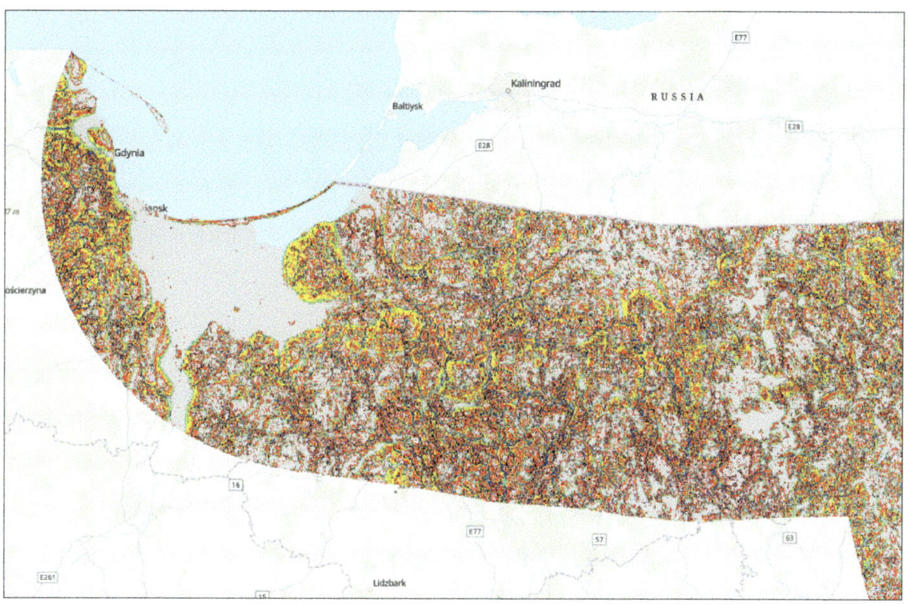

The next step will be to reclassify the output.

11. Open the Reclassify (Spatial Analyst Tools) tool.
12. For **Input raster**, select **frontier_landforms**. For **Reclass field**, select **Value**.

13. For **Reclassification**, use the following data to reclassify the values:

Landform	Value	New
Flat	1	0
Peak	2	10
Ridge	3	9
Shoulder	4	8
Spur	5	7
Slope	6	6
Hollow	7	4
Footslope	8	5
Valley	9	2
Pit	10	3
NO DATA	NODATA	NODATA

14. Set the name of the **Output raster** to frontier_landforms_rc.
15. Click **Run**.

    The symbology will be different because the **Unique Values** symbology will autogenerate a new random color palette.

    The next step is to use the **Reclassify** tool to reclassify **frontier_lc.tif.**
16. In the **Reclassify** tool pane, set the **Input raster** to select **frontier_lc.tif**. For the **Reclass field**, select **ClassName.**
17. For **Reclassification**, use the following data to reclassify the values:

ClassName	New
Rainfed Cropland	1
Herbaceous Cropland	1
Tree or Shrub Cropland	3
Mostly Cropland in a Mosaic with Natural Vegetation	5
Mostly Natural Vegetation in a Mosaic with Cropland	6
Closed to Open Canopy Broadleaved Deciduous Tree Cover	17
Closed Canopy Broadleaved Deciduous Tree Cover	18
Closed Canopy Broadleaved Evergreen Tree Cover	15
Mixed Tree Cover	16
Mostly Trees and Shrubs in a Mosaic with Herbaceous Cover	14
Mostly Herbaceous Cover in a Mosaic with Trees and Shrubs	13
Grassland	2

Sparse Vegetation	1
Fresh or Brackish Water Flooded Tree Cover	16
Flooded Shrub or Herbaceous Cover	18
Urban Areas	26
Bare Areas	1
Unconsolidated Bare Areas	3
Bodies of Water	24
NO DATA	NODATA

> **Tip:** If you get a warning that the field cannot be reclassified, click the Unique button at the bottom of the Reclassification stick.

18. Set the name of the **Output raster** to frontier_lc_rc and click **Run**.
19. Examine the results in the **Contents** pane.

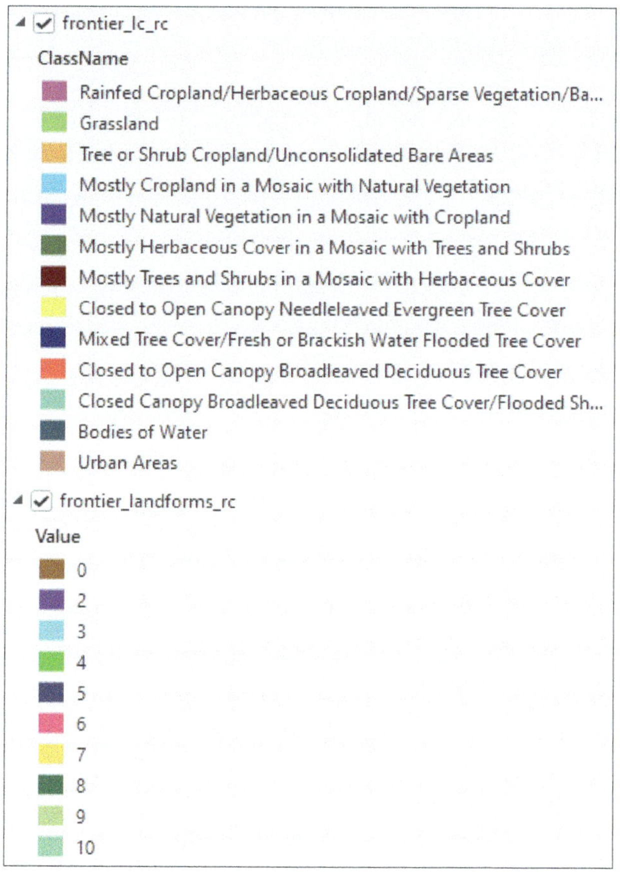

The symbology will change to a set of random colors, much like the landforms reclass raster. The weights have now been applied. Later in the exercise, you will update the symbology and apply color theory when preparing your final product.

20. Save your project.

Model weights according to Collins (1998)

- Urban areas receive the highest rating in this model of defensibility because urban combat disrupts unit cohesion, complicates control, blunts offensive momentum, and causes casualties to increase on both sides.
- Wetlands strongly compete for the title "Least Trafficable Terrain" because they are saturated with and partially, completely, perennially, or intermittently inundated by salty, brackish, or fresh water. They are given a higher rating in the defensive model.
- Comparing Bare Areas with Forests in the model depends on whether you are modeling offensive or defensive operations. Bare ground favors offensive forces, whereas forests favor defense.
- Crops, farms, and other fields in a defensive model should have a low rating.

4 Build a model in ModelBuilder and reclassify the data

This section uses ModelBuilder in ArcGIS Pro to build and customize a geoprocessing workflow. ModelBuilder allows you to develop reusable and reproducible Python toolboxes within a visual coding environment and to automate workflows to increase efficiency and accuracy. You were introduced to the data, tools, and outputs earlier in the chapter. Next, you will use ModelBuilder to document and automate the spatial analysis workflow. This next section is organized into three sections to help you execute the steps.

Customize the model in ModelBuilder

This section introduces the core inputs—elevation, land cover, and terrain analysis—and links key geoprocessing tools to prepare for further refinement. You will construct the foundation of the model.

1. To access ModelBuilder, click the **Analysis** tab. In the **Geoprocessing** pane, click **ModelBuilder**.

   A new **Model** window appears, for ModelBuilder.

2. On the **ModelBuilder** tab, in the **Model** group, click **Auto Save** to save your work.
3. In the **Contents** pane, locate the raster layers created earlier in this tutorial: **frontier_elev.tif** and **frontier_lc.tif.** Drag each raster from the **Contents** pane into the ModelBuilder workspace.

   The files appear as blue ovals labeled with their respective names.

   Note: In the View group on the ModelBuilder tab, click the Auto Layout button to organize the ModelBuilder workflow. Do this after adding a new tool or dataset to the workflow.

4. On the **ModelBuilder** tab, in the **Insert** group, click the **Tools** button and click the down arrow. Search for Focal Statistics (Spatial Analyst Tools).

5. Double-click the tool from the search results.

   The **Focal Statistics** tool will populate the workflow as a gray square with a gray oval for its output.

   Next, you will connect **frontier_elev.tif** to the **Focal Statistics** tool.

6. Click the **frontier_elev.tif** dataset and drag your cursor to create an arrow, connecting it to the **Focal Statistics** tool.

7. From the list of tool options, select **Input Raster**.

   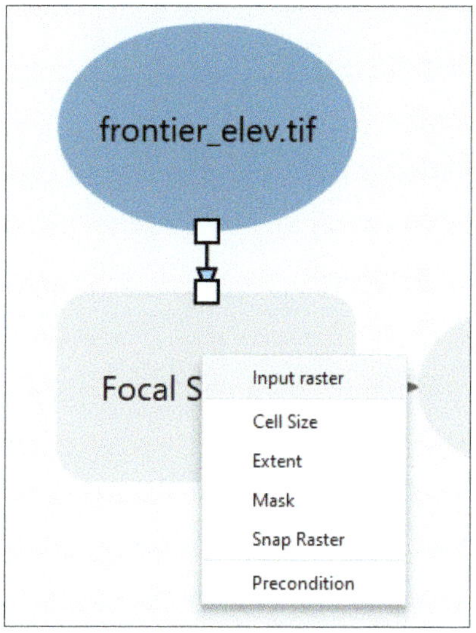

   After assigning the input raster, the **Focal Statistics** tool will activate and change from gray to yellow, indicating that it is ready for configuration.

8. Repeat the preceding steps to add the following tools to your model:

   - **Slope (Spatial Analyst Tools)**
   - **Geomorphon Landforms (Spatial Analyst Tools)**
   - **Reclassify (Spatial Analyst Tools)**
   - **Rescale by Function**

9. After adding the tools, click **Auto Layout**.

10. Connect the **frontier_elev.tif** dataset to the **Slope** and **Rescale by Function** tools as the **Input raster**. Connect the dataset to the **Geomorphon Landforms** tool as the **Input surface raster**.

11. Connect the **frontier_lc.tif** to the **Reclassify** tool as the **Input raster**.

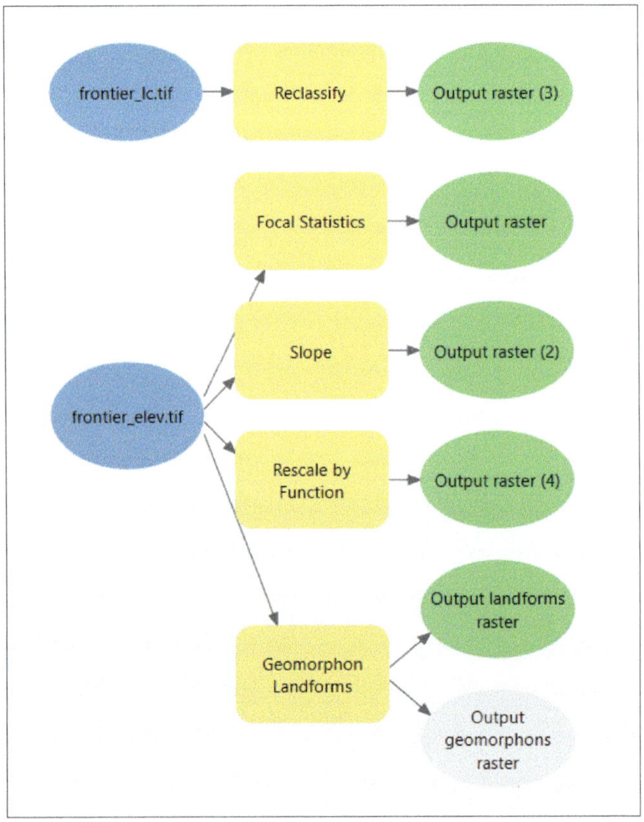

The first stage of ModelBuilder is now complete with essential data input and tools. Next, you will refine and customize these tools to generate output for the **Weighted Sum** analysis. Many of these geoprocessing tools will use the **Rescale by Function** tool, which transforms raster values, such as slope, elevation, or ruggedness, using a mathematical function to adjust their influence on the final analysis, and the **Reclassify** tool, which changes the classification of raster values based on defined ranges. A **Weighted Sum** is like raster math but with extra multiplication for specific datasets being weighted as more critical, more important, or more valuable for the final output. It adds up all the variables to produce a final output. Before running the final **Weighted Sum**, the next step is to refine and organize the model outputs.

Set up output rasters

The second part of the ModelBuilder process is to rename the output rasters and reclassify the data output so that the **Weighted Sum** tool can calculate the data correctly.

1. Right-click the **Slope** tool and open it. Apply the following settings:
    - **Input raster**: frontier_elev.tif
    - **Output raster**: Frontier_slope
    - **Output measurement**: Degree
2. Click **OK**. Check the green output raster oval. It should be named **Frontier_slope**. If not, right-click, click **Rename**, and update to the correct name.

3. Right-click the **Focal Statistics** tool and open it. Apply the following settings:
    - **Input raster**: frontier_elev.tif
    - **Output raster**: Frontier_range
    - **Statistics type**: Range
4. Click **OK** to save. Check the green output raster oval. It should be named **Frontier_range**. If not, right-click, click **Rename**, and update to the correct name.
5. Right-click the **Rescale by Function** tool and open it. Apply the following settings:
    - **Input raster**: frontier_elev.tif
    - **Output raster**: frontier_elev_rs
6. Click **OK** to save. Check the green output raster oval. It should be named **frontier_elev_rs**. If not, right-click, click **Rename**, and update to the correct name.
7. Right-click the **Geomorphon Landforms** tool and open it. Apply the following settings:
    - **Input raster**: frontier_elev.tif
    - **Output landforms raster**: frontier_landforms
8. Click **OK** to save. Check the green output raster oval. It should be named **frontier_landforms**. Update the name if needed.
9. Right-click the **Reclassify** tool and open it. For **Reclassification**, reclassify the land cover as you did earlier using the second data table. Apply the following settings:
    - **Input raster**: frontier_lc.tif
    - **Reclass field**: ClassName
    - **Reclassification:** Change **New** values to match the second data table earlier in the chapter.
    - **Output raster:** frontier_lc_rc
10. Click **OK** to save. Check the green output raster oval. It should be named **frontier_lc_rc**. Update the name if needed.
11. Add a second **Reclassify (Spatial Analyst Tools)** tool to the workspace.
12. Connect **frontier_landforms** (the output from the **Geomorphon Landforms** tool) to the second **Reclassify** tool as an **Input raster**.
13. Right-click the second **Reclassify** tool and click **Open**.
14. In the **Reclassify** tool, under **Reclassification**, click **Classify**, and a new pane appears. Update these settings and click **OK**:
    - **Method**: Defined Interval
    - **Interval size**: 1
    - **Classes**: 9

    The new **Reclassification** table has three sections: **Start**, **End**, and **New**. You will ignore the **End** section, but the **Start** and **New** categories will need to be modified.

15. In the **Reclassification** table, apply the following settings:

    - In the **Start** column, name the empty cell at the bottom 10.
    - Reclassify the values in the **New** column to match the first data table earlier in the chapter.
    - Ignore the **End** column.

Reclassification

Reverse New Values

Start	End	New
1	2	0
2	3	10
3	4	9
4	5	8
5	6	7
6	7	6
7	8	4
8	9	5
9	10	2
10	10	3
NODATA	NODATA	NODATA

Classify Unique

16. Change the name of the **Output raster** to frontier_landforms_rc and click **OK.**
17. Add two new **Rescale by Function** tools to the workflow.

    Next, you will connect the **frontier_range** and **frontier_slope** rasters to the new **Rescale by Function** tools.

18. Connect **Frontier_range** to one of the **Rescale by Function** tools as an **Input raster.** Open the tool, apply the following settings, and click **OK:**

    - **Input raster**: Frontier_range:1
    - **Output raster:** frontier_range_rs

19. Connect **Frontier_slope** to the other **Rescale by Function** tool as an **Input raster.** Open the tool, apply the following settings, and click **OK:**

    - **Input raster**: Frontier_slope:1
    - **Output raster:** frontier_slope_rs

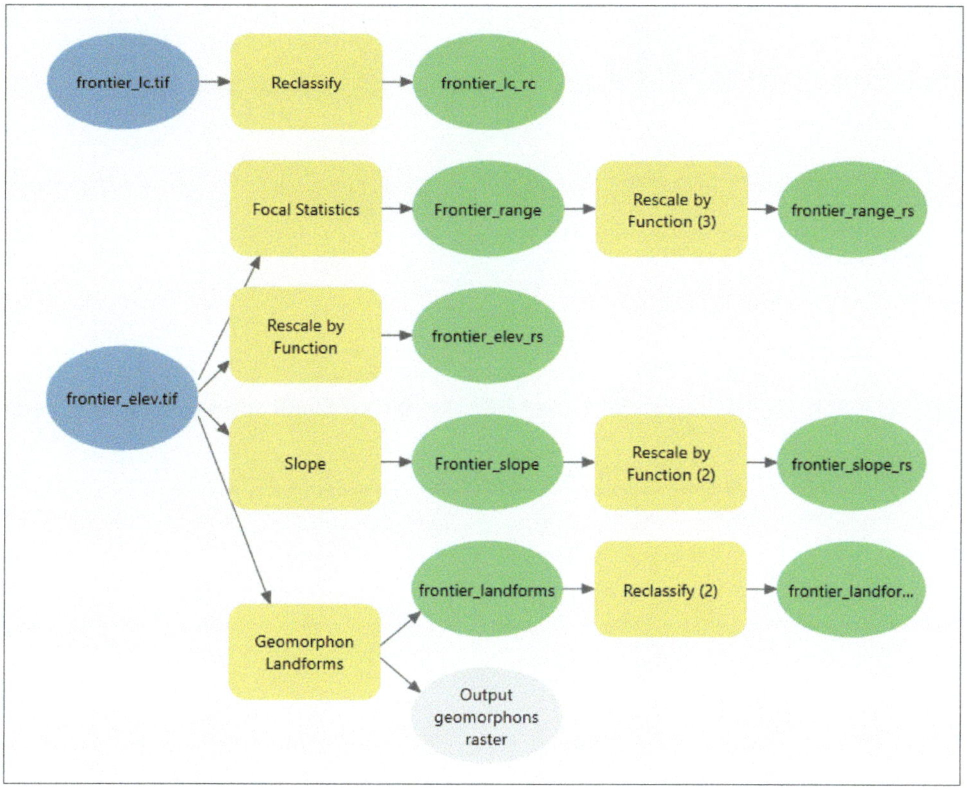

Integrate the tools and run the workflow

The third section of the ModelBuilder exercise is to integrate the **Reclassify** and **Geomorphon Landforms** tools into a final, comprehensive, predictive, defensive military geography model of Poland. By combining these processed rasters—representing land cover, ruggedness, elevation, slope, and landforms—into a weighted sum analysis, you will refine the model to highlight areas of strategic defensibility.

1. Add **Weighted Sum (Spatial Analysis Tools)** to the workflow.

   Next, you will connect each output raster to the **Weighted Sum** tool. Using ModelBuilder, you can process the GMTED file into ruggedness, landforms, elevation, and slope and then reclassify or rescale all the functions by your model weights.

2. Connect the following output rasters to the **Weighted Sum** tool as **Input rasters**:

   - **frontier_lc_rc**
   - **frontier_range_rs**
   - **frontier_elev_rs**
   - **frontier_slope_rs**
   - **frontier_landforms_rc**

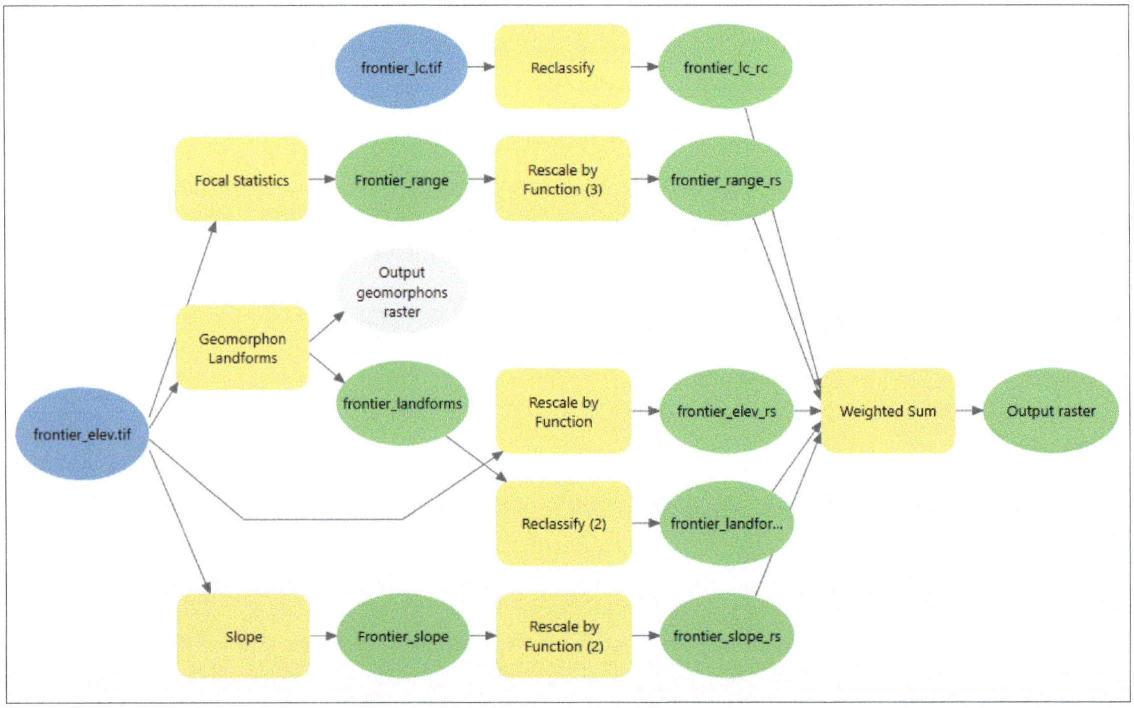

3. Right-click the **Weighted Sum** tool and open it. Keep the default values for the **Raster** and **Field** inputs. For the **Weight** value, assign weights to each model variable according to the values in this table.

Raster	Model Variable	Weight
frontier_lc_rc	Reclassified Global Landcover Raster	7
frontier_range_rs	Rescaled Ruggedness Raster (Range, Focal Statistics)	2
frontier_elev_rs	Elevation (World GMTED) Raster	2
frontier_slope_rs	Rescaled Slope Raster	2
frontier_landforms_rc	Reclassified Landform Raster	4

4. Change the name of the **Output raster** to Polish_Military_Geography and click **OK.**

   You have now created a new raster that shows which areas were the most defensible compared with the least defensible. Using ModelBuilder, you can process the GMTED file into ruggedness, landforms, elevation, and slope and then reclassify or rescale all the functions by your model weights.

5. On the **ModelBuilder** tab, in the **Run** group, click **Run.**

6. Once the model has finished running, right-click **Polish_Military_Geography** and click **Add To Display.**

   The new raster appears in the **Contents** pane and on the map.

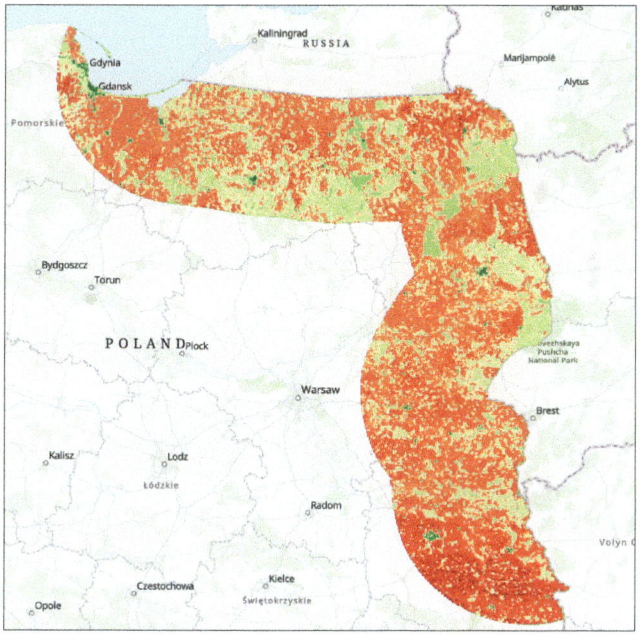

5 Customize symbology

Next, you will customize the symbology in the map.

1. In the **Contents** pane, right-click the **Polish Military Geography** layer and click **Symbology**.
2. In the **Symbology** pane, check the box for **Invert** so that red represents the most defensible areas (instead of the least defensible). The symbology now shows that red is the most defensible, yellow is dangerous, and green represents ease of trafficability for offensive forces.

> **Tip:** An easy way to test your map is to replace the topographic basemap with the human geography map to see whether cities are defensible. You can do this by clicking the Map tab, selecting the Layer group, clicking Basemap, and choosing the desired basemap. If you put the urban built environment in the land cover reclassification as value 26, the cities should be bright red.

6 Create a story in ArcGIS StoryMaps

The final part of this chapter is to create a story in ArcGIS StoryMaps to share the results of your work. ArcGIS StoryMaps lets you combine maps, 3D scenes, embedded content, and multimedia in an interactive narrative (a story). Stories can compellingly create awareness, influence opinion, and affect change.

1. In the **Contents** pane, right-click the **Polish_Military_Geography** layer and click **Sharing** > **Share As Web Layer**.
2. For **Summary**, type A map of defensible terrain in Poland along the borders of Russia and Belarus. For tags, type Ukraine War, geospatial modeling, land cover, and defensive operations. Type a comma between each tag.
3. In the **Share As Web Layer** pane, change the **Layer Type** to **Tile.**
4. Save this layer to the **SecurityFirst** folder.
5. You want your story to be viewable by everyone outside your organization, so change the **Sharing Level** to **Everyone (public).**

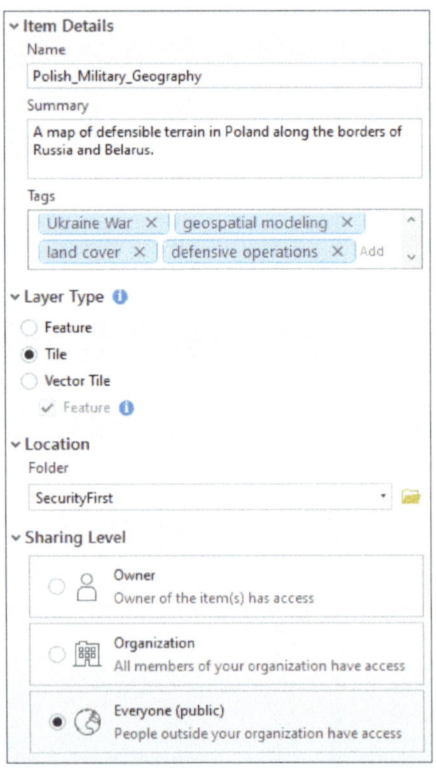

6. Click **Analyze** to confirm that the operation will run. Click **Publish.**
7. In your browser, navigate to www.arcgis.com and sign in to your ArcGIS organization.
8. In your **Content**, click the **Polish_Military_Geography** tile layer to open its item page, which contains various details. Click **Open in Map Viewer**.
9. On the **Contents** toolbar, click **Save and open** > **Save as.** Save your map as Polish Military Geography.
10. On the **Contents** toolbar, click **Create app.** Select **ArcGIS StoryMaps.**
11. For **Story title**, type Polish Military Geography.

Using this process, you can add multiple views of the same map throughout your story to emphasize different aspects of your findings. You can also create briefings—a slideshow version of ArcGIS StoryMaps, much like Microsoft PowerPoint, which emphasizes slides for the military, law enforcement, and first responders.

An intelligence analysis using the BLUF approach as the main content is recommended as an addition to this story, which uses multiple maps or images of the Polish borderlands.

Creating a story: The Military Geography of Poland

The authors reversed the color scheme to enhance visualization. In the uploaded map, red represents defensible areas in Poland, whereas green denotes less-defensible areas. The latter typically corresponds to flat farmland, highlighting a key geographic factor contributing to Poland's historical vulnerability to invasion.

Accuracy and uncertainty: This model does not incorporate temporal factors into the analysis. For example, as demonstrated during the Ukrainian War, seasonal weather conditions can significantly impact the terrain, turning soil into mud and hindering military movement and operational momentum. Including a soil raster dataset, segmented by seasonal variations and provided by the Polish government, would greatly enhance the model's accuracy and utility.

Interpretation of analysis

After completing this exercise, answer the following questions in detail. The answers to these questions, as well as the analysis results from the exercise workflow, will inform the intelligence brief.

- Based on this analysis, where are the areas within the map that are determined to be most defensible? Least defensible?
- Among the areas considered most defensible, are there strategic resources or assets that may make one of these areas most important to protect?
- What other variables might be used in this analysis to help further understand the military geography of Poland?
- How might your model in ModelBuilder be further developed or refined to improve the analysis on the military geography of Poland?
- What actionable information can be derived from this analysis? What efforts can this information support? To whom should this information be disseminated?
- What other non-GEOINT analyses might support this research?

Intelligence brief

Based on the analysis from the exercise workflow, produce a three- to five-page intelligence brief by responding to the following question prompts. Incorporate your geospatial outputs from the exercise to enrich and illustrate your conclusions.

- **BLUF:** List three to five primary observations from your analysis in bulleted-list format that summarize findings of this work.
- **Top-level analysis:** Report the primary data points supporting your BLUF.
- **Secondary analysis:** Provide the data sources and deeper analytic products from this analysis.
- **Impact:** Describe the impacts of your findings on the department or organization.
- **Confidence:** Report the level of confidence of your assessment based on available data, accuracy, and so on.
- **Bias:** Explain any background or information you have that could have affected your analysis.
- **Conclusions:** Determine what type of analytic conclusions there may be and how the data requirements, analysis, and brief writing may change based on what analytic framework you are following (descriptive, predictive, prescriptive).

Applying and extending your skills

Pick a specific country of interest and run the same analysis by exporting the raster for the whole country (usually, a country bigger than Poland will take too many hours of computing time). Examples include Taiwan, Singapore, and Kosovo.

If you wanted to repeat these steps for Taiwan, Vietnam, or another tropical or subtropical country, you would need to find a land cover dataset with a specific type of raster pixel for rain forest or jungle. The WorldCover dataset uses Closed to Open Canopy Broadleaved Evergreen Tree Cover for this pixel. You could then adjust the number for the weighted model when using jungle or rain forest land cover pixels.

Use the **Profile Plot Tool** to create a section of the country, such as the CIA military geography map of the former Yugoslavia. Add a chart at the bottom of the intelligence product to add context to the intelligence brief.

Because you have done this for a specific portion of a modern country or a historical example such as the former Yugoslavia, find a past country of interest (preferably a small polity) to see what your modern military geography models pull up. You can find GIS files for past empires and polities using the following websites:

- https://mosaic.ipums.org/historical-gis-datafiles
- https://worldmap.maps.arcgis.com/home/index.html

Additional resources

The following resources provide more details about military geography:

- American Association of Geographers Military Geography Specialty Group. www.aag.org /groups/military-geography/.
- CIA Map: *Military Geography of Former Yugoslavia*. Library of Congress. www.loc.gov /resource/g6841c.ct000964.
- Collins, John M. 2012. *Military Geography: For Professionals and the Public*. CreateSpace Independent Publishing Platform.
- Galgano, Francis A., and Eugene J. Palka, eds. 2010. *Modern Military Geography*. Routledge.
- Modern War Institute at West Point. n.d. Urban Warfare Project. https://mwi.westpoint.edu /urban-warfare-project/.
- Modern War Institute at West Point. n.d. Irregular War Initiative. https://mwi.westpoint.edu /irregular-warfare-initiative/.
- Spencer, John, and Liam Collins. 2022. *Understanding Urban Warfare*. Howgate Publishing.
- Winters, Harold A., Gerald E. Galloway Jr., William J. Reynolds, and David W. Rhyne. 2001. *Battling the Elements: Weather and Terrain in the Conduct of War*. Johns Hopkins University Press.

American Association of Geographers (AAG) Military Geography Specialty Group

The goal of the AAG Military Geography Specialty Group is to promote research and exchange information about military geography and advance scholarship that contributes to a deeper understanding of the subdiscipline. This subfield is defined as applying geographic information, tools, and techniques to solve military problems in peacetime or war. The subfield is broadly defined to include physical, cultural, political, historical, environmental, remote sensing, GIS, and other applications related to military or security issues.

About the authors

Logan Bolan, a graduate student, is also working on a master's in anticipatory intelligence at Utah State University. He has a bachelor of science in geography and GIS and a bachelor of liberal arts with a certificate in peacebuilding from Utah State University. He is proficient in GIS software, including ArcGIS Pro, ArcGIS Online, and ArcGIS Experience Builder. Logan is also fluent in Bosnian, Croatian, and Serbian and is a member of the American Association of Geographers (AAG) Military Geography Specialty Group.

Nathan Kozlowski is a graduate student working on a master's in anticipatory intelligence at Utah State University. He has a bachelor's degree in political science and a graduate geographic information science certificate from Utah State University.

Chapter 20

Processing advanced sensor imagery in ArcGIS Drone2Map for use in modeling and simulation intelligence activities

COL [R] Steven D. Fleming, *University of North Georgia, Institute for Environmental and Spatial Analysis*
Jason Knowles, *GeoAcuity*
Will Forker, *GeoAcuity*
Jonathan Hawes, *University of Southern California, Institute for Creative Technologies*

Overview

In this chapter, you will explore a variety of drone image types and learn how they can be processed using ArcGIS Drone2Map for intelligence analysis, modeling, and simulation. This software offers an intuitive, workflow-focused interface for converting drone-captured images into 2D and 3D products, including orthorectified imagery (orthomosaics), digital surface models (DSMs), point clouds, and textured meshes. These standard products are used for visualization and analysis in GIS and for modeling and simulation applications. In addition, you will explore other types of sensors, including multispectral and thermal infrared imagery that can help complete and properly segment datasets for modeling applications in intelligence activities.

Learning objectives

- Process RGB imagery using Drone2Map.
- Use datasets generated from RGB imagery in Drone2Map to measure volume and complete other analyses.

- Share layers created with Drone2Map in ArcGIS Online and create instant apps.
- Assess multiple methods to disseminate products and simulations to customers.

Technical components

- Software:
  - ArcGIS Online (at least Publisher level access with Creator user type)
  - ArcGIS Drone2Map v. 2024.2 (for system requirements, see https://doc.arcgis.com/en/drone2map/latest/get-started/arcgis-drone2map-system-requirements.htm)
- Dataset: Sample asphalt dataset

Prerequisites

- Basic understanding of ArcGIS Online and a fundamental understanding of imagery
- Technical skill level: Introductory GIS
- Time to complete chapter: Two to three hours (which includes an Esri Academy web course) without intelligence brief, four to six hours with intelligence brief.

Background and situational context

This exercise is a practical exploration of Drone2Map software. Using photos captured by uncrewed aerial systems (UAS), you will transform these images into comprehensive 2D and 3D output. Drone2Map is an Esri stand-alone solution for local photogrammetric image processing, contrasted with the company's cloud-based counterpart, Site Scan for ArcGIS. Drone2Map is particularly advantageous in environments with sparse internet access or when handling confidential data and provides all tools necessary to upload imagery, run photogrammetric reconstructions, and view outputs in 2D and 3D space. Conversely, Site Scan is useful in swiftly producing and distributing products for visualization, analysis, and simulation through cloud-based processing. Both systems produce output that integrates with ArcGIS Pro, supporting tasks such as terrain analysis, infrastructure examination, surveillance, geospatial intelligence, and various modeling and simulation endeavors. Drones are an ideal choice for these tasks because of their ability to capture high-resolution images on demand, unhindered by common limitations faced by traditional sensors, such as atmospheric conditions (including cloud cover), the demand or bandwidth limitations of space-based sensors, and the freshness of the data. These advantages are also valuable for simulation tasks where spatial accuracy, timeliness of data, and ease of data transfer are crucial.

Exercise workflow 20

1 Process RGB imagery with Drone2Map

Before you download the sample dataset, you will learn more about Drone2Map to get started.

1. Take ArcGIS Drone2Map Basics, an online course from Esri Academy, at links.esri.com/SF-D2MBasics.

   The Esri Academy course should take about two hours to finish.

2. For more practice loading projects into Drone2Map, take the free Esri tutorial Create 3D Products with ArcGIS Drone2Map at links.esri.com/SF-3DTutorial.

   This Esri tutorial should take about 45 minutes to finish. If you have time, you can take some of the other Esri tutorials on Drone2Map:

   - Inspect Infrastructure with ArcGIS Drone2Map
   - Use Catalog Datasets in ArcGIS Drone2Map
   - Create 2D Products with ArcGIS Drone2Map
   - Evaluate the Accuracy of Ground Control Points

   Next, you will download the sample dataset.

3. In a browser, go to links.esri.com/SF-D2MSampleData.

4. Under **Rapid**, download the **Sample Asphalt** dataset (courtesy of AirGon LLC).

Rapid

Sample asphalt	
General project information	Download
Data collected with AV-900 UAS by AirGon LLC.	
Location	Alabama, USA
Average Ground Sampling Distance (GSD)	1.66 cm / .65 in
Area covered	0.0129 km2 / 1.2935 ha / 0.005 sq. mi. / 3.198 acres
Image acquisition	
UAV	AV-900 UAS
Image acquisition plan	1 flight, grid flight plan
Camera	Sony NEX-5T
Images	
Number of images	21
Image size	4912 x 3264
Image geolocation coordinate system	WGS84

5. Extract the zip archive to the **SecurityFirst_ExerciseData/Chapter20** folder.

   The photos are located under the **Sample Asphalt** directory in the **Images** folder. Next, you will load the project template.

6. Open **ArcGIS Drone2Map** and sign in using your credentials.

7. Under **Processing Template**, select **Rapid**. For **Project Name**, title the project Ch20_SecurityFirst and save it in this chapter's folder.

   Note: The Rapid template is selected to load the photos. Later, you will add 3D processing options before starting the processing.

8. Click **Add images** to load the photos. Navigate to the **Images** subfolder and select all the images to add them.

   Once the images are added, you should be able to see the x,y,z information for each photo as well as the date and time they were collected. There should be 21 images.

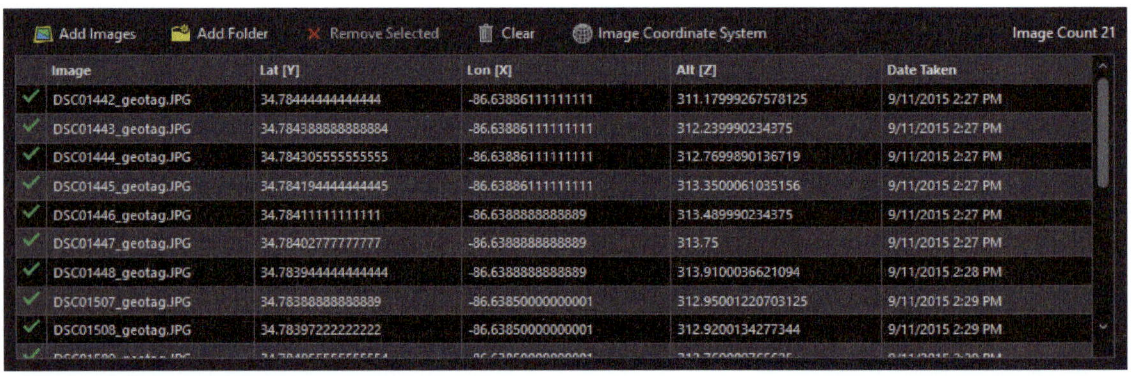

9. Click **Create** to create your project.

   In your new project, you will see the individual photos and the drone flight path on the map.

10. On the ribbon, click the **Home** tab. In the **Processing** group, click **Manage** to open the **Manage** pane on the right.

    The **Manage** pane helps you manage your output and products.

11. In the **Manage** pane, under **2D Products**, look for the list of four products (**Image Collection**, **True Ortho**, **Digital Surface Model**, and **Digital Terrain Model**). Check the box for each of them to enable all of them.

12. Under **3D products**, look for the list of three products (**DSM Mesh**, **3D Mesh**, and **3D Point Cloud**). Check the box for **3D Mesh** and for **3D Point Cloud** to enable and generate both of them.

> **Tip:** Because you chose the Rapid template earlier, no 3D products are automatically enabled.

13. On the **Home** tab, in the **Processing** group, click **Options**.

14. In the **Options** window, explore the default settings on each tab.

    Some tabs include settings for additional processing parameters and options for customizing the processing of your project to improve the quality of output or reduce processing time. For now, keep the default options, but remember that you can return to these options later to try different settings.

15. Click the **General** tab and review the **Hardware** section to ensure that your CPU resources are fully enabled. The slider should be at its maximum. If a CUDA graphics card is present, confirm that it is enabled, too.

    The computer will devote all its CPU processing toward the software during processing, including the graphics card.

16. Close the **Options** window.

> **Tip:** If the graphics card option is not available and you are certain that there is a graphics card in your computer, try updating the firmware on your graphics card. It may be out-of-date.

17. In the **Processing** group of the **Home** tab, click **Start** to start processing your project.

    The example dataset is a small project with only 21 photos and you are using rapid processing, so the processing should not take more than 30 minutes, depending on your computer and whether other processes are running.

> **Tip:** You can monitor the progress of your project at the bottom of the Manage pane.

18. When the processing is finished, view the list of output in the **Contents** pane. Review the **2D Map** and **3D Map** tabs. Toggle between the tabs to change views.

19. Save your project by clicking the **Project** tab > **Save Project**.

    You will now review the generated products and clean up your project.

20. Confirm that **2D Map** view is enabled. In the **Contents** pane, under **2D Map**, uncheck the box for the Project Data group layer to turn it off.

    Under the **Imagery Products** group layer, you should see the **True Ortho** (orthomosaic). Under the **DEM products** group layer, you should see the **Digital Surface Model** and **Digital Terrain Model** layers.

21. On the map, switch to **3D Map** view. In the **Contents** pane, under **2D Layers**, turn off **World Imagery**. Under **3D Products**, turn on the **3D Point Cloud** layer.

    You should now see the 3D map.

22. Practice navigating in 3D space.

    On the bottom left of the 3D map, you will find a white navigation ring with a compass arrow. Just above that ring is the option to expand to show all 3D degrees of navigation of the scene. This option can be helpful when not using a mouse.

    The orthomosaic allows you to take accurate horizontal measurements from your drone map. Similarly, generating the DSM allows estimates of volume. You have access to these capabilities in **2D Map** view in the central Drone2Map window.

23. Switch to **2D Map** view. In the **Contents** pane, select the **Digital Surface Model** layer.

24. On the ribbon, click the **Analysis** tab. In the **Measure** group, select the **Volume** tool.

25. Locate the three circular gravel stockpiles on the map. Start with the stockpile on the left. Click the edge of the stockpile to place a starting point. Trace the stockpile by adding multiple points along its edge.

26. Double-click to finish the trace of pile A.

27. Use the **Area** tool to trace pile B (the small stockpile on the right) and pile C (the large stockpile on the right).

    Assume that a ton of moderately damp gravel typically fills about 0.6 cubic meters and runs about $60 per ton. Using the **Volume** tool, you learn that pile A is 280 cubic meters, pile B is 345 cubic meters, and pile C is 2,303 cubic meters. You can then assume that this is approximately $292,800 worth of gravel (280 + 345 + 2303 = 2928 cubic meters of gravel total, 2,928/0.6 yields 4,880 tons of gravel, which multiplied by $60 per ton is $292,800). Results may vary on how you measure the piles.

28. In the **Contents** pane, hide the **True Ortho** layer. Assess the **Digital Surface Model** and the **Digital Terrain Model** layers. Toggle between the two DEM products and notice how the two are different.

> **Tip:** To view the digital terrain model (DTM), hide the Digital Surface Model layer.

The DTM does not account for everything present in the orthophoto because it is only an estimated bare-earth model. Bare-earth raster models can be crucial for various GIS analyses but can also work as "ground planes" for simulation and visualization purposes. Think about circumstances when a DSM would be more useful than a DTM, and vice versa (in other words, a DTM is better for hydrologic modeling).

29. Turn on the **True Ortho** layer and all **DEM Products** layers.

With your output models, further analysis can be done within and outside of Drone2Map. The **Analysis** tab in Drone2Map provides tools to inspect, measure, and interact with your newly created geospatial data beyond simply measuring in 2D or 3D. The **Profiles** tool allows for elevation to be analyzed across the landscape of the new elevation models, alongside global elevation datasets.

30. In the **Analysis** tool pane, in the **Tools** group, click **Profiles** to open the menu and select **Elevation Profiles.**
31. In the **Elevation Profile** pane, click the **Draw a new elevation profile** button.
32. On the map, draw a line that includes a flat area and an elevated area, such as the stockpiles. Double-click to finish drawing the line.

The elevation profile is instantly visualized as a linear chart. This format allows users to potentially uncover change, terrain patterns, and obscured objects in their area of interest previously unnoticed from a simple bird's-eye view.

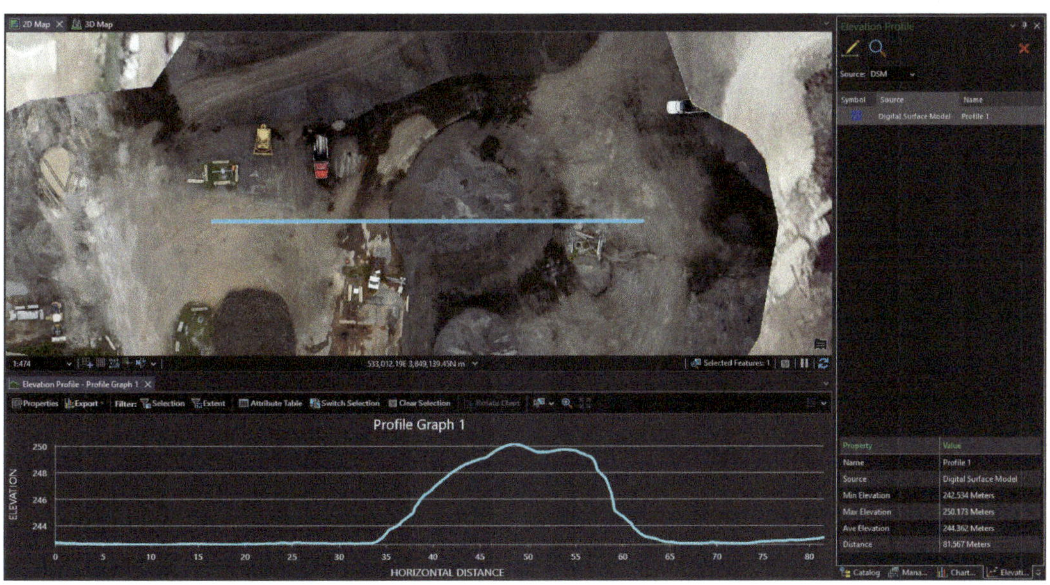

3D point cloud data allows for an effective combination of the visualization capabilities of an orthophoto with the 3D context that raster elevation models can offer.

33. Switch to **3D Map** view. In the **Contents** pane, turn on the **3D Point Cloud** layer and hide the **3D mesh** layer.

    You will see hundreds of thousands of colorized points that compose an interactive 3D model of the area captured by the drone images. These points contain x,y,z values and colors and can include classifications to segment buildings, ground, vegetation, and more.

34. Select the **3D Point Cloud** layer. On the ribbon, click the **Point Cloud Layer** tab. In the **Drawing** group, click **Symbology** > **Elevation.**

    The area of interest can now be understood and explored in 3D, giving analysts visual cues to the nature of the terrain.

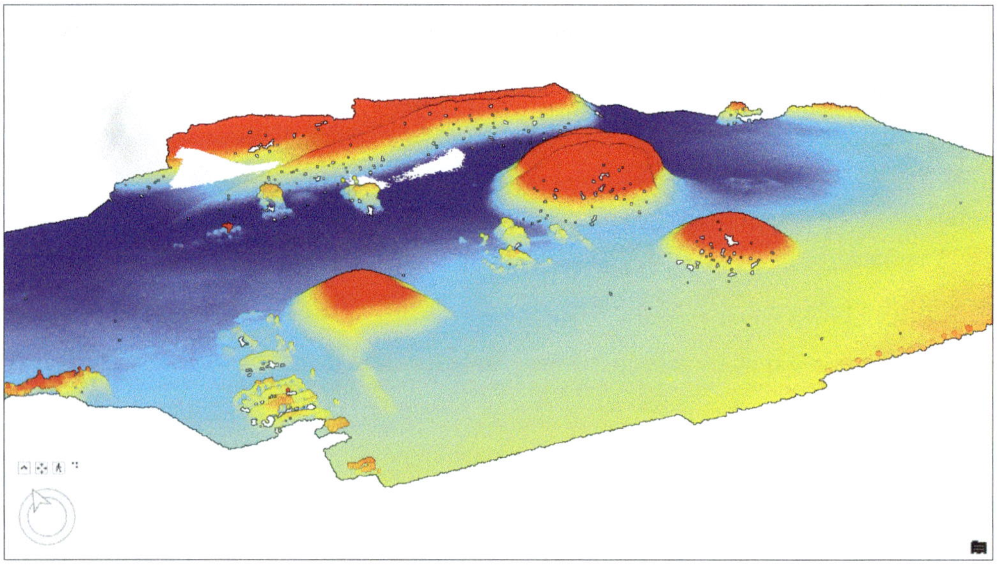

This workflow is an example of creating 2D and 3D geospatial datasets from true-color (RGB) UAS photographs. As seen from the sample analyses, the GIS products are greater than the sum of their parts. Although a well-trained image analyst may be able to derive significant insights from a collection of photos, having georeferenced orthophotos, elevation models, meshes, and point clouds allows for the quantification of change, the creation of high-resolution basemaps, and the visual segmentation of an area of interest. These products can be shared and exported for viewing in ArcGIS Pro, ArcGIS Online, and other GIS applications or web viewers, where they can be further visualized and analyzed. The following section is an example of how to disseminate your findings to a web viewer.

2 Disseminate your findings in ArcGIS Online

After processing your dataset to your desired specifications, you can share your visualizations and package your orthophotos, elevation models, meshes, and point clouds as tile packages and scene layer packages (TPKGs and SLPKs, respectively).

1. Confirm that you are still signed in to your ArcGIS account in Drone2Map.
2. On the ribbon, click the **Share** tab.

   In this example, you will share a scene layer to allow web users to perform 3D analyses.
3. In the **Share as** group, select the **Scene Layer** option.

   The **Share Scene Layers** pane opens.
4. For **Layers**, check the box for **3D Mesh.** For **Title**, set the name to 3DMesh_<your initials>. Optionally, you may share the dataset in a specific folder or leave the folders unchecked.

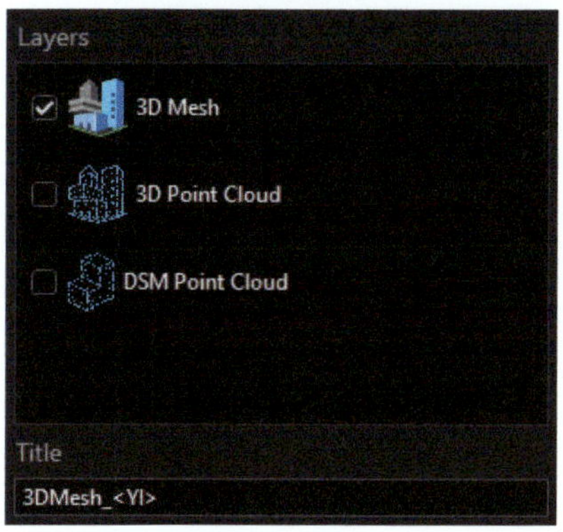

5. For **Groups**, choose the level of sharing. If you want to share this with the public, check **Everyone** and click **Share.**

   Once the scene layer is shared successfully, you can create a simple viewer app in ArcGIS Online to show your area of interest.
6. In a browser, go to www.arcgis.com and sign in.
7. Navigate to your **Content** page.

   You should see the item you just shared in the list as a **Scene layer package** and **Scene layer–integrated mesh (hosted).**
8. Select the **Scene layer–integrated mesh (hosted)** dataset and click **Open in Scene Viewer.**

   In the scene view, you can explore the 3D mesh the way you might explore it in Drone2Map. You can swap basemaps, add layers, adjust item heights, and more.
9. On the right toolbar, click the **Basemap** button. Change the basemap to **Imagery Hybrid.**

10. On the **Content** toolbar, click **Save.**
11. Save the scene as Sample asphalt data 3D mesh. Click **Save.**

    Once saved, you are given a variety of options to choose from to make apps and web interfaces for visualizing your data.

12. On the **Content** toolbar, click **Create app.**

    You will see the following app options:

    - **360VR Experience**
    - **Instant Apps**
    - **Experience Builder**
    - **StoryMaps**
    - **Dashboards**

13. Click **Instant Apps** and select **3D Viewer.**
14. Name your app Sample Asphalt Data 3D Viewer. Save it to the **SecurityFirst** folder. Click **Create app.**

    Express mode creates a viewer that allows users to explore a scene in their web browser. You will disable this feature and add app tools for viewshed analysis in your app.

15. Turn off **Express** mode to add more advanced analysis capabilities in the app.
16. On the **Content** toolbar, click the **Interactivity** tab.
17. Click **Modify.**
18. Turn on **Viewshed analysis**, which will add a widget directly to your app on the right.

19. In the app, click the **Viewshed analysis** widget. Click **Create viewshed**.

    Users can place a point on the mesh and estimate viewable areas from that point depending on the landscape around them. This capability can be useful for GEOINT, real estate, public safety, and other applications.

20. Use the app's **Explore** tools to view the viewshed in 3D.

21. Explore the tools that you can make available to your users through the app you are creating. When you have finished exploring the tools in the app builder, click **Publish** and confirm that you want to publish the app.

    You can now share this app with coworkers, colleagues, and the public. If the app is made public, any user with an internet connection and a browser can view and interact with your data.

    You have now completed the process of gathering, processing, analyzing, and disseminating a geospatially enabled UAS data collection using photogrammetry. This process not only allows for a larger audience to be able to view, understand, and analyze drone imagery but also places it in geospatial context, allowing any other geospatially enabled data to be added for viewing alongside it.

    Photogrammetry, however, is not a process limited only to sensors that capture imagery in the visible spectrum. The same workflows demonstrated in Drone2Map can be performed with image collections of multispectral sensors (such as infrared) as well as thermal sensors.

Interpretation of analysis

After completing this exercise, answer the following questions in detail. The answers to these questions, as well as the analysis results from the exercise workflow, will inform the intelligence brief.

1. What scenarios would require you to work with UAS-derived images and image products as opposed to those from aerial or space-based platforms?
2. What are some inaccuracies or artifacts you can identify in the output models from the Asphalt photoset? Based on the photos and their locations, what do you believe caused these anomalies?
3. If you were the drone pilot and wanted to enhance this dataset by collecting more images, what areas would you photograph and why?
4. You have demonstrated analysis and creation of a viewer for this data. Based on who you believe the audience of this analysis to be, what are some considerations for sharing and communicating your generated models and findings?
5. What GIS analysis tools do you think would be most useful for a web app viewer? Do they exist in ArcGIS Instant Apps?
6. What do you anticipate are potential use cases of photogrammetric workflows using images taken with multispectral or thermal sensors?

Intelligence brief

Based on the analysis from the exercise workflow, produce a three- to five-page intelligence brief by responding to the following question prompts. Incorporate your geospatial outputs from the exercise to enrich and illustrate your conclusions.

- **BLUF:** List three to five primary observations from your analysis in bulleted-list format that summarize findings of this work.
- **Top-level analysis:** Report the primary data points supporting your BLUF.
- **Secondary analysis:** Provide the data sources and deeper analytic products from this analysis.
- **Impact:** Describe the impacts of your findings on the department or organization.
- **Confidence:** Report the level of confidence of your assessment based on available data, accuracy, and so on.
- **Bias:** Explain any background or information you have that could have affected your analysis.
- **Conclusions:** Determine what type of analytic conclusions there may be and how the data requirements, analysis, and brief writing may change based on what analytic framework you are following (descriptive, predictive, prescriptive).

Additional resources

- See "Satellite vs. Drone Imagery in Vegetation Mapping" (March 24, 2020), a blog from AgEagle, at https://ageagle.com/blog/satellite-vs-drone-imagery-in-vegetation-mapping.
- See the ArcGIS StoryMaps story "One World Terrain: 3D Global Terrain Capabilities and Services to Support the Warfighter" (May 21, 2020) from USC Institute for Creative Technologies at links.esri.com/OWTStoryMap. Make sure to view the embedded videos within the narrative for a comprehensive understanding.

One World Terrain

Military and homeland security organizations have applications that span the nation and the globe. The One World Terrain (OWT) project is pivotal in advancing the collection and processing of space-based and drone-based imagery across challenging terrains, such as the Arctic snowscape, the dense Amazon rain forest, and vast deserts. It focuses on user-friendly approaches to transform the GEOINT field. These advances can and will revolutionize the way the Armed Forces collect, process, and disseminate intelligence.

An interactive application (figure 20-1) highlights a line-of-sight analysis using a drone image-derived mesh of the University of Southern California's Wrigley Marine Science Center on Catalina Island. Created from more than 500 images captured by a DJI Phantom Four Pro, the mesh was processed using a Bentley iTwin Capture Modeler. The application includes various modeling tools, such as line-of-sight simulators and weather and sunlight conditions, alongside measurement tools. It also features an elevation profile and a section slice tool, offering insights for practical applications, such as optimal solar panel placements. Applications such as the one shown in figure 1 can be quickly generated with mesh models uploaded to ArcGIS Online or Esri portals. These applications allow GIS data, models, and simulations to be shared with professionals not necessarily versed in those subjects, making interdisciplinary problem-solving increasingly possible.

Figure 20-1. A sample web application with a line-of-sight analysis. The spheres are observer points. The green lines show visibility from the central observer to four observers located at different points in 3D space. When the line is red, obstructions block the visibility (polygon viewshed).

GIS technology bridges the gap between large- and small-scale modeling and simulation. Integration of ArcGIS Runtime and SDK with Unity and Unreal Engine gaming development platforms exemplifies this advancement. The adoption of open-source standards, such as Cesium 3D Tiles, enhances the accessibility of detailed models for complex visualizations and simulations. The ongoing improvements in GPS, GIS, and drone technology, including interactive flight-planning tools such as Skydio and advanced GIS capabilities, hint at boundless future possibilities in efficient and effective simulation-based solutions. As GPS, GIS, and drone technology improve and benefit from integration with interactive flight planning with Skydio, more widespread access to real-time kinematic (RTK) and postprocessing kinematic (PPK) corrections, as well as cloud and Web GIS, the possibilities appear endless for quick and efficient simulations as solutions.

As with any innovative and emerging technology, 3D GIS, drone photogrammetry, modeling, and simulation all have drawbacks and room for improvement. However, many of these issues have workarounds and are being developed by user groups worldwide.

One of the most pressing issues with UAS-derived 3D mesh, point cloud, and elevation model data is assuring absolute accuracy. Models derived from images taken by UAS equipment often have unprecedented levels of relative accuracy: Resolution can be in millimeters, and features and peculiarities in the area of interest are captured with accurate measurements and internal relationships. However, as users of UAS platforms for gathering data to compare with satellite-derived data will find, consumer-level UAS accuracy does not match that of foundational data, such as data from the Shuttle Radar Topography Mission (www.earthdata.nasa.gov/data/instruments/srtm) or TanDEM-X (https://tandemx-science.dlr.de). This is observed most significantly in the z dimension, where drone data, when placed in 3D space alongside satellite elevation data, appears to float above or sink below the ground surface. For visualizations, ArcGIS allows for global vertical offsets of these 3D meshes, point clouds, or DEMs, but this causes problems when numeric analyses, such as change detection, are needed. Open-source coregistration methods, such as CODEM (https://github.com/NCALM-UH/CODEM) aim to solve this problem. Those with access to drones that allow for RTK or PPK corrections encounter these problems less frequently.

A solution is to supplement your collection with ground control points. The asphalt dataset in Drone2Map has available ground control points that can increase the absolute accuracy of the collection. To learn more about ground control points, see ArcGIS Drone2Map online documentation at https://doc.arcgis.com/en/drone2map/latest/help/ground-control.htm.

Problems in standardization are also prominent in 3D GIS, modeling, and simulation. For those experienced in game design, 3D graphics, or visual effects, the Wavefront OBJ format is considered an industry standard for displaying 3D meshes. However, an OBJ file is not inherently geographically enabled. Offsets and conversions must be run from the OBJ local coordinate system to work in 3D geographic space, such as WGS84. This complication makes it so that the Esri i3s and Cesium 3D tiles are more often used in situations in which geographically enabled 3D models are required. Both formats are open source, but both are not completely integrated into all 3D modeling spaces. Esri has created a solution for this by developing ArcGIS Maps SDK for Unreal Engine (links.esri.com/SDKforUnrealEngine) and ArcGIS Maps SDK for Unity (https://developers.arcgis.com/unity). For more advanced developers, this allows a pipeline for dynamically populating GIS data into industry-standard gaming engines, making geographically sound modeling and simulation applications possible.

It is important to remember to experiment and work with commercial and open-source solutions alike. Increasing support for customization in commercial solutions means that often the best of both worlds can be achieved, and solutions to many problems might be a few lines of code away. The solution that works is always the best solution, so whatever means of mission planning, capturing, processing, analyzing, and distributing drone-derived data works will be your best solution.

Data sources

- OpenStreetMap, Southern California (November 2023): https://openstreetmap.org/copyright
- Drone2Map video tutorials: https://mediaspace.esri.com/esearch/search?keyword=Drone2Map
- Drone2Map Esri tutorials: https://learn.arcgis.com/en/gallery/#?q=Drone2Map
- Drone2Map sample data: doc.arcgis.com/en/drone2map/latest/help/sample-data.htm
- “Satellite vs. Drone Imagery in Vegetation Mapping” (March 24, 2020), AgEagle blog: https://ageagle.com/blog/satellite-vs-drone-imagery-in-vegetation-mapping/
- “From Pest Management to Terrain Visualization: Drone Mapping with MicaSense Series Sensors” (April 20, 2020), AgEagle case study: https://ageagle.com/use-cases/from-pest-management-to-terrain-visualization-drone-mapping-with-micasense-sensors

About the authors

COL [R] Steven Fleming, PhD, is a professor of geospatial science and intelligence with the Institute for Environmental and Spatial Analysis at the University of North Georgia, where he teaches courses in geospatial intelligence, military geography, and the spatial sciences. From 2015 to 2022, he served as a professor at the University of Southern California's Spatial Sciences Institute and Institute for Creative Technologies. Before joining USC, he served on active duty for 30 years in the US Army, fulfilling Joint and NATO duties on multiple deployments in support of Operation Enduring Freedom. He culminated his military career at the US Military Academy, where he was an academy professor of geospatial information science. His research specialties focus on applications of geospatial technologies for national defense (military operations, homeland security, and disaster management).

Jason Knowles, PhD, is an adjunct associate professor with USC's Spatial Sciences Institute in the Dornsife College of Letters, Arts, and Sciences. Knowles has been a practitioner of geospatial sciences for more than 20 years, incorporating geospatial technologies into local municipal, commercial, military, intelligence, and academic communities. He is also the CEO and founder of GeoAcuity (geoacuity.com), a geospatial consulting firm that specializes in providing high-quality GIS, remote sensing, and GEOINT solutions to environmental, municipal, and federal organizations. His expertise includes skills in 3D modeling and visualization, pattern and predictive analytics, UAS data collection and analysis, complex geospatial modeling and design, spatiotemporal and multivariate analysis, cost-surface modeling, remote sensing (air photo and satellite imagery) analysis, cartographic visualization, field GPS data collection and support, GEOINT analysis, and enterprise GIS design and implementation.

Will Forker has been a GIS developer at GeoAcuity since 2021. From 2018 to 2022, Forker completed a bachelor of science in geodesign at the USC Spatial Sciences Institute, learning GIS, urban planning, and architecture. At USC, Forker developed an affinity for using GIS to solve complex problems and communicate realities of the world to spatial scientists and those unfamiliar with GIS alike. He is a Federal Aviation Administration (FAA) Part 107 licensed small UAS pilot and has put this certification to use collecting high-accuracy, high-fidelity drone imagery for use in spatial application on sites in the US and abroad. Merging the skills of a developer and data collector, Forker has found a specialty in the world of 3D GIS, having contributed to open-source algorithms for correcting point cloud and raster elevation data. Forker's work has engaged members of the public and private sectors relating to the vector of geospatial intelligence, environmental science, facilities management, government organizations, and research and development.

Jonathan Hawes, a senior geospatial data scientist at the USC Institute for Creative Technologies, applies his extensive academic background in geospatial intelligence to defense applications. He graduated summa cum laude with a master's degree in HSGI from USC, showcasing his academic dedication and expertise. In his military career with the US Army, Jonathan held various roles including Special Forces Intelligence sergeant and Communications sergeant. His diverse assignments ranged from leading tactical collection and integration for the One World Terrain project, serving as a liaison between the US Army Special Operations Command and the Army Futures Command, to training US Allied and Partner Nation Forces. His broad experience in these roles reflects a deep understanding and application of intelligence and operational planning in complex environments. As the director of emerging opportunities at RAVE Computer Association, Jonathan use his geospatial knowledge to spearhead business development and technological advancements, playing a key role in the company's growth in the geospatial technology sector.

Data credits

Chapter 1

WorldPop Total Population 2000–2020 100m, from WorldPop, CA Fire Perimeters 2018, from Fire & Resource Assessment Program FRAP, US National Grid from Federal Geographic Data Committee (FGDC). All data available from ArcGIS Living Atlas of the World.

Chapter 2

NOAA, National Hurricane Center, Humanitarian OpenStreetMap Team.

Chapter 3

Wildland Urban Interface/ArcGIS Living Atlas of the World; Los Angeles County and City Boundaries, Los Angeles County Fire Stations & Select Stations—County of Los Angeles Open Data Hub; Senior Well Being Boundaries—ACS/ArcGIS Living Atlas of the World; TIGER/Line Shapefile, 2018—Data.gov/US Census Bureau; Fire Hazard Severity Zones—CA.gov, California State Geoportal.

Chapter 4

Courtesy of Esri, MAXAR, Earthstar Geographics, and the GIS User Community; OpenStreetMap https://creativecommons.org/licenses/by-sa/4.0/.

Chapter 5

European Union, Copernicus Sentinel-1 IW 1SDV imagery, accessed on June 2, 2023, and June 9, 2023, from the Copernicus Open Access Hub. Ukraine Administrative Boundaries along with Sample Adjacent Urban/Populated Areas data court from the Humanitarian Data Exchange (OCHA), https://creativecommons.org/licenses/by/4.0/deed.en.

Chapter 6

Streets: 08/23, Nevada Department of Transportation; L2 EV Chargers: 08/23, Alternative Fuels Data Center, USDOE; Multi-Family Housing: 2017–2021 ACS 5-year.

Estimates, US Census; DACs: 05/22, USDOT/USDOE; Zoning: 08/23, Clark County (Nevada) GIS Management Office.

Chapter 7

Data courtesy of Leafmap.

Code sourced from: Wu, Q., and Osco, L. 2023. *Journal of Open Source Software*, 8 (89): 5663.

Chapter 8

FEMA Hazard Layer: https://msc.fema.gov/portal/advanceSearch.

MRLC LULC Dataset: https://www.mrlc.gov/data.

Census block group boundaries: https://www.census.gov/geographies/mapping-files/time-series/geo/cartographic-boundary.html.

American Community Survey API: https://www.census.gov/data/developers/guidance/api-user-guide.html.

Chapter 9

Parcel data from St. Lucie County Property Appraiser: https://www.paslc.gov/data-requests/.

Sea Level Rise Scenarios: https://coast.noaa.gov/slrdata/.

Shoreline data: https://geodata.floridagio.gov/.

Chapter 10

California Office of Environmental Health Hazard Assessment (2021), CalEnviroScreen 4.0, Environmental Protection Agency Greenhouse Gas Report (2021).

Chapter 11

US Census Bureau, Tree Survey Data courtesy of City of Phoenix.

Chapter 12

GeoEPR project; linguistic and political data from the Harvard University MAPA Digital Atlas of Ukraine: https://prod-mapa.drupalsites.harvard.edu/.

Chapter 13

Esri, Microsoft, European Commission, European Space Agency

Mariupol Updated Building Damage Assessment Overview Map: https://unosat.org/products/3300, courtesy of Unitar.

Chapter 14

Datasets from Global Fishing Watch and Miller, Nathan A., Aaron Roan, Timothy Hochberg, John Amos, and David A. Kroodsma. 2018. "Identifying Global Patterns of Transshipment Behavior." *Frontiers in Marine Science* 5 (July). doi.org/10.3389/fmars.2018.00240.

Chapter 15

Daily and monthly AIS data from satellites (SAT–AIS) and from vessel traffic services (T–AIS); Land Administrative boundaries: https://connectivity.esa.int/satellite-%E2%80%93-automatic-identification-system-satais-overview.

Direção-Geral de Recursos Naturais, Segurança e Serviços Marítimos (DGRM); Food and Agriculture Organization (FAO)—Global Administrative Areas (GADM).

Chapter 16

Courtesy of National Snow and Ice Data Center (nsidc.org).

ArcGIS Living Atlas of the World, Arctic Ocean Base: Esri, GEBCO, NOAA, Garmin, and other contributors.

Chapter 17

Patrick Kenney, Whitespace.

Chapter 18

Land Use Land Cover data from Sango Bay in East Africa. Created through funding by the NSF (awards numbers 0119821), BE/CNH: Climate and Land Use Change. Courtesy of Samuel Mugisha.

Processes in East Africa (0308420), BE/CNH: An Integrated Analysis of Regional Land–Climate Interactions; Citation for one article that used these products: Olson, J. M., Alagarswamy, G., Andresen, J. A., Campbell, D. J., Davis, A. Y., Ge, J., and Wang, J. (2008). Integrating diverse methods to understand climate–land interactions in East Africa. *Geoforum* 39 (2), 898–911.

Chapter 19

ArcGIS Living Atlas of the World: Poland state boundaries, Russian Federation Raioni Boundaries, and Belarus state boundaries; Esri, Michael BauerResearch GmbH 2023, Central Statistical Office, Eurostat.

Global Land Cover: Esri, ESA. Land Cover CCI Product User Guide v. 2. Tech. Rep. 2017. Available at maps.elie.ucl.ac.be/CCI/viewer/downloaLC-Ph2-PUGv2_2.0.pdf.

Danielson, J. J., and D. B. Gesch. 2011. Global multiresolution terrain elevation data 2010 (GMTED2010): US Geological Survey Open-File Report 2011–1073.

Chapter 20

Sample Asphalt Dataset collected with AV-900 UAS by AirGon LLC. Courtesy of Esri.

About Esri Press

Esri Press is an American book publisher and part of Esri, the global leader in geographic information system (GIS) software, location intelligence, and mapping. Since 1969, Esri has supported customers with geographic science and geospatial analytics, what we call The Science of Where. We take a geographic approach to problem-solving, brought to life by modern GIS technology, and are committed to using science and technology to build a sustainable world.

At Esri Press, our mission is to inform, inspire, and teach professionals, students, educators, and the public about GIS by developing print and digital publications. Our goal is to increase the adoption of ArcGIS and to support the vision and brand of Esri. We strive to be the leader in publishing great GIS books, and we are dedicated to improving the work and lives of our global community of users, authors, and colleagues.

Acquisitions

Stacy Krieg
Claudia Naber
Alycia Tornetta
Jenefer Shute

Product Engineering

Craig Carpenter
Maryam Mafuri

Editorial

Carolyn Schatz
Mark Henry
David Oberman

Production

Monica McGregor
Victoria Roberts

Sales & Marketing

Eric Kettunen
Sasha Gallardo
Beth Bauler

Contributors

Christian Harder
Matt Artz

Business

Catherine Ortiz
Jon Carter
Jason Childs

Related titles

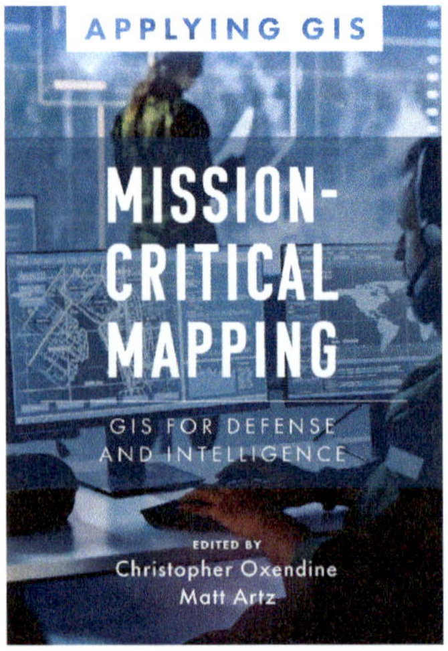

Mission-Critical Mapping

Christopher Oxendine & Matt Artz (eds.)

9781589487994

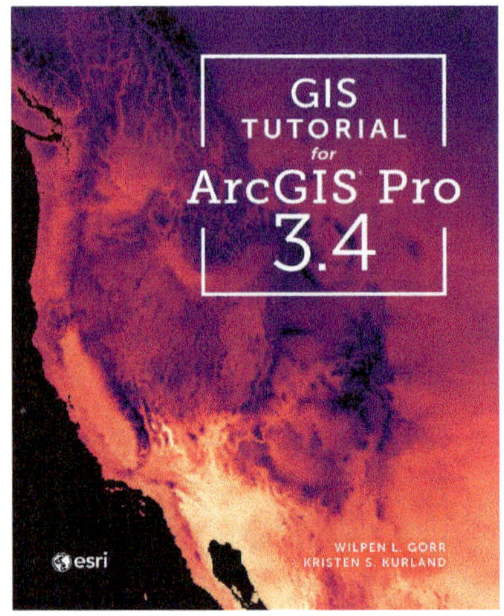

GIS Tutorial for ArcGIS Pro 3.4

Wilpen L. Gorr & Kristen S. Kurland

9781589488151

Top 20 Essential Skills for ArcGIS Online

Craig Carpenter, Jian Lange & Bern Szukalski

9781589487802

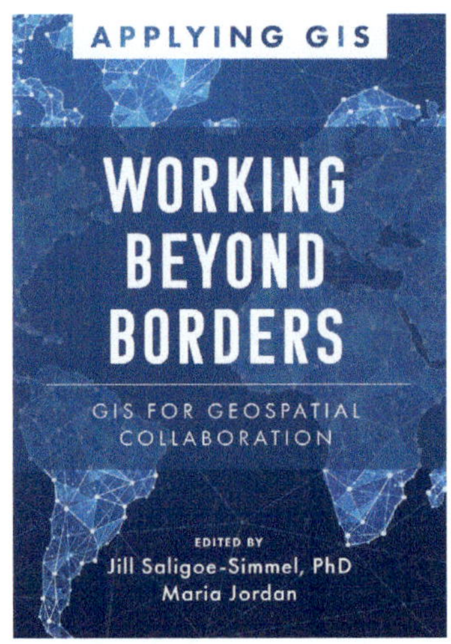

Working Beyond Borders

Jill Saligoe-Simmel & Maria Jordan (eds.)

9781589487628